KU-330-544

Pollutant Transfer and Transport in the Sea

Volume II

Editor

Gunnar Kullenberg
Professor of Physical Oceanography
Institute of Physical Oceanography
University of Copenhagen
Copenhagen, Denmark

CRC Press, Inc.
Boca Raton, Florida

UNIVERSITY LIBRARY
- 8 FEB 1984
LANCASTER

Library of Congress Cataloging in Publication Data
Main entry under title:

Pollutant transfer and transport in the sea.

Includes bibliographical references and index.
1. Marine pollution. 2. Ocean circulation.
I. Kullenberg, Gunnar.
GC1085.P63 628.1′686162 81-4500
ISBN 0-8493-5601-6 (v. 1) AACR2
ISBN 0-8493-5602-4 (v. 2)

This book represents information obtained from authentic and highly regarded sources. Reprinted material is quoted with permission, and sources are indicated. A wide variety of references are listed. Every reasonable effort has been made to give reliable data and information, but the author and the publisher cannot assume responsibility for the validity of all materials or for the consequences of their use.

All rights reserved. This book, or any parts thereof, may not be reproduced in any form without written consent from the publisher.

Direct all inquiries to CRC Press, Inc., 2000 N.W. 24th Street, Boca Raton, Florida, 33431.

© 1982 by CRC Press, Inc.

International Standard Book Number 0-8493-5601-6 (Volume I)
International Standard Book Number 0-8493-5602-4 (Volume II)

Library of Congress Card Number 81-4500
Printed in the United States

83 004470

PREFACE

The distribution of pollutants in the sea is influenced by physical, chemical, and biological transfer processes. These processes interact, and in order to understand the pollution problems and implications for the marine environment it is necessary to take this into account. The aim of this book is therefore to give a presentation of various physical, chemical, and biological transfer processes which includes an account of our present understanding of relevant aspects of the marine systems, discussions of the state of the art as regards parametrization and modeling of transfer processes, and an account of applications to some specific examples.

The subject areas are covered by different authors and the editor has not attempted to really integrate the various chapters. Rather, the aim has been to cover those subject areas which we now consider essential for the transfer and transport of pollutants in the marine environment, obviously without claiming complete coverage. The basic message is that a proper understanding of the transfer of pollutants in the marine environment requires an interdisciplinary approach and due consideration of physics, chemistry, biology, and geology.

The individual chapters by and large stand on their own. A list of notations has been included in the individual chapters when required, but it should be noted that all notations have not been unified. That was simply not feasible.

The first two chapters of Volume I deal with physical processes in the ocean and predictive modeling of transfer of pollutants. Chapter 1 is an attempt to present recent ideas and observations of the physical conditions in the ocean thought to be relevant to the pollution problem in general. Chapter 3 gives a very brief account of techniques commonly used to investigate experimentally the physical spreading of pollutants on small to mesoscales. Chapter 4 gives an account of the air-sea exchange of pollutants, where much progress has been made in recent years.

Processes and transfer related to biological conditions, suspended matter, and sediments are considered in Volume II, Chapters 1, 2, and 3. In all these subject areas the research is moving fast and several new results are presented here.

In Chapter 4 estuaries and fjords are discussed separately since these coastal zones are considered to be of special interest. Likewise in Chapter 5 many aspects of the spreading of oil from oil spills are considered, including processes and modeling with a discussion of findings from several recent spills, since oil spills constitute a serious problem in many areas. It is hoped that this publication will stimulate discussions and interdisciplinary research relevant to pollution problems as well as serve as an educational reference book.

G. Kullenberg

THE EDITOR

Gunnar E.B. Kullenberg received a Ph.D. in Oceanography from Göteborg University, Göteborg, Sweden, in 1967.

Since completion of his doctoral work, he has held positions as Associate Professor of Oceanography at the Institute of Physical Oceanography, University of Copenhagen, 1968-1977; Professor of Oceanography at Göteborg University, 1977-1979; and his current position since 1979, Professor of Physical Oceanography, Institute of Physical Oceanography, University of Copenhagen.

Dr. Kullenberg's main research interests include turbulent mixing, air-sea interaction, optical oceanography, problems related to marine pollution, and coupling between physical and biological processes.

His participation in several international programs has led to work not only in northern European waters but also in the Mediterranean, central and eastern subtropical Atlantic, Southern, and Pacific Oceans.

Approximately 50 papers have been published since 1968, including review papers and presentations at several international symposia and workshops.

Dr. Kullenberg is engaged in the following international organizations: ICES (International Council for the Exploration of the Sea), SCOR,IAPSO (International Association for the Physical Sciences of the Ocean), GESAMP (the U.N. Technical Agencies Joint Group of Experts on Scientific Aspects of Marine Pollution), and in the work of some U.N. Agencies directly. He is chairman of the joint ICES and SCOR WG on the study of the pollution of the Baltic, chairman of ICES Advisory Council on Marine Pollution, and chairman of the GESAMP WG on the health of the ocean. He is engaged in the IAPSO Committee on dispersion problems and inthe WG on ocean optics. In addition to these international engagements, he is involved in the work of Danish and Swedish national committees including the subcommittee on earth sciences of the Swedish Natural Science Council.

CONTRIBUTORS

R. Chester, Ph.D.
Reader in Oceanography
Department of Oceanography
University of Liverpool
Liverpool, England

E. K. Duursma, D. Sc.
Director
Delta Institute For Hydrobiological Research
Yerseke, The Netherlands

Scott W. Fowler, Ph.D.
Head, Biology Section
International Laboratory of Marine Radioactivity/IAEA
Musée Oceanographique
Principality of Monaco

Herman G. Gade, D. Philos.
Head, Department of Oceanography
Geophysical Institute
University of Bergen
Bergen, Norway

Gunnar Kullenberg, D.Sc.
Professor of Physical Oceanography
Institute of Physical Oceanography
University of Copenhagen
Copenhagen, Denmark

Stephen P. Murray, Ph.D.
Assistant Director and Professor
Coastal Studies Institute
Louisiana State University
Baton Rouge, Louisiana

Maarten Smies
Environmental Toxicologist
Shell International Research Maatschappij BV
Group Toxicology Division
The Hague, The Netherlands

Dr. G. C. van Dam
Head of Physics Division
State Public Works
Directorate of Water Management and Research
The Hague, The Netherlands

Michael Waldichuk, Ph.D.
Senior Scientist
Pacific Environment Institute
Fisheries and Marine Service
Canada Department of Fisheries and the Environment
West Vancouver, B.C.
Canada

TABLE OF CONTENTS

Volume I

Volume II

Chapter 1

BIOLOGICAL TRANSFER AND TRANSPORT PROCESSES

Scott W. Fowler

TABLE OF CONTENTS

I. INTRODUCTION

The ability of marine organisms to accumulate pollutants is well documented; however, one aspect not so well understood is how and to what degree they affect the distribution of these substances in the marine environment. As discussed in the following chapters, pollutant distribution strongly depends upon current and water mass movements, eddy diffusion, and sedimentation processes. Movements of pollutants associated with biological material are also subject to physical and geological transport processes but, in addition, are affected by bioaccumulation, retention, and subsequent food chain transfer, horizontal and vertical migration of many species, and passive sinking of biodetritus. Clearly, the relative importance of these biological transport mechanisms compared to physical and chemical processes will be a function of the oceanic biomass at any given location.

In the following sections each of these biological processes will be discussed with respect to four major classes of marine contaminants: heavy metals, artificial radionuclides, chlorinated hydrocarbons, and petroleum hydrocarbons. Obviously, much more is known about biological transport for some types of contaminants than others, and this is reflected by their treatment in this chapter. Furthermore, the literature abounds with examples of baseline studies of pollutant levels in marine species and their tissues, biochemically mediated pollutant transformations, and physiological and environmental factors which control pollutant bioaccumulation. Given the scope of this review, these types of data are included only in so far as they are considered pertinent to an understanding of the processes that govern biological transfer and transport of pollutants through the marine environment.

II. MECHANISMS

A. Accumulation from Water

The fact that most marine species can accumulate pollutants to levels many times those in sea water suggest that they are instrumental in redistributing these substances in the sea. Uptake from water occurs by either adsorption of the substance onto body surfaces or absorption across body surfaces such as gill and gut walls, or a combination of both. For heterotrophs the alternative mode of accumulation is through the ingestion and assimilation of contaminated food. The relative ability of organisms to concentrate pollutants is often expressed by a concentration factor, defined as the ratio of the amount of substance per unit fresh weight of tissue to that in an equal weight of seawater. Since these ratios take into account only the amounts of pollutants in water and the organism, they give no information on the relative importance of the different routes of uptake. It also must be kept in mind that concentration factors are not constant but vary considerably. Concentrations in marine species are in a state of dynamic equilibrium and are the net result of both pollutant uptake and elimination processes occurring simultaneously. These dynamic processes are controlled by exposure time, the physico-chemical form of the pollutant, salinity, temperature, competitive effects with other substances, life cycle of the organism, physiology, feeding habits, etc. In this respect concentration factors are best viewed in terms of ranges rather than absolute values.

1. Inorganic Pollutants

Several comprehensive reviews present ample evidence indicating the strong affinity of metals and radionuclides for marine organisms.[1-5] Depending on the chemical species and the element or radionuclide, concentration factors range from roughly 10^0 to 10^6. Concentration factors such as these attest to the organism's ability to concentrate elements from sea water. However, biological concentration processes become even more striking when it is considered that recent advances in sea water metal analyses[6,7] have shown that certain metals are present in sea water at levels as much as 3 orders of magnitude lower than previously thought. This means that many concentration factors may be underestimated by the same order of magnitude.

It is evident that the elements with the highest concentration factors will be those most readily transported by biological means. These elements are in many cases not those which are biologically essential. For example, in many marine species, especially the smaller ones, Al, Pb, Ru, Zr and certain lanthanides are normally concentrated more than physiologically important elements such as Zn, Cu, and Co.[2,3] This may be because many of these nonbiologically essential elements are mainly particulate in sea water and, thus, are more apt to sorb to surfaces, especially those of small organisms like phytoplankton and zooplankton with high surface area to volume ratios.[3,8] Regardless of the uptake mechanisms which lead to high concentrations of these elements, plankton is probably the most important vector in biological transport since these organisms not only constitute the largest biomass in the sea, but they represent the first point of entry for pollutants into marine food chains.

Phytoplankton, because of its large surface area to volume ratio, quickly takes up radionuclides and metals and reaches extremely high concentration factors. The biphasic process involves rapid sorption to the cell surface, perhaps by cation exchange, followed by slower diffusion across the cell membrane and subsequent binding within the cell.[9,10,329] Equilibration times are generally short (minutes to hours), and there is some evidence in the case of Hg,[11-13] Cd,[12] Zn,[14] and plutonium[329] that uptake is a passive process. Controlled experimentation in the laboratory has shown that uptake of metals like Hg and Cd depends on the metal concentration in sea water, the length of exposure, and the algal species.[12] Furthermore, for the same algal species metal availability varies widely; for example, Sick and Windom[12] found that Hg uptake was one to two orders of magnitude greater than that of Cd for all phytoplankton species tested.

Likewise zooplankton generally displays high radionuclide and metal concentration factors and can absorb these elements directly from water. While members of the zooplankton community are extremely diverse and represent several phyla, most of the experimental work on bioaccumulation has been with crustaceans, such as copepods and euphausiids. Direct uptake from sea water occurs both by adsorption onto body surfaces and absorption of the elements across surfaces, such as gills or gut wall. Once across the cellular boundaries the elements are translocated to other organs and tissues by either active or passive processes where they are either stored or eventually eliminated. Uptake rates strongly depend on the element, with reported equilibration times ranging from several hours[15,16] to several days.[17-21] Typical curves for the absorption of certain metals and radionuclides from water by euphausiids are shown in Figure 1.

In the case of crustacean zooplankton, the amount of element absorbed from water is strongly affected by molting. For example, bioaccumulation experiments with ^{65}Zn,[17] ^{237}Pu,[20] and ^{141}Ce[8] showed that euphausiid body burdens of these elements were reduced on the average by 41, 53, and 99%, respectively, when the animals molted. The significance of this observation is twofold: (1) individual crustacean zooplankton probably never reach a state of equilibrium with the pollutant in its environment, and it is therefore necessary to consider uptake and elimination rates averaged over an entire population of individuals to estimate the relative bioaccumulation potential of the species; and (2) the release of a large fraction of the organism's pollutant load with the molt profoundly affects the biological transport of the substance, an aspect which will be discussed in detail in a later section.

Macroalgae, macroinvertebrates, and fish also absorb metals and radionuclides

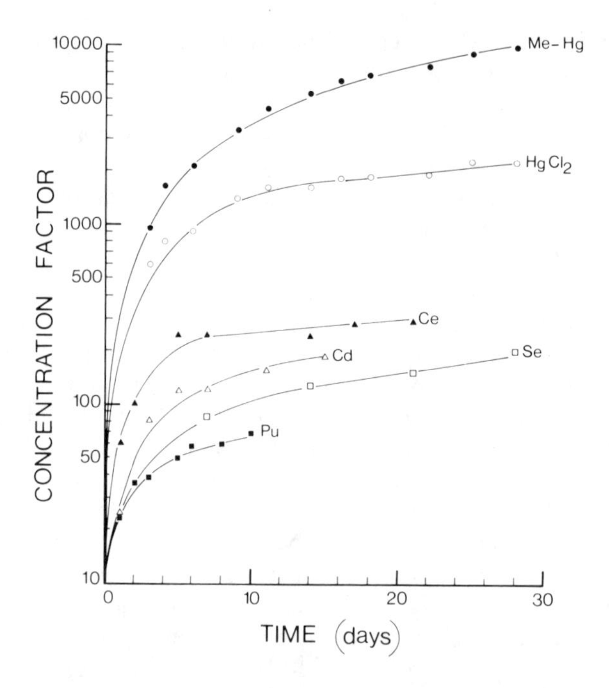

FIGURE 1. Accumulation from water of various metals and radionuclides by the euphausiid *Meganyctiphanes norvegica*. Data are taken from radiotracer experiments utilizing Hg-203,[19] Ce-141,[8] Cd-109[18], Se-75[21] and Pu-237.[20]

from water although the degree of uptake is usually much less than that of smaller organisms since the role of surface area in total accumulation is of far lesser importance in larger species. Numerous examples of the accumulation of these elements by a myriad of species under a wide variety of laboratory and field conditions are given in several recent reviews[4,5,22,23] and will not be reiterated here. As noted for plankton, uptake is generally nonlinear and often biphasic with an initial rapid component representing surface adsorption followed by a slower rate of accumulation as the element is incorporated into internal tissues. The uptake rate generally decreases until a steady state is reached between the element in the water and the organism's tissues. Because in larger organisms internal tissues are often relatively isolated from the surrounding sea water, equilibration times based on element absorption from water are normally much longer (days to weeks) than those observed with plankton.

The importance of the initial component of uptake depends to some extent on the surface characteristics of the organism. Hardshelled, calcareous animals may deposit much of the element in the shell during growth. Indeed substantial concentrations of radionuclides and metals are present in mollusc shells[24,25] and exoskeletons of crustaceans[24,26-28] and echinoderms.[29] Soft bodied organisms with no hard, external covering are able to equilibrate their internal tissues more rapidly. The surface coating of mucus on many of these species, including fish, plays a primary role in the initial complexing of the element.[30-32] For example, the epidermis of tuna, representing about 0.25% of the fish's weight, contains about 52% of the total lead content.[31] Somero et al.[33] suggest that the rapid turnover of lead in mucous-covered tissues of teleosts results from the sloughing off of the mucous layer containing complexed lead. Once the element has diffused through the epithelium, the blood or haemolymph circulation in invertebrates is the principal vector for translocating contaminants to the various tissues.[4,317] The degree of pollutant build-up in these tissues depends on the chemistry of the element, the number of binding sites, potential for detoxification, retention time in a tissue, sexual cycle, and general physiology of the organism. Although not always, liver and kidney of both invertebrates and vertebrates often contain the highest concentrations of metals and radionuclides accumulated from water.[4,34-38,317] Muscle, on the other hand, normally concentrates inorganic contaminants absorbed from water to a much lesser extent[5,35] except, for example, in the case of methylmercury[39,40] and As.[41,42]

The accumulation of metals in the tissues of marine organisms often depends on the size of the animal. Both negative and positive correlations between concentration and size have been reported.[43-45] Negative correlations observed in small planktonic crustaceans[44] are most likely related to the predominance of surface sorption because of the relatively greater surface to volume (weight) ratios in smaller individuals. Higher concentrations noted in the soft tissues of smaller-sized organisms[45] may result from relatively higher metabolic rates in these individuals. For example, laboratory studies with oysters have demonstrated that smaller individuals accumulate Hg from water more rapidly than larger ones.[46] The classic example of a greater increase of metal accumulation with size is Hg in marine fish muscle.[43,47] Cross et al.[43] estimated the Hg concentration factors in bluefish muscle to range from 1000 in young fish to 6000 in the largest individuals. These authors suggested that uptake exceeds elimination leading to a slow buildup of Hg residues throughout the lifespan of the fish. This may be particularly true if most of the Hg accumulated is methylated. However, the size relationship noted with Hg cannot be considered a simple function of metal absorption from water since feeding habits are instrumental in achieving Hg body burdens in fish (see Section II.E.1).

The degree of metal and radionuclide uptake from water largely depends on the physical and chemical form of the element. Davies[48] has shown that iron hydroxide particles readily sorb to the surface of diatoms; this mechanism may enhance the uptake of other elements since metal hydroxides are known to act as scavengers for other elements in sea water.[9] Several studies with phytoplankton give indirect evidence for a reduction in bioavailability of metals and radionuclides which are organically bound or chelated.[49-52] From a review of the literature, Davies[9] concluded that only inorganic ionic species or possibly only the free ions are available for uptake by phytoplankton. The response of zooplankton to different chemical forms of an element resembles that of phytoplankton. Lowman and Ting[53] reported that a pelagic macruran crustacean absorbed four times more ionic ^{58}Co than ^{57}Co — cobalamine from sea water containing equal amounts of the two chemical forms. Small[54] presented experimental evidence suggesting that phytoplankton released exudates which chelated ^{65}Zn, rendering it unavailable for uptake by euphausiids. The situation is less clear for larger invertebrates. For example, experiments with mussels have shown that chelation or complexation greatly reduces the uptake of Zn[55] but enhances the accumulation of Cd[56] and Fe.[57] On the other hand, alkyl mercurials such as methylmercury, with covalent metal-carbon bonds have a much higher affinity for marine organisms than inorganic mercury.[19,39,40,58] An example of the relative bioavailability of methyl and inorganic Hg is shown in Figure 1.

Although relatively few studies have addressed the question of how chemical forms of inorganic pollutants which are naturally present in sea water affect uptake processes, it is evident from the literature that this single factor may largely control the initial transfer of the element from water to organism. Clearly, this is an important area for future research.

The uptake of metals and radionuclides in many species is proportional, or nearly proportional, to the concentration of the element in the surrounding sea water. This is particularly true for small plankton[59,60] macroalgae[4] and for certain elements such as Cu and Cd in polychaetes,[61,62] Cr in clams,[63] and Pb,[64] Cd,[65] and Se,[66] in mussels.

On the other hand, some species show little or no increase in element concentration in response to elevated levels in their surroundings. It is often difficult to discern whether this is a net result of the physiological processes (uptake, storage, and elimination) or simply that uptake is reduced by the saturation of binding sites on or within the organism. However, there is strong evidence from both laboratory and field studies indicating that for essential metals such as Zn, certain crabs,[67] polychaetes,[61,62] and fish[43] can effectively regulate the content of this metal in their tissues.

Differences in the oxidation state of metals also strongly affect their uptake characteristics. Clams accumulate more ^{51}Cr in the hexavalent form than as the trivalent ion.[63] Selenite has an affinity for mussel tissue approximately four times over that of selenate when the ions are absorbed from water.[66] Ruthenium-106 chloride complexes are more available for uptake than ^{106}Ru nitrosyl-nitrato forms in mussels[68] and euphausiids.[69]

As mentioned above, several environmental factors play a role in controlling the bioaccumulation of inorganic pollutants from sea water. Of these, temperature and salinity probably exert the strongest effect. Generally speaking, metal and radionuclide uptake rates correlate positively with temperature in a variety of species, including phytoplankton,[59] zooplankton,[17] benthic crustaceans,[65,70,71] mussels,[40,66,72] and fish.[33,73] However, there are some notable exceptions with crustaceans[42,66] and bivalves,[22,65] indicating that temperature has little or no effect. In the case of crustaceans, one explanation is that at elevated temperatures crustaceans molt more frequently and thus eliminate more rapidly the accumulated element. Hence, the net effect is lower levels than those in animals exposed at lower temperatures.

On the other hand, uptake rates of these elements in marine species generally show an inverse correlation with salinity.[4,72,74-78] This effect, usually attributed to lesser amounts of competing ions in low salinity waters, is thus most pronounced in estuaries and coastal waters receiving runoff. One exception appears to be monovalent ions such as ^{134}Cs which concentrate to higher levels in fish living in high-salinity waters.[73]

A third environmental factor affecting the uptake of a given inorganic pollutant is the coexistence of several inorganic elements in the medium. Studies on this aspect are few, and Phillips[22,75] points out that uptake responses are highly variable and largely dependent on the combinations and relative concentrations of the elements involved.

The pH of sea water may also affect element absorption to some degree. Gutknecht[79] found that the ^{65}Zn uptake rate in several species of macroalgae increased when the pH was raised from 7.3 to 8.6. Although of little or no consequence in open ocean waters where the range is extremely narrow, pH may come into play in estuaries where mixing with fresh water takes place.

2. Organic Pollutants

Because of the generally low solubilities of chlorinated and petroleum hydrocarbons in water, these compounds can be expected to become associated with surfaces such as bacteria, plankton, and other marine species. Furthermore, these classes of compounds differ from metals and radionuclides in two respects: (1) they are lipophilic and often accumulate in fatty tissues, (2) certain compounds can be degraded *in situ* by microbial activity or in vivo by enzymatic processes within the tissues of marine species. Nevertheless, certain aspects of the uptake pattern and tissue distribution of inorganic and organic pollutants are similar.

Knowledge of the uptake kinetics for chlorinated hydrocarbons in lower trophic levels is sparse. Bacteria have been shown to accumulate various compounds to levels 10^2 to over 10^4 times the concentration in water presumably by direct lipid/water partitioning.[80] The uptake of DDT and PCB compounds by phytoplankton is thought to be controlled mainly by specific surface area and to consist of an initial adsorptive step followed by absorption into the cell.[81,82,306] Rapid uptake, with typical equilibration times ranging from 0.5 to 3 hr,[81,82,306] indicates that adsorption is most likely the predominant process. Equilibrium concentration factors following uptake from water are typically 10^4 to 10^6 and agree well with the values calculated from *in situ* measurements.[82] Furthermore, recent evidence[83] indicates that not only dissolved PCB but those associated with suspended microparticulates can be rapidly taken up and incorporated into algal cells. In the case of DDT, once it is accumulated, several phytoplankton species can metabolize this compound to the more persistent metabolites DDD and DDE.[84]

Information on petroleum hydrocarbons in marine phytoplankton is somewhat more limited. There is circumstantial evidence that aromatic hydrocarbons are rapidly adsorbed to diatom frustules.[85,333] Whether or not adsorption is the sole mechanism of uptake is not known; however, some insight can be gained from freshwater phytoplankton studies which show that naphthalene is continually accumulated by *Chlamydomonas angulosa* during a 7-day period[86] and that the cells apparently do not metabolize the compound.[87] An overview of hydrocarbons in marine phytoplankton, both biogenic and those derived from petroleum, has recently been given by Corner.[88]

Chlorinated hydrocarbons are accumulated to relatively high levels in marine zooplankton.[89-91] Concentration factors for PCB and DDTs as high as 10^5 are not uncommon, although the values depend on size, with smaller organisms attaining higher concentrations.[89,92] The uptake of DDT and PCB by copepods and euphausiids is rapid and concentration dependent.[92,93,290,309] Temperature has little or no effect on the process.[92] Equilibration times are rapid (order of days) and appear to be positively correlated with animal size.[92,290,309] The uptake process for organochlorines is thought to involve an initial surface adsorption step, with subsequent absorption across the gill and/or through the integument into the lipid reserves. Thus, the final concentrations reached will largely depend on the lipid content of the species.[94] It appears that copepods living in colder waters cannot metabolize DDT absorbed from water;[95] thus, DDT metabolites which have been measured in copepods and other zooplankton probably come from their food.

Petroleum hydrocarbons behave somewhat similarly to organochlorine compounds in zooplankton. Rapid uptake of naphthalene has been observed in copepods with steady state being reached after 1 to 3 days.[96,97] Net uptake depended both on the naphthalene concentration in sea water and the lipid content of the copepods. The results from these studies suggest that surface adsorption is only a minor factor in the accumulation of this aromatic compound. Unlike organochlorines, marine zooplankton can metabolize absorbed petroleum hydrocarbons.[88,96]

Most uptake studies with larger invertebrates have dealt with benthic organisms. The processes governing uptake from water by zooplankton generally apply to larger invertebrates except that equilibration usually takes longer (days to weeks). Following a relatively long exposure to PCB, concentration factors typically range from 10^3 to 10^5 in a variety of invertebrates.[98-103] Some experiments[99] show that lower chlorinated isomers are taken up more readily than the higher chlorinated isomers, suggesting that bioavailability is related to their solubility in water (i.e., the lower the chlorine content, the greater the solubility). However, Vreeland[100] found the reverse, with uptake being proportional to the chlorine content; thus, the exact nature of the relationship appears to be both species and isomer dependent. Surficial mucous secretions may not sequester chlorinated hydrocarbons from water as effectively as they do inorganic elements.[103] Following absorption across the gills, the highest concentrations of organochlorines generally appear in the hepatopancreas, a lipid-rich organ which likely regulates biotransformation and distribution of chlorinated hydrocarbons in many invertebrates.[104]

The uptake of petroleum hydrocarbons by benthic invertebrates is rapid and dependent upon their concentrations in the water.[105-107,291] These compounds are absorbed across the gill membrane and most likely accumulate in the hepatopancreas.[105,108,111] Hydrocarbon absorption appears to depend upon partitioning of the individual hydrocarbons between the water and the tissue lipids.[106,108] Furthermore, aromatics such as naphthalene, and mono-, di-, and trimethyl derivatives are accumulated preferentially over the alkanes with bioaccumulation factors increasing with the molecular weight of the compound.[108,315] In general, whole body concentration factors of 10^1 to 10^3 appear to be somewhat less than those for chlorinated hydrocarbons.[109,110] Whereas benthic crustaceans readily metabolize hydrocarbons absorbed from water,[111] bivalve mollusks, like mussels, appear to have only a limited capability to do so.[105] Very little is known about how environmental variables affect uptake processes. Studies with clams indicate that lower temperatures increase uptake while salinity has little or no effect.[112] Furthermore, the uptake of both saturated and aromatic fractions of petroleum hydrocarbons by clams appears to be inversely related to the dissolved organic matter concentration in sea water.[307]

Addison[84] has recently reviewed several aspects concerning chlorinated hydrocarbon accumulation in fish. Uptake of organochlorines from water by fish occurs primarily by absorption across the gill membrane. To what extent mucus on fish skin hinders or facilitates transdermal absorption is unknown. Following absorption these compounds are presumed to enter the blood and are subsequently transported to the various tissues, especially those rich in lipids. The efficiency of absorption generally follows the sequence PCB $>$ DDT $>$ dieldrin $>$ lindane. There are data to show that DDT compounds absorbed from water can be metabolized in vivo by fish; however, similar convincing evidence for PCBs is lacking.

When fish are exposed to complex mixtures of petroleum hydrocarbons in water, resultant concentration factors vary considerably among the different organs and tissues as well as among fish species.[310] Polycyclic aromatic hydrocarbons are absorbed from water via the gills, metabolized by the liver, and transferred to the gall bladder where parent compounds and metabolites are both stored.[113] The gall bladder has the most hydrocarbons with lesser but significant amounts in the brain and liver.[108,113] Furthermore, high uptake of petroleum hydrocarbons in some fish tissues does not necessarily correlate with high lipid content.[310] Unlike bivalve mollusks, fish can rapidly metabolize incorporated petroleum hydrocarbons.

Temperature, salinity, sex, age, size, reproductive state, etc. can all affect the uptake of individual petroleum hydrocarbons in many of the organisms discussed above. Trends for the effects of these variables are not often consistent, but appear to depend strongly upon species, compound, and experimental conditions. Their interaction with hydrocarbon uptake has been reviewed in detail by Varanasi and Malins[110] and Neff.[325]

B. Accumulation from Food

The absorption of pollutants from food takes place in the gut with transport to the various tissues via the circulatory system. Tissue accumulation of these pollutants depends on the assimilation efficiency and the subsequent metabolism and retention of the contaminant. Once assimilated from the gut, many of the same factors affecting pollutants absorbed directly from water will determine the fate of ingested contaminants.

1. Inorganic Pollutants

Absorption and tissue distribution of ingested metals and radionuclides depend on the bioavailability of the element. Biologically essential elements such as Zn, Fe, and Mn are rapidly absorbed across the gut and assimilated into tissues of euphausiid crus-

Table 1
TISSUE DISTRIBUTION OF ^{60}CO IN THE CLAM *SCROBICULARIA PLANA* FOLLOWING A 1-MONTH EXPOSURE TO LABELED SEA WATER AND 13 SEPARATE RATIONS OF LABELED PHYTOPLANKTON GIVEN OVER A 45-DAY PERIOD[117]

Uptake route	% total ^{60}Co					
	Shell	Hepatopancreas	Muscle	Mantle	Siphon	Foot and remainder
Water	54.3	21.8	1.0	5.4	1.3	16.3
Food	2.5	77.6	2.0	11.5	1.2	5.3

taceans.[60,114] The localization of ingested zinc is similar to that following absorption from water, although there are marked quantitative differences.[114] For example, after several days of exposure to ^{65}Zn, muscle to exoskeleton ^{65}Zn ratios were three times higher in euphausiids accumulating the radioisotope from food than in those individuals which absorbed ^{65}Zn from water. These results suggest that equilibration times between metals in the organism and the source are not immediate and that the eventual tissue distribution of an inorganic pollutant can be a function of its mode of entry, especially in the case of an acute contamination. On the other hand, nonessential elements or those displaying enhanced biological availability, like Ru,[69] Ce,[8] and Pu,[20] are poorly absorbed by zooplankton and are excreted with the feces.

When metals are ingested by larger crustaceans, in a matter of hours there is a rapid depletion from the stomach fluid and concomitant increase in blood, hepatopancreas, and urine.[115,116] Manganese thus absorbed can also be transported to and deposited in the calcified portion of crustacean shell.[116] As with zooplankton, metal absorption through the gut of benthic invertebrates can lead to quite different tissue distributions of the metal than when absorbed directly from water. Table 1 shows that even after long periods of exposure to ^{60}Co via the two separate uptake pathways, clams maintain distinctly different tissue distributions of the radioisotope.

Uptake of metals and radionuclides from food is important to the levels of contamination reached in fish.[5,41,73,118-120] The liver appears to be the principal organ of accumulation, with elasmobranch livers accumulating relatively more of the absorbed dose than those of teleosts.[5,119,121,122] The relative distribution of an ingested metal or radionuclide in fish tissues has been shown to depend on fish species, food quality, chemical form of the element, and the time elapsed following ingestion.[5,73,119,120] The relative build-up of ingested metals in fish tissues also depends on the rate of elimination; this aspect is discussed in detail in a later section. Nevertheless, where excretion is slower than absorption, accumulation will increase with repetitive ingestion of contaminated food. This was shown to occur in flounder[123] which ingested periodic rations of ^{65}Zn-labeled brine shrimp.

Absorption efficiencies of certain metals and radionuclides ingested with food are given for several species in Table 2. Although experimental conditions varied considerably, it is clear that absorption into tissues depends both on the organism and element.

2. Organic Pollutants

The lipophilic nature of hydrocarbons facilitates their absorption through the gut with the lipids of the prey and subsequent transport to the fat reserves of the consumer. Hence, the temporal build-up of hydrocarbons following ingestion is probably largely limited by the size of the consumer's lipid pool.

Table 2
THE FRACTION[a] (%) OF INGESTED METAL OR RADIONUCLIDE ABSORBED FROM CONTAMINATED FOOD AND ASSIMILATED INTO TISSUES OF MARINE ORGANISMS

Organism	Food	Element																Ref.
		Ag	Zn	Cd	Mn	As	Fe	Cu	Se	Co	Hg[b]		V	Ce	Ru	Pu	Tc	
											I	O						
Zooplankton																		
Psuedodiaptomus coronatus	Phytoplankton			13—68														16
Euphausia pacifica	Phytoplankton		59—71															54
Meganyctiphanes norvegica	Brine shrimp			10					66		36	97		≃0.1—1	≃50			8, 18, 19, 21, 69
Benthic crustaceans																		
Orchestia gammarella	Macroalgae							66										124
Lysmata seticaudata	Artemia					78							25			≃15	26—79	32, 42, 320, 331
Crangon crangon	Mussel		≃60							≃20								71
Carcinus maenas	Worms, clams, brine shrimp		58	10			35	31		≃99			38			20—60		4, 125—127, 320
Benthic mollusks																		
Mytilus galloprovincialis	Phytoplankton												7					313
Scrobicularia plana	Phytoplankton									83								125
Starfish																		
Coscinasterias tenuispina	Mussel															≃70—90		29
Fish																		
Raja clavata	Worms or crab hepatopancreas	49		17		85					14	97				≃0.1—1		41, 119, 120, 122, 293
Pleuronectes platessa	Worms or crab hepatopancreas	4	36	5	40	10	32			4	6	91				<0.01		41, 73, 118—121

[a] Certain values have been estimated from the original data. In several cases mean values have been computed.
[b] I = inorganic mercury, 0 = methyl mercury.

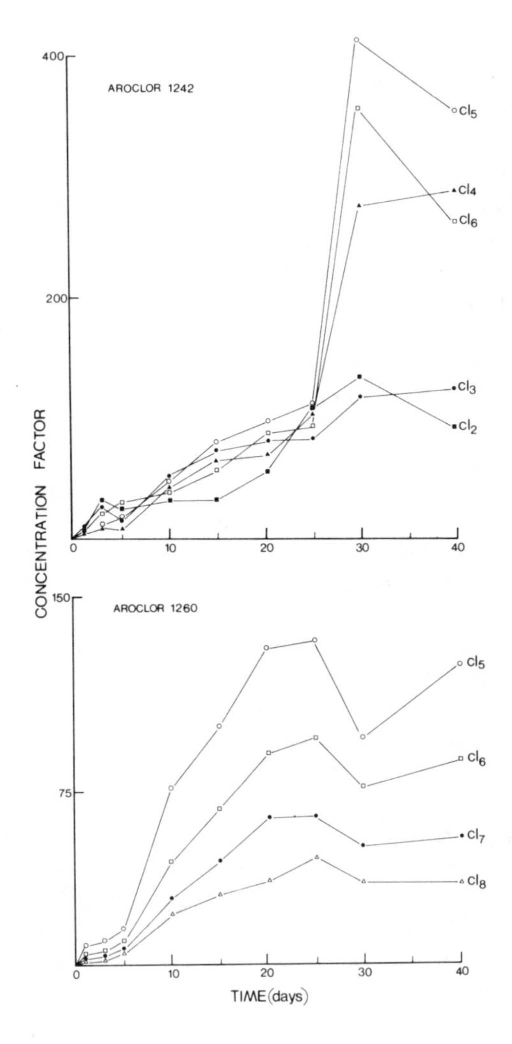

FIGURE 2. Accumulation of PCB homologues by the bivalve *Macoma balthica.* Cl subscripts refer to di-, tri-, tetra-, penta-, hexa-, hepta-, and octachlorobiphenyls. (From Langston, W. J., *Mar. Biol.*, 45, 265, 1978. With permission.)

Small planktonic euphausiids can accumulate DDT residues from their food with assimilation efficiencies (60 to 80%) comparable to those reported for carbon in the same organism.[92] Zooplankton ingest petroleum hydrocarbons both as particles and droplets of crude oil[128,129] or as individual compounds incorporated in their food.[88,97,333] These studies demonstrate that particulate oil is poorly assimilated and passes through zooplankton relatively unaltered with the feces. On the other hand, water-soluble compounds such as naphthalene incorporated in phytoplankton are assimilated more by copepods ($\simeq$ 60%) than are the dietary nutrients N and P.

PCBs fed to shrimp are assimilated and accumulate in the hepatopancreas.[102] Bivalves are also effective in removing PCBs from ingested particulates.[130,131] Langston[131] has shown that after ingestion of PCB-contaminated particulates, isomers of five chlorine atoms per molecule were accumulated by bivalves, whereas isomers of very low and very high molecular weights were taken up at much slower rates (Figure 2). Likewise, more pentachlorobiphenyl was assimilated than di- and trichlorobiphenyls by worms (*Nereis virens*) which had ingested contaminated food for 3 weeks.[132] These studies indicate that the chlorine content of the biphenyls can strongly affect food chain transfer rates.

Crude oil deposited on the bottom apparently can be ingested by a variety of benthic invertebrates.[133] For example, under laboratory conditions, shrimp readily ingested oil which remained in the animal's foregut until it molted. Benthic crustaceans also accumulate petroleum hydrocarbons incorporated in their food by routes similar to those for chlorinated hydrocarbons. Crabs were found to assimilate from 2 to 10% of a variety of paraffinic and aromatic hydrocarbons ingested with their food.[111] The same study also showed that crab hepatopancreas was the principal site of hydrocarbon accumulation and metabolism. Little information exists on food chain uptake of petroleum hydrocarbons in mollusks. The apparent assimilation of ingested hydrocarbons was demonstrated in one study in which mussels (*Mytilus edulis*), filtering kaolin particles contaminated with diesel fuel, reached tissue levels over 1000 times ambient.[134] It is not certain to what degree mollusks can metabolize ingested hydrocarbons; however, bivalves like *Mytilus* appear to have the metabolic capability to eventually detoxify these compounds, albeit at extremely slow rates.[328] This may partially explain results of field studies which indicate a lack of petroleum hydrocarbon metabolism in mussels chronically exposed to these compounds in both their food and water.[135] In contrast to crustaceans and mollusks, there is clear evidence both in the laboratory[136] and field[137] suggesting that assimilation of petroleum hydrocarbons from food or ingested sediment is very low in benthic polychaetes.

Most evidence for the assimilation of ingested organochlorines in fish comes from freshwater species, although a few feeding experiments have been performed with marine fish.[84,138,139,292] Chlorinated hydrocarbons absorbed in the gut enter the blood and are redistributed to other tissues. Information on absorption efficiencies for these compounds in marine species is scant. Sole (*Solea solea*) fed various concentrations of radiolabeled DDT for 3, 28, and 57 days retained 72, 60, and 43%, respectively, of the ingested dose at the end of the experiment.[292] Following ingestion brain, liver, and gastrointestinal tract contained the highest DDT concentrations while skeletal muscle displayed the lowest tissue levels. Information on absorption efficiencies for DDT and other chlorinated hydrocarbons has also been derived from studies with fresh-water fish;[84] values of roughly 10, 24, and 68% retention for DDT, dieldrin and PCB, respectively, appear to be typical. Once assimilated, PCBs accumulate in lipid active tissues such as adipose, spleen, and lateral line muscle. Furthermore, when individual PCB compounds were chronically ingested, hexachlorobiphenyl accumulated more in fish tissues than did tetrachlorobiphenyl.[139] While metabolism of ingested PCB is currently viewed as an extremely slow process, DDT and aldrin taken in with the food are readily converted to DDE and dieldrin, respectively, although a portion of the conversion process may be carried out by intestinal microorganisms.[84] In contrast to PCB and DDT, higher molecular weight compounds like chlorinated paraffins ingested with food may be little, or not at all, assimilated by fish.[138]

Visual evidence of tar balls in the stomach of the epipelagic fish (*Scomberesox saurus*) demonstrates that fish do ingest these residues; however, there are relatively few data available for assessing the fate of the oil particles in tissue following ingestion.[140] Analysis of codling (*Gadus morhua*) fed a squid diet contaminated with topped Kuwait crude oil indicated that the liver discriminated against n-alkanes in the chain length range C15 to C33.[141] Although the liver is probably the center of lipid metabolism and may control deposition in tissues, in codling very little of the alkane mixture was assimilated into muscle. On the other hand, plaice ingesting topped Kuwait crude oil accumulated n-alkanes (C15 to C22) in the muscle, which quickly decreased to background levels when ingestion of oil ceased.[142] This difference may be because plaice store significant amounts of lipid in muscles, whereas the lipid content in cod muscle is very low.[143] Ingestion studies with cod[143] using the individual radiolabeled hydrocarbons, hexadecane, and benzo[*a*]pyrene demonstrated that, once absorbed, these compounds build up in stomach, liver, and gall bladder. A strong preference for benzo[*a*]pyrene over hexadecane was noted in gall bladder bile. Once in the bile, these compounds are subsequently released into the gut in response to the passage of food. Roubal et al.,[144] feeding benzene, naphthalene, and anthracene to coho salmon, showed that these compounds accumulated principally in liver and brain, in the following order: anthracene > naphthalene > benzene. This suggests that the accumulation of ingested aromatic hydrocarbons in fish tissue relates to the number of benzenoid rings. In addition, this study as well as others[110,143] clearly demonstrates the marked ability of fish to metabolize ingested petroleum hydrocarbons.

C. Relative Importance of Uptake from Food and Water

It is clear from the foregoing discussion that marine organisms can accumulate pollutants both directly from water and from contaminated food. What is not clear is which pathway prevails under a given condition of the natural environment. Obviously, many biological, chemical, and environmental factors act in concert to regulate the relative importance of the two pathways. However, two basic conditions are of primary importance in governing which uptake pathway will predominate: (1) the relative concentrations of the pollutant in water and food and (2) the relative abundance of food biomass available for ingestion. Considering both the nonhomogeneous distribution of food organisms in the sea and the wide range of concentration factors for different pollutants, it is highly probable that no one pathway always predominates but that the relative contribution of each route depends upon the prevailing ecological conditions and, thus, varies in time and space. Because of the inherent difficulties in delineating these pathways, few studies have addressed this question in a rigorous manner. Most evidence has been derived from either indirect estimates based on theoretical calculations or short-term experiments carried out under ideal feeding conditions; hence, our ability to extrapolate these findings to the natural environment is limited.

1. Inorganic Pollutants

Under certain conditions absorption of metals and radionuclides from water can contribute significantly to the body burden of these elements in small organisms like zooplankton. As discussed in Section II.A.1, uptake from water occurs largely by adsorption leading to contamination of the organism's external tissues.[8,16,17,20,54,145] With acute exposures, the water pathway may predominate in these species. Sick and Baptist[16] demonstrated that in 24 hr the copepod *Pseudodiaptomus coronatus* took up significantly more cadmium from water than from contaminated algal suspensions at cell concentrations of 5×10^3 and 10^4 $m\ell^{-1}$. Only when the cell concentration was raised to 10^5 $m\ell^{-1}$ did uptake from food equal or surpass the rate of Cd uptake from water. The authors concluded that, over the short-term and at phytoplankton concentrations normally found in the environment ($\simeq 10^3 - 10^4$ cells $m\ell^{-1}$), uptake of metals from water may represent a primary and biologically significant mechanism for trace metal accumulation in zooplankton. Their study underlines the effect of food density on the concentration of an element in an organism. There is some evidence from laboratory experiments[54] that during bloom conditions, soluble organic exudates can chelate metals, thereby reducing their bioavailability from water. However, as Davies[9] points out, this effect only occurs at cell concentrations much higher than are normally found throughout the year; thus, a reduction in the importance of the water pathway by this mechanism would take place only rarely or under a special set of oceanic conditions.

Long-term experiments[18,21,146,147] in which isotopic equilibrium between elements in water and organism has been considered, indicate that the food pathway is essential for attaining total body burdens of an element over relatively long periods of time. Using radiotracer techniques Nassogne[146] showed that when ^{65}Zn was absorbed from

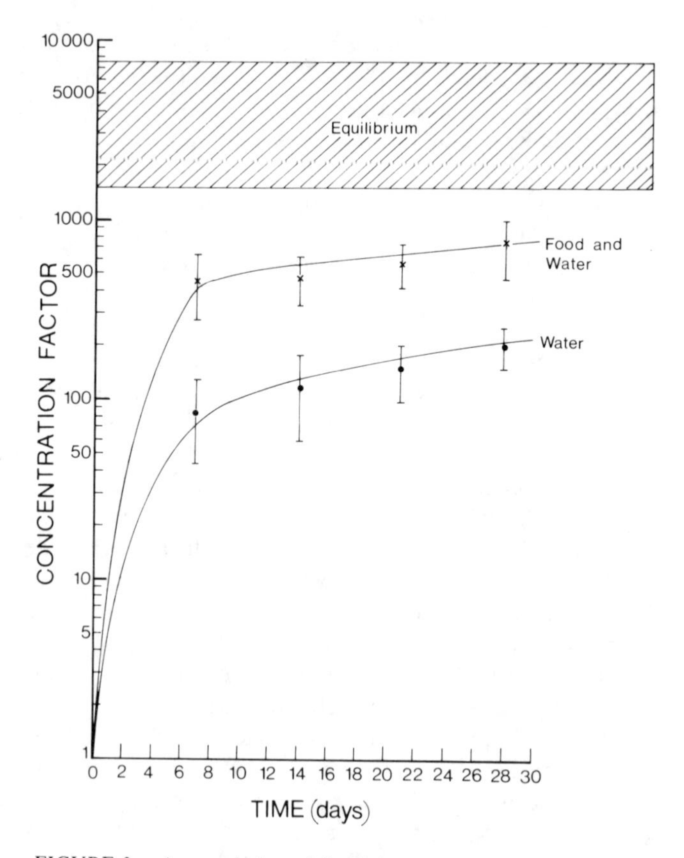

FIGURE 3. Accumulation of Se-75 from water and from food and water together by the euphausiid *Meganyctiphanes norvegica*. Shaded area represents range of equilibrium concentration factors based on stable selenium concentrations in euphausiids and water. (Reprinted from Fowler, S. W. and Benayoun, G., *Mar. Sci. Commun.*, 2, 43, 1976 by courtesy of Marcel Dekker, Inc. With permission.)

water by the copepod *Euterpina acutifrons*, after 4 days much of the stable zinc within the animal was not freely exchangeable with the radiotracer in the sea water. However, when the copepods ingested phytoplankton contaminated with ^{65}Zn for 5 days, greater than 95% of the stable zinc had exchanged by that time. These results suggest that zinc uptake from water only involves sorption or ion exchange processes at or near the body surface and that zinc input through the food is required for the metal to become fixed within the tissues. Similar studies with euphausiids also imply the importance of the food pathway in the long-term accumulation of metals.[18,21] A typical example is given in Figure 3 in which euphausiids (*Meganyctiphanes norvegica*), accumulating Se from both food and water, reached concentration factors much closer to steady-state values than did animals which took up the element only from water. The difference between the two uptake curves after 4 weeks indicates that under *ad libitum* grazing conditions, food chain transfer accounted for roughly 70% of the total Se accumulation by the euphausiids. For metals and radionuclides with high assimilation efficiencies like Se (66%), the contribution of the food pathway can be significant; however, for elements which are poorly assimilated, absorption from water may be the primary mechanism for accumulation. For example, plutonium uptake rates from water and the combined food and water pathway are similar in euphausiids due to poor assimilation of this radionuclide via the food chain.[20] Cerium ingested by zooplankton behaves similarly and appears to be taken up primarily by absorption from water.[8]

In an environment where chronic exposure to a metal or radionuclide takes place, organisms can establish internal and external equilibrium with the contaminant. Assuming that metal concentration in adult euphausiids results from steady state between uptake and loss, Small et al.[147] estimated the flux of zinc through *M. norvegica* by measuring the rates of zinc ingestion, excretion, and assimilation under various conditions. They found that zinc normally ingested by euphausiids was sufficient to account for both the total amount of metal excreted and that assimilated during growth. Similar studies on cadmium[18] and selenium[21] gave an equally good balance between the amount of element ingested with the food and that assimilated and excreted; however, mercury ingestion rates[19] were far too low to account for losses and tissue incorporation, implying that mercury is taken up by euphausiids mainly from water. This heuristic approach, while not measuring accumulation from either pathway directly, has nevertheless furnished good insight into the importance of food vs. water as a source of metals to this particular planktonic species.

Several attempts have been made to delineate the two pathways in larger invertebrates and fish, but usually results are inconclusive because of difficulties involved with simulating natural ingestion rates and relative pollutant concentrations in the food and water.[127,148] Baptist and Lewis[149] carried out an elaborate experiment to assess the relative importance of food and water in the accumulation of ^{65}Zn and ^{51}Cr at each trophic level (phytoplankton → *Artemia* → post larval fish → mummichog). Although realizing certain limitations of their experimental design, they concluded that the food chain was generally the more efficient pathway for uptake of the two radionuclides by all higher trophic levels except the second. Enhanced uptake from water by the small secondary grazers (*Artemia*) was attributed to their relatively large surface area for adsorption. Using ^{65}Zn and ^{59}Fe to measure separately uptake from water vs. that from food, Young[150] calculated the inputs to both water and food metal pools in the winkle (*Littorina obtusata*) grazing on macroalgae. Extrapolating from the conditions of metal concentrations and feeding regime used in the experiment, it was concluded that in the environment the food chain was the major pathway for accumulation of these two metals in *L. obtusata*. On the other hand, in experiments in which mussels were chronically exposed to lead in their food and water,[64] it was found that the organisms accumulated equal amounts of the metal from both sources. In a similar study with fish, Hoss[123] demonstrated that flounder reached higher ^{65}Zn concentration factors from food (*Artemia*) than from water; however, again these results were obtained under conditions not necessarily encountered in the natural environment.

A different approach was used by Renfro et al.[151] in which shrimp, crabs, and small fish were allowed to accumulate ^{65}Zn for 3 months either from water alone or from the combined food and water pathway. Despite the continuous ingestion of radioactive food, shrimp and crabs in both treatments reached the same ^{65}Zn level at the end of the experiment. In contrast, with fish the food pathway accounted for 2.5 times more ^{65}Zn than that obtained from water alone. These results suggest that for zinc, food chain uptake may be of lesser importance for certain crustaceans which turn over the ingested metal rapidly. Furthermore, examination of specific activities in organisms and water as well as the organisms' radioactive and stable zinc concentration factors indicated that isotopic equilibrium had not been reached after 3 months of exposure. The authors concluded that there are zinc pools in the adult which exchange slowly, if at all, with zinc absorbed from either water or food. If these results are typical for inorganic pollutants in general, only those individuals spending a significant portion of their life cycle in contaminated waters might be expected to thoroughly equilibrate with the pollutant.

Several other radiotracer experiments with crustaceans[35,42,65] and mollusks[34,35,65,66] have also demonstrated that concentration factors for radionuclides absorbed from water fall far short of those for the corresponding stable elements. Similar conclusions have been drawn from studies with fish. Pentreath[5,119,120,152] presented convincing evidence that for several heavy metals and radionuclides only a small fraction of the element content in teleost and elasmobranch tissues can exchange with the radionuclide taken up from water. Only in the case of monovalent ^{137}Cs[152] and inorganic mercury[118] does it appear that the water input into fish is significant (≃ 50%). By and large, all these results implicate the importance of the food chain in affecting the total turnover of inorganic pools within marine species.

Owing to the complex interaction of ecological and chemical parameters in nature, it is indeed difficult if not impossible to draw general conclusions about the relative importance of the two pathways in the natural environment. Nevertheless, several trends can be discerned which could help predict the behavior of an inorganic pollutant under certain conditions. Accumulation from water tends to be more important in smaller organisms than larger forms principally because of the greater relative surface area for adsorption in the former. In an acute contamination situation, where most of the pollutant is in the water and not in the food chain, absorption from the water will predominate. If the contamination becomes chronic, more and more of the pollutant will be passed along the food chain. Then the input of the pollutant via the food chain will depend on the food supply and feeding rates. Contamination of highly productive areas will result in increased importance of food chain transfer, especially at the lower trophic levels. Food chain transfer is enhanced by the following: (1) high concentration factors in the prey, (2) high assimilation efficiencies in the predator, and (3) strong retention of the assimilated fraction by the predator. For example, certain fission products,[152] lanthanides,[8] and actinides,[20,121,153,334] which in general are poorly assimilated and weakly retained, are mostly derived from the water, whereas metals which are readily assimilated[18,21,54,147,152] or those like organic mercury with long biological half-times of the assimilated fraction,[19,118] are most likely transferred via the food chain.

Evidence suggests that in the short term neither pathway alone will lead to total equilibration of inorganic contaminants in the organism's tissues. This implies that uptake into discrete element pools in an animal is governed primarily by one or the other pathway. Future work should concentrate on relating the input pathway to a given pollutant pool within the organism.

2. Organic Pollutants

Very few studies have specificially examined the relative importance of food and water in the accumulation of hydrocarbons by marine species. Although the corresponding data base for these compounds is smaller than that for inorganic pollutants, many of the same principles discussed above for metals and radionuclides will hold for chlorinated and petroleum hydrocarbons. However, for hydrocarbons the picture is somewhat more complicated since the relative importance of an uptake pathway may change with time as the parent compounds are degraded.

In a recent review of the fate of organochlorine compounds in aquatic organisms, Addison,[84] drawing principally upon fresh-water studies, concluded that small invertebrates accumulated most of their chlorinated hydrocarbons from water, whereas ingestion was the predominant uptake pathway in larger invertebrates. Recent evidence with marine species[94,154,155] supports the theory that uptake from water predominates in lower trophic level heterotrophs. Scura and Theilacker,[155] using a laboratory three-step food chain, showed that in a contaminated system unfed anchovy larvae accumulated the same amount of PCBs as fed larvae and that the final concentration in the tissues was related to the PCB concentration in the water. Another set of experiments,[154] designed to measure the uptake of dieldrin from food and water by crab

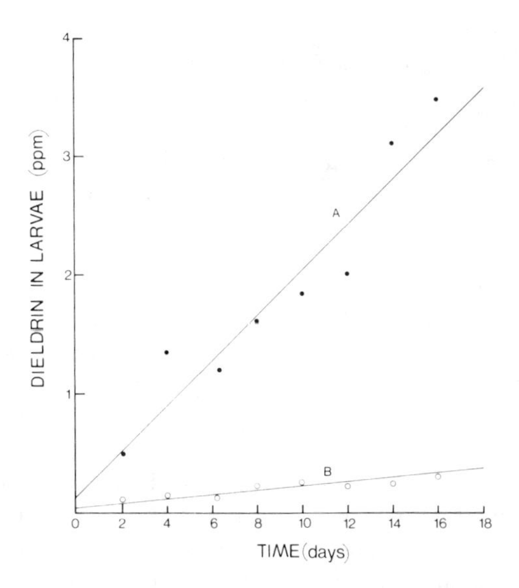

FIGURE 4. Uptake of ^{14}C-dieldrin by crab larvae *Leptodius floridanus* from 0.5 ppb dieldrin in water. (A) 0.213 ppm in food, (B) from the time of their hatching (time zero). (From Epifanio, C. E., *Mar. Biol.*, 19, 320, 1973. With permission.)

larvae, demonstrated that the pesticide was accumulated about 8000 times more readily from water than from food (Figure 4). Using this ratio and dieldrin concentrations in food and water normally found in the environment, it was calculated that the larvae under these conditions would accumulate the pesticide 1.23 times as quickly from water as from their food. In a field study, Clayton et al.[94] found that lipid-water concentration ratios of PCB in zooplankton were uniform over a wide range of spatial and temporal regimes, species composition, and lipid content. They concluded from their study that water is the predominant pathway for uptake due to surface properties of zooplankton which favor the establishment of rapid exchange of PCB between internal lipid pools and the aqueous medium.

However, despite the well-documented importance of equilibrium partitioning as a mechanism for chlorinated hydrocarbon uptake by zooplankton, dietary input cannot be overlooked as a pathway leading to enhanced organochlorine body burdens, in particular, following short-term exposures to these pollutants. For example, in a carefully controlled laboratory study Wyman and O'Connors[309] found that the herbivorous estuarine copepod *Acartia tonsa* fed PCB contaminated phytoplankton accumulated more PCB relative to unfed copepods, despite similar PCB levels in the water. It may be significant that their experiments employed herbivores, whereas the aforementioned food chain studies[154,155] involved carnivorous feeding modes. Clearly, future work should examine the aspects of food type and food concentration as they affect the relative importance of the dietary pathway in chlorinated hydrocarbon accumulation by lower trophic level heterotrophs.

There is evidence that the water pathway may also be more important to larger invertebrates and fish. Nimmo et al.[102] exposed benthic shrimp to PCBs in either their food or water and compared the resultant tissue distributions of PCBs with those of feral shrimp from a contaminated bay. The PCB distribution in the tissues of shrimp exposed in water most closely resembled those of the naturally contaminated organisms leading the authors to conclude that the contaminated feral shrimp obtained most of the PCB from water. Bahner et al.[156] followed the accumulation of Kepone along a three-step food chain composed of plankton (*Artemia*), mysids (*Mysidopsis bahia*), and fish (*Leiostomus xanthurus*). They demonstrated that the insecticide was rapidly taken up from both water and food; however, it took 3000 times as much Kepone in food as in water to produce similar residue levels in fish after one month.

Experimental data on the relative importance of both pathways in large marine fish are sparse. However, there is enough evidence from field data [334] and analogous fresh water fish studies[84] to suggest that food chain transfer predominates in larger marine species.

Although surface adsorption has been cited as one mechanism by which petroleum hydrocarbons are taken up by small zooplankton (Section II.A.2), experimental evidence so far indicates that assimilation from food may be more important for zooplankton in nature. In one recent study Harris et al.[97] exposed the copepod *Calanus helgolandicus* simultaneously to naphthalene in sea water alone and sea water with phytoplankton, *Biddulphia sinensis,* containing hydrocarbons. Even at relatively low cell concentrations (2×10^4 l^{-1}), it was found that the naphthalene level in sea water alone would have to be 2 to 3 orders of magnitude greater than in the food to give the same increase in copepods. The authors concluded that even if suspended matter in the sea contained only small amounts of hydrocarbon relative to those dissolved in sea water, these particulates could be a more important source of hydrocarbons to copepods than the water. On the other hand, some recent experimental evidence from studies on the food chain transfer of benzo[*a*]pyrene[333] suggests that under natural conditions in which concentrations in food would be expected to be higher than in the surrounding waters, uptake from water and food may be of equal importance in the accumulation of polycyclic aromatic hydrocarbons by filter-feeding bivalve larvae.

Other indirect evidence implies that petroleum hydrocarbons are efficiently transferred through zooplankton up the food chain. Teal[157] noted that copepods of the genus *Calanus,* which synthesize pristane, are eaten by certain predatory copepods. These predators contain an order of magnitude more pristane than the herbivorous species, which presumably obtain small amounts from ingested phytoplankton. The author concluded that the low pristane level in herbivorous species indicates the absence of any major uptake pathway (e.g., water) other than food. In further support of his conclusion, Teal[157] used pristane concentrations in sea water in conjunction with concentration factors obtained from fresh water studies to estimate the maximum amount of pristane that copepods could accumulate from water alone. These values were 2 to 3 orders of magnitude less than pristane concentrations normally found in copepods, again suggesting the importance of the food pathway in hydrocarbon uptake.

In contrast to the strong evidence for the predominance of food chain transfer in the accumulation petroleum hydrocarbons in zooplankton, less detailed data for a variety of other species seems to point to the importance of direct absorption from water. Burns and Teal[158] found that contaminated *Sargassum* contained significant amounts of natural hydrocarbons in addition to the petroleum compounds, whereas animals associated with the weed contained mostly petroleum hydrocarbons. Assuming that natural hydrocarbons only enter the animals through food while petroleum compounds are available from both food and the water, they concluded, in light of the very high nonnatural to natural hydrocarbon ratio in the associated fauna, that these animals obtained petroleum hydrocarbons from water rather than from food. A later study of the same *Sargassum* community[159] also failed to find significant correlations between natural and petroleum hydrocarbons (in the biota) and similarly concluded that petroleum hydrocarbons were absorbed from water, primarily across respiratory surfaces.

Water also appears to be of major import in the uptake of petroleum hydrocarbons by certain benthic infauna. The polychaete *Neanthes arenaceodentata* rapidly accumulated radiolabeled naphthalene from water, reaching concentration factors of 40 after only 24 hr.[136] Worms were then fed *ad libitum* for 16 days with contaminated organic detritus and allowed to clear their gut before analysis. Although the worms fed actively (as evidenced by radioactivity in their fecal pellets), naphthalene could not be detected in their tissues during the experiment, indicating an extremely low availability of ingested petroleum hydrocarbons. Although these findings suggest that the food pathway is unimportant for hydrocarbon uptake by detritivores, in view of results from feeding studies with other invertebrates (Section II.B.2), far more work is needed utilizing a wider variety of petroleum hydrocarbon compounds before any general conclusions can be drawn about detritus feeders in general.

Besides absorbing petroleum hydrocarbons through the gills, marine fish can also take up these compounds while drinking and feeding; however to date insufficient work has been carried out to assess the relative importance of these pathways. Nevertheless, considering the small gill surface area in larger fish as well as similarities in hydrocarbon profiles in environmental samples of fish and their prey,[143] it appears that most fish obtain petroleum hydrocarbons primarily from their diet.

Many aspects of petroleum hydrocarbon uptake by marine species remain to be examined. Hydrocarbon partitioning between tissue lipids and water is a major uptake mechanism for many organisms; nevertheless, where direct comparisons of the two pathways have been made under well-simulated conditions, food chain transfer was the major mechanism of accumulation. The contrasting results for detritivore worms are unique; thus, additional and more detailed studies with infauna may prove to be particularly instructive. A further consideration is that most conclusions on the relative importance of food and water pathways are based on experiments using a single compound, e.g., naphthalene. In view of the many petroleum hydrocarbon compounds in the environment and since each may vary in its degree of assimilation and persistence, extrapolating results with a single compound to petroleum hydrocarbons as a group is somewhat tenuous.

D. Elimination

Pollutants in or on marine organisms are eliminated in either soluble or particulate form. Depending on the element or compound, loss of the soluble fraction can occur passively by simple desorption and ion exchange processes, or actively through metabolic excretion. Alternatively, organisms can rid themselves of incorporated pollutants by releasing contaminated particulate products such as fecal material, molts, mucus, and reproductive products. Factors influencing the relative importance of each of these pathways in a given species include the nature of the pollutant, route and duration of accumulation, metabolic pathway, and life cycle, to mention only a few. In any event, release of soluble and particulate materials is the prime mechanism by which marine organisms redistribute pollutants in their environment.

Despite the multitude of parameters affecting the rate of elimination from aquatic biota, most studies indicate that pollutants are lost more slowly than they are accumulated. Generally, when organisms are transferred from a contaminated to a clean environment, there is an initial rapid drop in pollutant concentration followed by a slower loss. The initial rapid loss can be due to desorption of loosely bound pollutants, defecation of unassimilated materials, release of nonmetabolized compounds, etc., whereas the slower depuration rates generally reflect loss from pools within the organism where the contaminant is more tightly bound to tissue components. The shape of the curve describing depuration of the pollutant with time is a function of all these processes working in concert.

Various models involving linear, exponential, and power functions have been used to describe depuration processes. These mathematical representations have proven useful for comparing pollutant kinetics between different organisms and between different contaminants, although it is evident that for a given pollutant or species, no one model can be expected to hold under all circumstances. Biological half-time (the time for one half of the pollutant to be lost from an organism or a pool within the organism) is a useful tool to quantify elimination processes. However, as mentioned above, the loss rate is rarely constant over the entire period of depuration, thus both fast and slow components with different half-times are normally observed.

1. Inorganic Pollutants

There is little information on the loss of metals and radionuclides from contaminated phytoplankton. Measurements are complicated by the fact that actively dividing cells cause a "dilution" of the pollutant concentration by an increase in the number of cells. However, experimental evidence[9,14,160] points to rapid elimination, with an almost instantaneous desorption from extracellular binding sites followed by a much slower release from intracellular pools of the element. Half-times for these processes are on the order of hours to a few days, and can be expected to decrease for cells in water containing organic chelating agents.[14,160,329] There is also evidence[10,329] that longer exposures lead to greater metal and radionuclide retention, at least over the short term. However, normally the most probable way in which contaminated phytoplankton effectively reduce their pollutant concentration is by "biological dilution" (i.e., 50% reduction in concentration at each cell division).[160]

Several investigations have examined metal and radionuclide excretion kinetics in marine zooplankton.[8,17-21,60,145,147,161-163] Keunzler[161,162] measured loss of radioactive I, Co, Fe, and Zn from many different zooplankton species, obtaining rates ranging from 1 to 22%/hr. These rapid loss rates reflect both the short exposure time and the fact that loss was only measured for a few hours. More recent evidence[18,21,60,147,163] indicates that long-term depuration from zooplankton does not follow a single exponential rate but takes place from both fast and slowly exchanging compartments. Hence, the rates mentioned above probably reflect loss from a rapidly exchanging pool which may represent only a small fraction of the organisms' total pollutant load. When zooplankton are chronically exposed to inorganic contaminants, subsequent depuration can take much longer. Several examples of depuration over a relatively long time are given in Figure 5 using euphausiids exposed to radionuclides in their food or in both water and food for 2 to 4 weeks. Rough estimates of the biological half-times, computed for the slow component, range from a few days or less for elements with low assimilation efficiencies like Ce, Cs, and Pu, to several months for methylmercury which is tightly bound within the organism. Interestingly, methylmercury is lost 10 times more slowly than the inorganic form. The transition metals and selenium are intermediate. Despite the small size and rapid metabolic turnover rates characteristic of zooplankton, the exposure time to a pollutant is critical in determining the subsequent depuration rate. For example, euphausiids which either spend more time in contaminated water,[8,69] or receive contaminated food for a longer period[18] lose the incorporated metals more slowly. As element pools within organisms equilibrate with the contaminant, subsequent excretion rates are relatively unaltered by further exposure to the pollutants unless some change in chemical form of the element takes place. One example might be in vivo methylation of inorganic mercury, although recent work[19] indicates that methylation probably does not occur in planktonic crustaceans.

Several studies have documented the importance of particulate excretion in zooplankton. Molting of planktonic crustaceans is an effective mechanism for eliminating incorporated pollutants, particularly when the organism is briefly exposed to a waterborne metal or radionuclide with a high affinity for surfaces.[8,17,20,28,60,145] Experiments with euphausiids have shown that depending upon the amount of residual ^{65}Zn in the

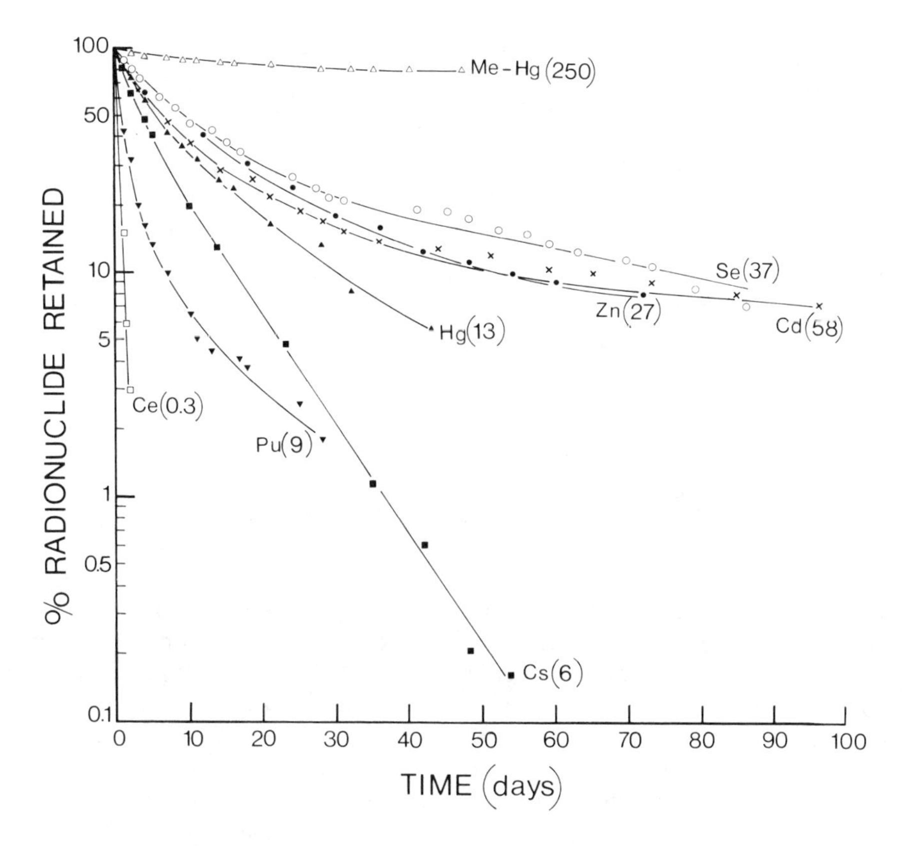

FIGURE 5. Long-term depuration of different elements from euphausiids chronically exposed to radiotracers in food and water. *Meganyctiphanes norvegica:* CH_3-^{203}Hg and inorganic ^{203}Hg from food and water;[19] ^{75}Se (selenite) from food and water;[21] ^{109}Cd from food and water;[18] ^{65}Zn from food and water;[147] ^{237}Pu from food and water.[20] *Euphausia pacifica:* ^{137}Cs and ^{144}Ce from food.[163] Numbers in parentheses are biological half-times in days which have been estimated for the slow component of loss.

water, these organisms lose roughly 20 to 40% of their ^{65}Zn body burden at molt.[145] For elements like Ce and Ru which have no biological function and are strongly sorptive, these percentages increase from 70 to nearly 100%.[8,69] When the contaminant is either in the food alone or in combination with the water, the fractions lost at molt are much lower[18,60,163] due to the relatively larger amounts bound to internal tissues.[114] It is interesting to note that even following absorption from water, Hg and Se accumulate in internal tissues so that only small fractions (3 to 6%) are associated with cast molts.[19,21] Temperature strongly affects the loss of pollutants from pelagic crustaceans, since the release rate of molts is directly related to temperature.[164,165] The total depuration of pollutant is further enhanced by the inherent rapid molting rates (every few days) in many of these small crustaceans.[164]

Evidence from radiotracer studies with planktonic crustaceans indicates that once metals have been assimilated into tissue, very little is transferred back into the gut and excreted with the fecal pellets.[18,19,21,60] This suggests that in the absence of pollution, these small organisms eliminate the majority of their contaminants by soluble excretion or direct exchange across body surfaces. This differs somewhat from zooplankton living in continued contact with pollution. For example, these organisms may approach an equilibrium with the surrounding element and, while soluble excretion still plays an important role in eliminating the assimilated contaminant from the tissues, the major fraction taken up with food often moves through the animal with the feces.[18,21,26,27,147]

Table 3
RELATIVE DISTRIBUTION OF ELEMENT EFFLUX[a] THROUGH THE EUPHAUSIID *MEGANYCTIPHANES NORVEGICA* BY MOLTING, DEFECATION, AND SOLUBLE EXCRETION

Element	Molting (%)	Defecation (%)	Soluble excretion (%)	Ref.
Zn	1.1	92.6	6.3	147
Cd	3.3	84.5	12.2	18
Se	2.4	54.4	43.2	21
Hg (inorg.)	2.5	29.1	68.4	19
$^{239,240}Pu$	0.8	98.6	0.6	27

[a] Calculations based on euphausiids grazing under sufficient food conditions during the non-egg-laying period.[147]

This occurs because much of the pollutant entering with the food remains associated with the unassimilated fraction and thus passes out with the fecal pellets in a more concentrated state. While fecal excretion may not include the fraction of the pollutant assimilated into tissue, nevertheless, it is in this way that zooplankton process, repackage, and eliminate the pollutant in a more concentrated form than was originally ingested. The relative importance of defecation compared to other elimination vectors is shown for four trace elements and $^{239,240}Pu$ in Table 3. The role played by zooplankton defecation in pollutant transport and redistribution is discussed in a later section.

A review of the literature shows that most of the zooplankton excretion studies have dealt with microcrustaceans. While microcrustaceans usually comprise the largest fraction of the zooplankton biomass, at times members of other taxa become dominant (salps, coelenterates, pelagic mollusks, etc.) and, yet, almost nothing is known about routes or rates of excretion in these organisms. In order to verify the applicability of microcrustacean excretion models to the zooplankton community as a whole, some effort should be expended on examining excretion kinetics of metals and radionuclides in noncrustacean zooplankton.

There is considerable information on metal and radionuclide elimination from macrobenthos. The loss of incorporated inorganics from macrobenthos varies widely with species. For example, macroalgae lose heavy metals at different rates, depending on circumstances. Several elements, including iron, are readily lost from contaminated macroalgae removed to clean environments[150,166] whereas little, if any, Zn is lost from the same species under similar conditions.[4,150,166] The excretion mechanisms in macroinvertebrates are generally the same as those in zooplankton; however, the loss rates are usually much slower commensurate with lower metabolic rates in larger species. Elimination patterns for many elements in a variety of species suggest two or more loss processes, each with a simple negative exponential clearing rate.[40,42,71,167,168] Biological half-times range from days to months depending upon the element and the way in which it is accumulated. Bivalves typically retain metals and radionuclides for a long time; biological half-times of roughly 1 year or more have been reported for some elements (Table 4). Crustaceans generally lose inorganic contaminants much more rapidly than mollusks. This is partly due to the large fraction of the pollutant lost when crustaceans molt.[32,35,42,65,66,71] Defecation has also been cited as an important mechanism for the elimination of heavy metals from benthic crustaceans.[4,173]

The manner in which crustaceans and other species are contaminated strongly influences elimination rates.[35,42,174] Loss of arsenic and plutonium from benthic shrimp

Table 4
BIOLOGICAL HALF-TIMES FOR ELIMINATION OF ELEMENTS FROM MUSSELS FOLLOWING ACCUMULATION FROM WATER AND FOOD; RADIONUCLIDES WERE USED TO FOLLOW WHOLE BODY LOSS

Species	Element	Form of radionuclide or metal used	Uptake route	Loss conditions	Total loss (%)	Tb½ (days)	Ref.
Mytilus edulis	Zn	^{65}Zn	Field	Field 14—22°C	100	76	169
	Zn	$^{65}ZnCl_2$	Water	Lab 14—17°C	≃25—70	48—60	170
	Co	$^{60}CoCl_2$	Water	Lab 14—17°C	30—70	57—72	170
Mytilus galloprovincialis	V	$^{48}VOCl_2$	Water	Field 13—15°C	40	103	313
	Cd	$^{109}CdCl_2$	Food & water	Field 15—19°C	78	307	65
	Se	$Na_2\,^{75}SeO_3$	Food & water	Field 18—23°C	22	81	66
	As	$Na_2H\,^{74}AsO_4$	Food & water	Field 20—22°C	23	32	72
	Hg	$^{203}HgCl_2$	Food & water	Field 13—16°C	51	82	40
	Hg	$CH_3\,^{203}HgCl$	Food & water	Field 13—16°C	99	63	40
	Pu	$^{237}Pu\ (+4, +6)$	Water	Field 13—24°C	≃30	190	171
	Am	$^{241}Am\ (+3)$	Water	Field 13—24°C	≃25	480	171
	Np	$^{237}Np\ (+3)$	Water	Field 13—24°C	≃50	81	172

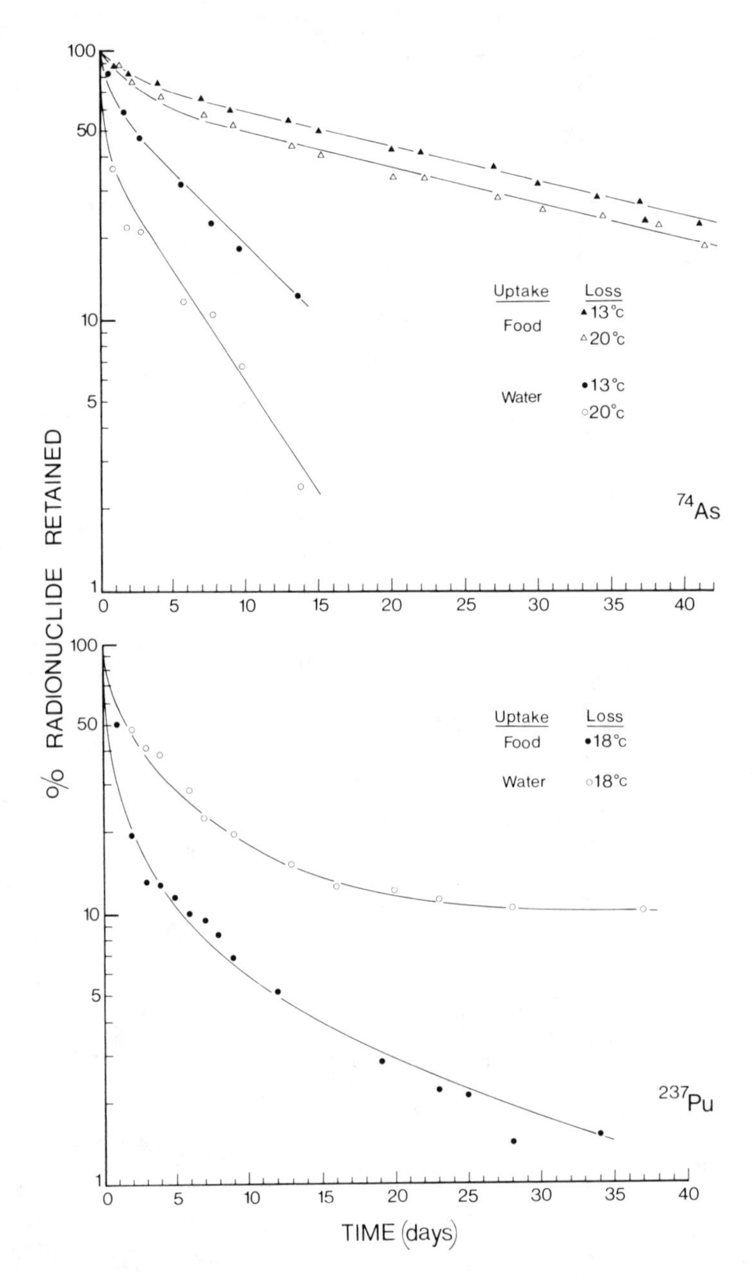

FIGURE 6. Elimination of As-74[42] and Pu-237[32] from the benthic shrimp *Lysmata seticaudata* following uptake from either food or water. (After the data of Fowler et al.[32] and Fowler and Ünlü[42])

(*Lysmata seticaudata*) which had accumulated the elements from food and water is shown in Figure 6. Arsenic, which is readily assimilated and accumulates in internal tissues, is retained to a much greater degree when uptake is through the food chain whereas plutonium, with its low assimilation efficiency, is lost more rapidly after similar uptake. The data for As (Figure 6) also illustrate how temperature can accelerate the loss of an element under certain conditions. This effect has also been observed for Co and Zn in shrimp[71] and can be explained in part by more rapid molting at higher temperature. However, some metals do not behave in this manner; for example, Cd and Se are lost from shrimp at rates which are independent of temperature and molting.[65,66]

Besides temperature, exposure time, and uptake pathway, other variables affect the elimination of inorganic pollutants. The total body burden,[4,46] chemical form of the element,[40,53,175] salinity,[176] and size[46] have all been cited as factors which affect metal flux in macroinvertebrates.

Biological half-times in fish are highly variable and depend on many of the same factors known to affect loss rates in other marine species.[5] Although single, double, and triple exponential models have been used to describe loss of radionuclides and metals from fish, elimination commonly follows a double rate function.[5,36,39,118-120,122] Loss from fish is rapid following ingestion of elements with low assimilation efficiencies such as Pu,[121] and slow for metals like methyl mercury which are assimilated and bound in interval tissue.[118] Methyl mercury is lost from the plaice, *Pleuronectes platessa*, much more slowly than inorganic mercury.[36,39,118] Following uptake from food the biological half-time for methyl Hg averages 163 days as opposed to only 33 days for the inorganic form. Likewise, when the metal is accumulated from water, half-times average 275 and 135 days for methyl and inorganic Hg, respectively. One of the critical factors determining loss in fish may be species. For example, Pentreath[39,118,120-122,293] has shown that the thornback ray, *Raja clavata L.*, retains more metals and radionuclides than plaice regardless of the uptake pathway. Silver-110m ingested by plaice is poorly retained and eliminated with a half-time of about 12 days, whereas rays assimilate relatively large fractions of the metal and lose it very slowly ($Tb_{1/2} \simeq$ 315 to 1386 days).[120] Similar differences have also been noted for Pu, an element which generally is poorly assimilated.[121,122] These differences may be typical for elasmobranchs and teleosts in general, however, more data from other species within each group are needed for confirmation.

2. Organic Pollutants

Little is known about clearance rates for either chlorinated or petroleum hydrocarbons in phytoplankton, but they are likely slower than accumulation rates since internal binding is involved.[81,83,333] Metabolites of the accumulated compound might also give rise to different elimination rates depending upon the compound. Most probably in living cells the greatest reduction of incorporated hydrocarbon residue levels takes place during cell division by dilution with new cells.[86]

Biological half-times on the order of a few days appear to be typical for chlorinated hydrocarbons in small zooplankton. Following a 2-week exposure to DDT in sea water, loss of the residues from copepods was described by a single exponential loss rate with a biological half-time of about 14 days.[93] The euphausiid *Thysanoessa raschii* (13 to 19 mg dry), similarly exposed to DDT but for only 3 hr, subsequently eliminated the compound at almost the same rate ($Tb_{1/2}$ = 16 days).[290] However, in another study[92] smaller euphausiids (*Euphausia pacifica;* 3 to 10 mg dry) exposed to DDT for 2 hr lost the residues much more rapidly than either the copepods or *Thysanoessa*. In addition, the smaller *Euphausia* excreted DDT at higher rates than larger individuals of the same species. Although the species used in these experiments were different, the results taken as a whole suggest that clearance rates in zooplankton depend on both size of individuals and the duration of exposure. Furthermore, there is evidence that overall metabolic health of the organism is critical in determining organochlorine depuration rates.[290]

Elimination of chlorinated hydrocarbons from planktonic crustaceans will also be enhanced by molting. Chitinous molts of euphausiids contain high concentrations of PCBs relative to levels in the whole animal.[178] Thus, with a short intermolt period of about 1 week,[164] the total elimination of these compounds from euphausiids and presumably other planktonic crustaceans will be accelerated. The length of the euphausiid intermolt period depends on the size of the individual, with smaller euphausiids of a given species molting more frequently than larger individuals.[164] Elimination of DDT by molting could partially explain the observation of rapid DDT loss rates in smaller euphausiids cited above. A second important route of chlorinated hydrocarbon excretion is through the release of fecal pellets. While zooplankton feces contain significant concentrations of PCBs,[178] it is unclear whether these levels arise from residues remaining with undigested food or result from direct excretion from the organism's tissues into the gut. Nevertheless, since the amount of fecal material is greater in these animals than that produced from molts, defecation would appear to be more important than molting in eliminating organochlorine body burdens.

Petroleum hydrocarbons appear to be relatively persistent in zooplankton. Loss rates for aromatic hydrocarbons in zooplankton are not constant but diminish with time.[88,96,333] Although biological half times for major pools of incorporated aromatic hydrocarbons may be only 2 to 3 days, even after several weeks of depuration some residues still remain with the zooplankton.[96,97] Furthermore, the depuration rate is slower when aromatics are accumulated via the food chain than following uptake from sea water alone.[88] Since zooplankton are known to metabolize petroleum hydrocarbons[88,96] their loss from zooplankton should increase at higher temperatures, principally because of the commensurate increase in rate of hydrocarbon metabolism.[97] In addition, unfed zooplankton would tend to lose incorporated hydrocarbons more rapidly due to enhanced utilization of lipid reserves under conditions of starvation.[97] No information exists on the importance of molting in eliminating petroleum hydrocarbons from planktonic crustaceans; however, it may not be as important as with chlorinated hydrocarbons since recent studies suggest that surface adsorption is not a major factor in the retention of aromatic hydrocarbons.[96,97] On the other hand, zooplankton fecal excretion does appear to be of prime importance in the elimination of both oil particles and the water-soluble hydrocarbon fractions of crude oil.[88,97,128,129] Some studies have shown that copepods can eliminate as much as 40% of ingested naphthalene in fecal pellets;[88,97] however, it is not clear how much, if any, of the assimilated fraction is excreted in this manner.

As with zooplankton, chlorinated hydrocarbons are lost from benthic invertebrates at slower rates than they are accumulated.[84] Furthermore, loss rates measured over a long duration are not constant but vary both with time[98,132,312] and initial concentration in the organism.[103,179] Depuration half-times range from days to weeks depending upon the compound and the organism. Oysters contaminated with Kepone lost 35% in 1 day and all of the insecticide within 7 to 20 days, whereas shrimp eliminated only 30 to 50% in 24 to 28 days.[156] Some studies indicate that PCBs are more persistent in marine biota than certain pesticides. For example, oysters and clams can eliminate 50 to 90% of incorporated DDT residues in one week.[180] Biological half-times of about 3 days have been reported for dieldrin and endrin in similar species.[179] In contrast, PCBs are lost from benthic worms,[103,132] shrimp,[102,181] and mollusks[98,182,312] much more slowly, with half-times on the order of weeks. Loss does not always proceed at similar rates from the different tissues. PCBs appear to depurate mainly from shrimp hepatopancreas, whereas the remaining tissues lose little of the residues.[181] On the other hand, lobster hepatopancreas was found to eliminate PCBs more slowly ($Tb_{1/2}$ = 45 days) than other tissues examined.[183] Similar experiments with contaminated clams have shown that PCBs are effectively lost from foot muscle but not from the visceral mass.[182]

Of particular importance in the elimination of PCB compounds is the ability of some species to preferentially excrete individual homologues of different chlorine content. Figure 7 shows the elimination of di- , tri- , and pentachlorobiphenyl from the worm *Nereis virens* following 3 weeks of ingestion of contaminated food. The higher chlorinated isomer was far more persistent than either the di- or trichloro compounds;

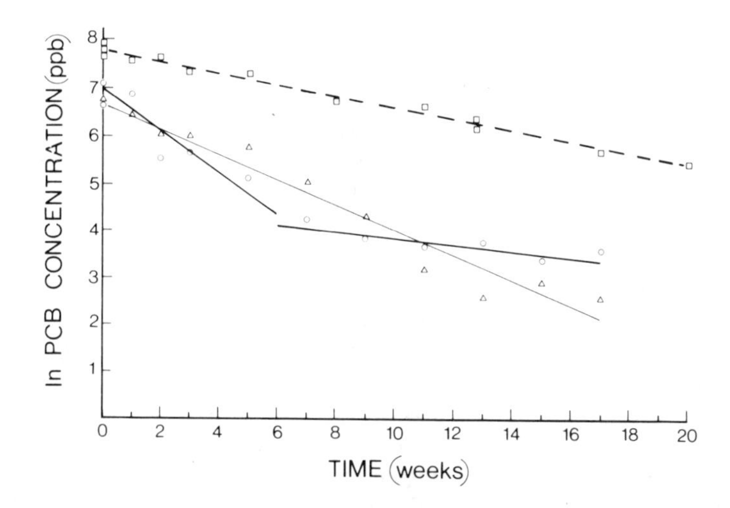

FIGURE 7. Elimination of PCB compounds from the worm *Nereis virens* following ingestion of contaminated food for 3 weeks. □: 2,4,6,2′,4′-pentachlorobiphenyl; ○: 2,5,4′-trichlorobiphenyl; △: 2,2′-dichlorobiphenyl. (From Goerke, H. and Ernst, W., *Chemosphere*, 6, 551, 1977. With permission.)

biological half-times for elimination were 3 and 8.7 weeks for the di- and pentachlorobiphenyl, respectively. Loss kinetics of the trichlorobiphenyl compound followed a double exponential rate with half-times of 1.8 and 10 weeks for the fast and slow component. Similar studies with bivalves[184] have demonstrated virtually no elimination of incorporated hexachlorobiphenyl, whereas dichlorobiphenyl is lost relatively rapidly, more so at higher (15°C, $Tb_{1/2}$ = 5 days) than at lower temperatures (8°C, $Tb_{1/2}$ = 7 days).

Fecal excretion is considered a primary route for the elimination of chlorinated hydrocarbons from many benthic species.[182,183,185] Clams can excrete higher chlorinated PCB compounds in their feces, while retaining the lower chlorinated isomers in their tissues.[182] Similar studies[185] have shown that the worm *N. virens* can considerably degrade ingested PCBs and excrete the metabolites in their feces, with the higher chlorinated pentachlorobiphenyl being defecated more efficiently than either the di- or trichloro compounds.

Generally speaking, a large fraction of incorporated petroleum hydrocarbons is rapidly lost from contaminated benthic invertebrates when transferred to clean environments.[109,110,291,315,325] This fraction will probably be a maximum following acute exposures. Bivalves contaminated with petroleum hydrocarbons for only a few hours were observed to rapidly eliminate these compounds with half-times on the order of 1 to 5 days.[105,107,315] However, it is clear that these elimination rates depend upon the absolute concentrations of petroleum accumulated as well as the concentration differential between levels in the organism and the water.[109] The situation differs following chronic contamination, as there is a tendency for slower depuration rates after longer exposure periods.[109,134,135,186,318] In this case, generally an initial rapid discharge with a short biological half-time is followed by a much slower loss from the fraction of petroleum hydrocarbons associated with a more stable compartment within the organism. For example, oysters,[106] clams,[318] and mussels[134,186] chronically exposed to hydrocarbons retain substantial fractions which are not significantly eliminated over a period of several weeks. Benzo[*a*]pyrene is lost from environmentally contaminated mussels with a biological half-time of 16 days.[135] Furthermore, this polycyclic aromatic appears to be much more persistent in bivalves than the naphthalenes.[108] Biological half-times for loss of various petroleum hydrocarbons from bivalves under different conditions have been recently summarized by Lee.[109] Values range from 2 to 7 days following short-term exposure and reach as high as 60 days for bivalves chronically exposed. Normally, paraffins are discharged from mollusks at a faster rate than aromatics, and within the latter group, naphthalene is eliminated first followed by methyl-naphthalene and then the dimethyl- and trimethyl-isomers.[108,187] In some cases, the rapid loss of paraffins may be linked to their incorporation in feces or pseudofeces and subsequent defecation.[108,187] These differences indicate that factors such as molecular weight and polarity can also affect the elimination rate of petroleum hydrocarbons in these species. Regardless of the route of excretion these compounds are eliminated from bivalves essentially unchanged, since these mollusks apparently have only limited ability to metabolize petroleum hydrocarbons.[105,135]

Crustaceans tend to eliminate petroleum hydrocarbons faster than other species in part because they can metabolize them.[108,109,111] Lee et al.[111] have shown that crabs which accumulate hydrocarbons from either water or food rapidly lose these compounds, with defecation providing the major pathway for elimination. Loss rates vary for different crustacean tissues; shrimp exposed to water-soluble aromatics excrete naphthalenes most rapidly from muscle and exoskeleton.[108] In comparison, elimination is relatively slow from shrimp hepatopancreas, the tissue with the highest concentration of aromatics after contamination. Similar results were noted for crabs, whereby 20 days after exposure the benzo[*a*]pyrene was almost completely eliminated except for that in the hepatopancreas.[111] In view of available information on organochlorine compounds, molting is a likely mechanism by which benthic crustaceans can rapidly reduce whole-body petroleum hydrocarbon levels; however, data are needed to confirm this pathway.

Benthic worms also can metabolize petroleum hydrocarbons and, consequently, eliminate them fairly rapidly.[109,326] Polychaetes[136] and sipunculids[188] exposed to naphthalenes eliminated virtually all the compounds within 2 weeks. Because assimilation of ingested petroleum hydrocarbons in these worms is relatively poor, fecal excretion is probably a major pathway for the elimination of these compounds.[136,188]

Elimination rates of organochlorine compounds in fish are generally slower than accumulation rates.[84] Biological half-times for clearance vary from hours[189] to weeks[156,190,292] depending upon the species, the contaminated tissue pool within the animal, and the compound. For example, studies[189] of skates injected with PCBs show that the 2- and 4-chlorobiphenyl isomers are eliminated from plasma much more slowly than the 3-isomer, the latter compound being excreted into the water with half-times of only minutes. In contrast, fish chronically exposed to kepone in water eliminate this insecticide only very slowly, e.g., 30 to 50% in 24 to 28 days.[156] Evidence from fresh-water fish indicates that organochlorines tend to be lost in the order lindane > dieldrin > DDT > PCB,[84,190] suggesting that loss is largely governed by the water solubility of the compound. Furthermore, biliary and fecal excretion appear to be predominant pathways for their elimination. Fish clearly can metabolize DDT to DDE but metabolism is not a prerequisite for loss, since fish tissues and feces contain the parent compound as well as the metabolite.[84,292] Less information exists on the ability of marine fish to metabolize PCB; however, indirect evidence suggests that metabolism occurs to a limited degree in certain species.[84]

There is evidence[113] that marine fish, like crustaceans, metabolize petroleum hydrocarbons and excrete them as water-soluble metabolites. In addition, elimination rates are slower than uptake rates.[108,110] Among the aromatic compounds, naphthalenes are eliminated from fish more rapidly than benzo[*a*]pyrene and its metabolites.[113] Fish tissues also lose these compounds at different rates. Following acute exposure of killifish to the water-soluble fraction of No. 2 fuel oil, naphthalenes were slowly lost from

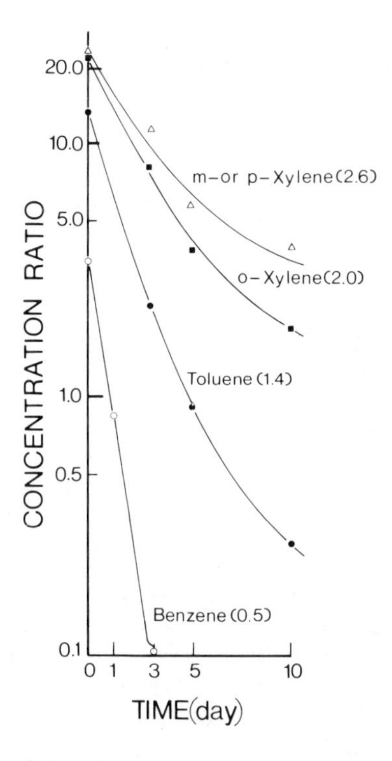

FIGURE 8. Depuration of aromatic hydrocarbons from the eel (*Anguilla japonica*) following a 10-day exposure in sea water. Numbers in parentheses are biological half-times (days) based on calculations from semilogarithmic plots of the data. (From Ogata, M. and Miyake, Y., *Water Res.*, 12, 1041, 1978. With permission.)

gall bladder and brain but were rapidly eliminated from gill and muscle.[108] Similar experiments with eels[191] demonstrated the rapid loss of incorporated benzene, followed by toluene, m- or p-xylene and o-xylene, in that order (Figure 8). Excretion of aromatics takes place via the gills,[191] from fish urine,[113] and feces,[109]

Little information exists on the metabolism and excretion of n-alkanes in marine fish; however some evidence suggests that selective excretion of individual hydrocarbons can occur.[141] Chronically contaminated fish may lose these compounds from compartments with slow turnover rates. For example, during migration naturally contaminated mullet eliminated n-alkanes (C9 to C13) with a relatively long half-time of 18 days.[192]

In most long-term elimination studies with fish and other organisms, loss rates for petroleum hydrocarbons diminish with time. While loss rates derived from acute exposures are useful for describing the kinetics of physiological pools with rapid half-times, they may be completely unrealistic in determining total depuration times for chronically contaminated marine species. For this reason, biological half-times from long-term studies using organisms contaminated from both food and water are necessary to determine the total residence time of these compounds in biota.

E. Food Chain Magnification

Pollutant levels measured in marine organisms are the net result of accumulation, metabolism, storage, and elimination processes acting in concert. Regardless of the many physiological and environmental conditions which regulate the pollutant content within an animal, contaminants are often visualized as moving along a food chain from prey to predator with concentrations either decreasing or increasing (biomagnification) in the organism's tissues at each successive trophic level. Biomagnification might be expected to occur if a pollutant is readily assimilated from food and subsequently retained for long periods in the predator's tissues. This situation could lead to a net increase in pollutant body burdens with time, resulting in a concentrated source of the contaminant being passed on to top predators such as man. The basic assumption here is that pollutants are passed only from prey to predator but as was shown in Section II.C, this is not always the case. A test of the biomagnification hypothesis is difficult to carry out experimentally and, therefore, most evidence on accumulation up the food chain is derived from comparisons of pollutants in organisms from different trophic levels. Several aspects pertaining to the biomagnification concept are highlighted below.

1. Inorganic Pollutants

Levels of heavy metals[2-4] and radionuclides[1-3,5,152,334] have been comprehensively examined in a wide variety of feeding groups, but there is little evidence for biomagnification at higher trophic levels. For contaminants with elimination rates commensurate with accumulation rates, biomagnification normally cannot occur. Following ingestion, a combination of factors such as incomplete absorption across the gut, rapid excretion via feces, and dilution in tissues like muscle which accounts for a large fraction of the organism's total weight results in lower pollutant levels in the whole predator per unit mass than in its food. Despite the vast number of studies showing a decrease in radionuclide or metal concentration along the food chain, there are exceptions in specific food chains where predators do contain higher levels than their prey.

Cesium-137 is often cited as a radionuclide which preferentially accumulates in higher trophic level fish.[1,5] It has been shown that although roughly 50% of the ^{137}Cs body burden of certain fish is obtained from water,[152,294] the high degree of assimilation of ^{137}Cs from prey results in an overall accumulation up the food chain. This effect is thought to be due to the high percentage of body weight (> 50%) represented by muscle coupled with its relatively high cesium concentration.[5]

Peden et al.[193] surveyed metals in a variety of estuarine species and noted that dog whelks contained 3 times more cadmium and 4 times more zinc than the limpets upon which they were feeding. Subsequent depuration experiments showed that the metals were strongly retained by the whelks which suggested that, in this particular mollusk, cadmium and zinc uptake exceeded elimination. However, a recent laboratory study[319] in which dog whelks were fed contaminated barnacles failed to confirm a metal magnification trend for either zinc or iron in this food chain.

In another field study[194] starfish feeding upon the mussel *Mytilus* contained four times more plutonium than their prey. Similar ^{238}Pu to ^{239,240}Pu ratios in both organisms implied transfer via food, and it was concluded that plutonium was biomagnified along this simple food chain. This hypothesis was further strengthened by laboratory experiments[195] which showed that starfish ingesting mussels contaminated with ^{237}Pu assimilate and retain large fractions of the radionuclide. However, more recent ^{237}Pu studies[29] have revealed that starfish also rapidly take up plutonium from water to relatively high levels and eliminate it very slowly. Furthermore, the resultant tissue distribution of ^{237}Pu following uptake from water closely resembles the natural distribution of ^{239,240}Pu in tissues of starfish contaminated by fallout. These results, implicating water as an important source of plutonium for starfish, illustrate the inherent difficulties in drawing conclusions on food chain biomagnification without prior knowledge about the uptake pathway.

Table 5
METHYL MERCURY AS PERCENTAGE OF TOTAL MERCURY IN THE MUSCLE OF FIN FISH FROM THE SOUTHEASTERN U.S.[197,198]

Trophic level	Common name	Species	Methyl mercury (%)
Secondary consumer	Fringed flounder	*Etropus crossotus*	10
	Spot	*Leiostomus xanthurus*	80
	Atlantic croaker	*Micropogon undulatus*	66
	Star drum	*Stellifer lanceolatus*	54
			$\overline{X}$ = 53 ±30
Tertiary consumer	Spotted sea trout	*Cynoscion nebulosus*	100
	Silver perch	*Bairdiella chrysura*	85
	Ocellated flounder	*Ancyclopsetta quadracellata*	89
	Bay whiff	*Citharichthes spilopterus*	94
			$\overline{X}$ = 92 ±6

Only in the case of mercury do the data appear to clearly support the biomagnification hypothesis. It is now generally accepted that top level predators such as large piscivorous or omnivorous fish contain the highest levels of mercury, most of which is methylated. Unlike chlorinated hydrocarbons, methyl mercury in fish is bound to protein rather than dissolved in lipids. Consequently, mercury concentrations in fish are not lipid-dependent but are more closely related to trophic level and fish size or age.[196] This is borne out by fish studies which show an increase in the methylated fraction of total mercury with trophic level (Table 5). Using a trophic level approach, Ratkowsky et al.[199] examined total mercury levels in fish as a function of their diet and found the highest concentrations in vertebrate predators with lesser amounts in invertebrate predators and herbivores, in that order. Assuming that much of the mercury is in the methyl form and that it is indeed obtained via the fish's food (fish cannot methylate inorganic mercury in vivo[36,295]), it is easy to envisage progressive accumulation along the food chain given the preferential absorption and extremely long biological half-times for methyl mercury in fish.[4,118] However, it should be borne in mind that methyl mercury is also readily absorbed from sea water by fish although the extent to which this process occurs in nature is not known.[39,197] Cross et al.,[43] noting a positive correlation between size and mercury content, considered that age rather than trophic position influences the mercury levels in fish. They hypothesized that mercury uptake, by whatever route, is more rapid than elimination which would result in older fish containing higher Hg levels. As the older, larger fish are often also the top-level carnivores, the end result of high Hg concentrations in older fish is similar to that found when mercury levels in fish are ranked strictly according to trophic level. Which of these two processes — age/size or diet — predominates in nature is only speculation at this point; however, it is highly likely that both contribute to the mercury levels observed in fish.

Examination of many organisms from lower trophic levels has failed to turn up comparable evidence for biomagnification of mercury.[200,201,297] However, using a different approach, Windom et al.[198] calculated mercury transfer efficiencies from phytoplankton through tertiary consumer fish and found that mercury was more efficiently transferred to predators at each successive link up the food chain. While this method holds promise for examining certain aspects of biomagnification along food chains, its basic assumption that mercury input occurs only through food is questionable in light of recent studies[19,202] indicating that uptake from food may not be the principal pathway for mercury accumulation by small pelagic species.

Thus, at the trophic levels where biomagnification may occur, the observed increased concentrations of mercury almost certainly result from transfer of the methylated form. This appears to be the case for fish but virtually no information exists for lower trophic level species. More detailed information on ratios of methyl mercury to total mercury in organisms other than fish from different trophic levels would help clarify this aspect. Furthermore, if it is shown conclusively that methyl mercury does, in fact, biomagnify at the higher trophic levels, it would prove instructive to examine the behavior of other toxic elements (e.g., arsenic, selenium, and lead) which are also known to occur as organic compounds in the environment.

2. Organic Pollutants

Perhaps the most convincing evidence put forth in support of the pollutant biomagnification hypothesis comes from aquatic studies on the fate of organochlorine compounds. Interest in the possibility of food chain magnification has arisen from numerous reports of DDT and PCB concentrations increasing up the food chain and correlations between reproductive failure in fish-eating birds and high levels of these organochlorine compounds in their tissues.[203-205] However, as was pointed out in previous sections, there are two views regarding how marine organisms accumulate organochlorine compounds: (1) uptake through their diet and (2) partitioning of the compound between lipid and water phases. Thus, any discussion of organochlorine biomagnification in food chains must consider the relative importance of these uptake mechanisms at various points in the food web.

Where higher trophic level species such as fish, mammals, and birds have been examined, definite increases in PCB and DDT concentrations (dry weight, wet weight and lipid weight basis) from prey to predator have been noted.[204,334] Furthermore, there is evidence[139,204] that the percentage of PCB compounds with higher degrees of chlorination increases up the food chain from prey to predator. This suggests that lesser chlorinated PCBs are metabolized or more rapidly excreted than the higher chlorinated compounds. If partitioning between lipids and water were primarily responsible for uptake, one would expect residue concentrations on a lipid weight basis to remain fairly constant. The fact that organochlorine residue concentrations in lipids[204-206,296] show an increase, although not as much as concentrations based on fresh or dry weight, suggests that phase partitioning is not the sole mechanism involved in bioaccumulation. Certainly in fish there is clear evidence[84,156] that organochlorine residue uptake from food is more rapid than elimination; hence, residue burdens tend to increase with time and selectively accumulate at higher trophic levels.

The situation is less clear in organisms from the lower trophic levels. Harvey et al.[90] surveying a large number of pelagic species from the Atlantic ocean, found no conclusive evidence for food chain magnification. They noted, however, that PCB/ΣDDT ratios decreased from about 30 in water and plankton to 3 in high predators, indicating that despite similar molecular structure, DDT and PCB compounds move differently in marine food chains. Their survey was broad in both time and space, and thus simple comparisons of residue levels in organisms from different trophic levels may be confounded by spatial and temporal variations in the environment. Other studies, which tried to eliminate these variables by intensive sampling at a single station or studying specific food chains, have given conflicting results. Baird et al.[207] analyzed pelagic species from the Gulf of Mexico and at one station found that organochlorine levels in zooplankton were one to two orders of magnitude less than those in mesoplagic predators. In contrast, a similar study in the Mediterranean[91] found 10 to 100 times less PCBs in myctophid fish than their zooplankton prey. These studies assumed that uptake at each level is principally through food, whereas recent experiments[94,155] suggest that water is the primary source of chlorinated hydrocarbons, at least for small zooplankton. If this is generally true for small pelagic organisms, then the observed differences noted in the studies cited above may be due in part to local variations in the atmospheric input of organochlorines. These variations would be rapidly reflected in zooplankton but would take longer to appear in higher trophic levels.

Until the relative importance of the pathways for organochlorine uptake are clarified, it will be difficult to draw any definite conclusions regarding their biomagnification. For the moment we can speculate that plankton and small invertebrates, which accumulate substantial amounts of these compounds from water, will contain levels closely reflecting those in their surroundings. Vertebrates, which are generally larger and have relatively less surface area for absorption, tend to accumulate most of their organochlorines from food, and therefore their levels depend more on trophic level than on water content.

Unlike the case for the more persistent chlorinated hydrocarbons, there is no convincing evidence that petroleum hydrocarbons biomagnify in marine food webs.[110,158,159,208,325] Lack of biomagnification is probably a result of two important factors. First, much evidence suggests that direct uptake from sea water across respiratory surfaces is the predominant pathway for petroleum hydrocarbon accumulation in most marine organisms, including higher fish. Hence, when certain top predators have higher hydrocarbon contents than their prey,[159] the apparent food chain magnification may result more from the species' ability to absorb these compounds from water than their relative positions in the food web. Second, although elimination generally proceeds at a slower rate than uptake, biological half-times for petroleum hydrocarbons in most species tested are extremely short with near total depuration often occurring in a few hours or days. In addition, the fact that top predators such as benthic crustaceans and fish usually metabolize and eliminate petroleum hydrocarbons rapidly further reduces the likelihood that these compounds will concentrate at the higher trophic levels. Nevertheless, data for assessing food chain transfer and magnification of petroleum hydrocarbons are considerably fewer than for other pollutants. Considering the many individual compounds and metabolites involved, further study is prudent before drawing general conclusions on the behavior of the petroleum hydrocarbons.

III. ECOLOGICAL ASPECTS

Pollutants put into the ocean by land runoff, atmospheric fallout, or dumping rapidly associate with the biota. As a result of different physiologies and ecologies of marine species, the pollutants are subsequently distributed throughout the marine environment in a variety of ways. Figure 9 outlines schematically the principal biological pathways of pollutants in the sea. In the following sections these pathways are discussed with respect to their importance in the biogeochemical cycles of various types of pollutants.

A. Horizontal and Vertical Migration

Large mobile organisms like fish moving in and out of polluted waters can transport pollutants away from their point of incorporation. For example, it is easy to envisage contaminated anadromous fish leaving a polluted estuary and being ingested by some predator or man hundreds or thousands of kilometers away from the polluted area. This pathway of biological transport is most important for pollutants with long biological half-times; however, to what extent this process is really effective has been little studied. Pollutants which rapidly associate with sediments and benthic organisms are, in the short term, probably little affected by horizontal migration. For example, Aarkrog,[209] studying an area near Thule, Greeland that had been accidentally contaminated

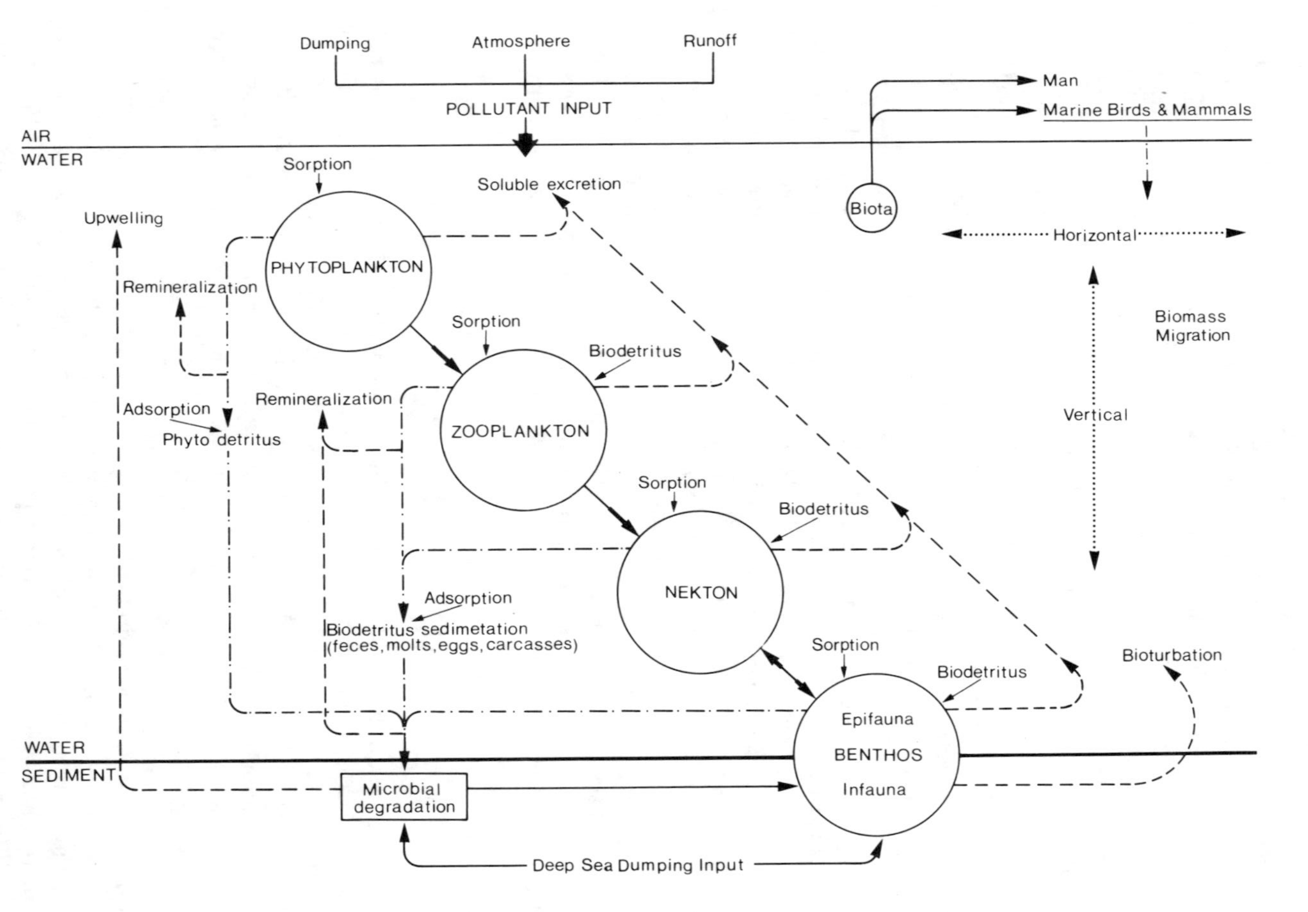

FIGURE 9. Scheme of marine pollutant transfer and transport by biological processes.

with plutonium, found contaminated biota and sediments as far as 15 km from the point of impact a few months after the accident. Two and six years later the radius of biological contamination had spread to approximately 30 and 40 km, respectively. While some highly mobile contaminated fish may have migrated much further, by and large dispersion of plutonium in sediments by horizontal movements of organisms appears to be a relatively slow process. On the other hand, contaminants normally associated with the aqueous phase and hence more available for bioaccumulation may be more readily dispersed by moving organisms. For example, high concentrations of cadmium (100 to 200 $\mu g\ g^{-1}$ dry) have been found in the epineustonic seaskater, *Halobates,* which lives on the pollutant-enriched surface microlayer.[327] It is thus conceivable that many wind-driven epineustonic forms such as these insects, *Sargassum* weed, and jellyfish may be able to transport pollutants over considerable distances. However, it seems clear that for either active or passive horizontal movement to be an effective biological mechanism for pollutant dispersion vis à vis physical processes, a large portion of the total biomass must be involved. Since present information indicates that only a relatively small fraction of the biomass — mainly certain fish species — migrate long distances, horizontal migration will be of only limited importance in pollutant dispersion, except perhaps under certain unique circumstances.

Contrary to the case of horizontal migration, many species of zooplankton and nekton, constituting a large fraction of the oceanic biomass, migrate vertically. Vinogradov[210] visualizes these migrations as a series of overlapping "ladders," whereby animals feeding in the surface layers at night descend to depth and are eaten by organisms from deeper waters which, in turn, descend to still greater depths where they become the prey of organisms inhabiting the deepest layers. If this is correct, pollutants introduced into the surface layers could thus be actively transported downwards by predation between overlapping vertical migrants to depths of at least 4000 to 5000 m. Although diel vertical movements do occur in all planktonic phyla, and in some cases to relatively great depth, some authors[3,211] feel that this mechanism has been given undue emphasis since migratory species only constitute a small fraction of the total planktonic biomass. Longhurst,[211] after a thorough review of the literature, concluded that there was little evidence for diel vertical migration of epipelagic zooplankton below roughly 1000 to 2000 m. He further noted that information on deep vertical migration was inadequate to draw any firm conclusions about its importance. It was suggested that in the absence of any large-scale upward vertical migration in deep waters, abyssopelagic species obtain most of their nourishment from sinking biodetritus, even though some upward movement of larger organisms into the base of the pycnocline to feed on settling phytoplankton probably occurs. Lowman et al.[3] attempted to evaluate the relative effectiveness of vertical transport of elements upward from deep waters by both biological and physical processes. Reworking earlier data[212] in light of new information, they calculated that concentration factors in deep-dwelling zooplankton would have to be about 10^4 to 10^5 for biological transport upward to equal physical transport. As concentration factors rarely exceed 10^4, it was concluded that physical transport is generally more effective than biological transport in moving radionuclides from depth to the surface.

Although it is difficult to attribute pollutant transport to extreme depth solely to vertical migration, one can readily envisage organisms contaminated in the surface layers descending across physical boundaries such as a thermocline or pycnocline in the upper few hundred meters where vertical migrations do occur. Studies to quantify pollutant translocation across these interfaces are difficult and, accordingly, are few. Pearcy and Osterberg[213] collected macrozooplankton and micronekton from different depths off Oregon in an area contaminated by radionuclides from Columbia River runoff. Based on ^{65}Zn concentrations in migrants and nonmigrants in the 0- to 1000-

m water column, they estimated that roughly 40% of the ^{65}Zn in the organisms sampled was transported twice daily through the base of the pycnocline at 150 m. The euphausiid *Euphausia pacifica,* one of the principal migrators in this region, contained relatively high levels of ^{65}Zn throughout the year, leading Osterberg et al.[214] to suggest that biological transport of zinc across the pycnocline by these organisms was more important than by physical processes. These authors noted that because of vertical migration, a major loss of ^{65} Zn from the mixed later would be caused by predation on euphausiids below the pycnocline. These suggestions led others[54,145] to examine the mechanisms by which *E. pacifica* excreted ^{65}Zn above and below the pycnocline. Comparable data for hydrocarbons are not available; however, it is likely that vertical migration will move major fractions of these materials across discontinuity layers since biological half-times of compounds like DDT in euphausiids[92] are long (days) relative to the time spent migrating (hours).

Kuenzler[161,162] studied biological transport in the eastern Pacific Ocean where a low-yield nuclear test contaminated surface waters with radioactivity. He found that only about one fourth of the zooplankton biomass in the top 500 m migrated out of the mixed layer carrying down and eliminating cobalt below the thermocline. At this rate, the quantity of cobalt in the top 100 m would be removed in about 56 years. This biological transport rate was about the same as the computed rates for vertical eddy diffusion, thus substantiating an earlier hypothesis[212] that migrating zooplankton were instrumental in transporting material downward through physical boundaries. Significantly, these studies[161,162,213,214] not only considered that zooplankton descending through the discontinuity layer could transfer their pollutant load by soluble excretion or as prey, but also recognized the mechanisms of particulate release (e.g., fecal pellets, molts, eggs, and carcasses) as being of importance in the transport process. More recent data now allow assessing the relative importance of the individual biological processes in the downward transport of a number of pollutants by zooplankton.

B. Particulate Transport

While active migration is undoubtedly important in transporting pollutants vertically in the upper water layers, it probably cannot account for the presence of inorganic and organic pollutants[215-220] found in organisms at great depth. The rapid sinking of large biogenic particles such as fecal pellets, crustacean molts, eggs, carcasses, and shells from zooplankton in the upper water column, offers an attractive hypothesis to explain the influx of pollutants to depth.

1. Direct and Indirect Evidence

Osterberg et al.[215] found that sea-cucumbers at 2800 m in the northeast Pacific Ocean contained the short half-lived fission products ^{95}Zr-^{95}Nb and $^{141,144}Ce$. The presence of these isotopes at great depth was impossible to explain solely on the basis of the Stokesian settling of fallout particles, and the authors hypothesized that zooplankton grazing in the contaminated surface layers packed the radionuclides in fecal pellets which, because of their greater size and density, rapidly sank to depth. Differences in ^{95}Zr-^{95}Nb concentrations in deep sea-cucumbers and those from shallow depths indicated a transit time of 7 to 12 days from the surface to 2800 m. Later work[221,222] showed that natural fecal pellets from larger copepods and euphausiids sink at rates similar to those calculated by Osterberg et al.[215]

Zooplankton molts have also been proposed as effective carriers to depth of heavy metals,[28,223,224] radionuclides,[8,20,27,145] and chlorinated hydrocarbons.[92,178] This process is most important for small pelagic forms which molt frequently throughout their life span.[164] The release of eggs is another means by which zooplankton can redistribute pollutants; however, its importance relative to defecation and molting is limited since egg production is normally a discontinuous process.[147] A final pathway for the downward biological transport of pollutants is the sinking of dead organisms. The relative importance of fecal pellets, molts, eggs, and carcasses in moving pollutants to depth depends on production rates, sinking rates, pollutant concentrations in the materials, and rates of pollutant remineralization as the debris sinks.

Evidence in support of sinking biogenic debris as an important mechanism in the vertical transport of contaminants comes from several studies. Coupling estimates of feeding rates, trace-metal assimilation efficiencies, and trace-metal concentration factors in zooplankton and phytoplankton with previously published data,[161,162] Lowman et al.[3] estimated the role of zooplankton and their particulate products in the downward vertical transport of several elements and radionuclides under different oceanic conditions. Using iron as an example they estimated that fecal pellets, molts, other excreted particulates, and dead organisms accounted for over 90% of the iron transported across the thermocline in the northeastern Pacific. This suggested that $^{55,59}Fe$ in surface waters would be transported out of the upper mixed layer in about 7.8 years by sinking particulates and carcasses but would take 90 years by vertical migration alone. Calculations for several other radionuclides gave similar results. Although these transport times appear long relative to physical transport, the computations were for an open ocean area with a relatively low standing crop of zooplankton (0.04 cc m^3). The authors pointed out that in coastal regions and areas of upwelling, where typical biomass estimates were 0.44 and 1.0 cc m^3, respectively, biological transport rates become much more rapid and represent a far greater proportion of the total transport. Davies[9] subsequently used these data to calculate the fraction of the elements removed annually by biological processes from the upper layers of the oceanic regions studied by Lowman et al.[3] and compared these values with fractions of the same elements lost each year by geochemical sedimentation. Comparisons suggested that Mn, Fe, Zn, Co, Zr, and Pb would be removed from the upper mixed layer more by sinking biodetritus than by either vertical migration or geochemical processes.

Although good first approximations, the estimates of Lowman et al.[3] are somewhat limited by lack of basic data on production rates of biogenic particulates, element concentrations in particulates, and the effects of metabolism on element uptake and excretion. Later works[18,19,21,27,147] considerably refined the estimates of Lowman et al.[3] by examining the flux of several elements through a single euphausiid species, *Meganyctiphanes norvegica.* These studies used element concentrations in the various particulate products along with production rates of the particulates to arrive at estimates of the relative importance of each mode of release for a given element. The key to this approach was obtaining measurements of pollutants in fecal pellets, molts, and eggs. Table 6 gives the concentrations of elements and compounds measured in pelagic crustaceans and their particulate products. The most striking feature is the generally high concentrations in the biogenic particulates, especially fecal pellets, relative to the whole animals or their prey.

As discussed in Section II.D.1, the high elemental concentrations (Table 6) and high rate of production often combine to make fecal pellets the dominant mode of element excretion from zooplankton. Fecal pellets sink relatively rapidly,[221,222,304] disintegrate only slowly at low temperatures,[224] and certainly have the potential to transport their pollutant load to great depths. Several studies highlight the importance of fecal pellets in the vertical flux of materials in the sea. McCave,[227] reviewing the literature on vertical flux of particles, presents an excellent synthesis of evidence to suggest that even though most material in suspension throughout the water column is very fine, the principal mode of transport to depth is by relatively rare, rapidly sinking large particulates such as fecal pellets. His hypothesis has been supported by recent work which directly measured the flux of sinking biogenic particulates. Using a large volume *in*

Table 6
TRACE ELEMENTS, RADIONUCLIDES, AND PCB IN PELAGIC CRUSTACEANS, THEIR PARTICULATE PRODUCTS AND MICROPLANKTON[27,28,178,225,226,236]

Sample	μg g⁻¹ dry										
	Ag	Cd	Co	Cr	Cu	Fe	Mn	Ni	Pb	Zn	Ce
Meganyctiphanes norvegica											
Whole euphausiid	0.71	0.74	0.18	0.85	48	64	4.2	0.66	1.1	62	0.21
Molts[a]	2.9	2.1	0.80	5.3	35	232	11.7	6.7	22	146	1.2
	(31)	(22)	(34)	(48)	(6)	(28)	(21)	(78)	(≈100)	(18)	(44)
Fecal pellets	2.1	9.6	3.5	38	226	24000	243	20	34	950	200
Eggs	0.96	0.58	0.80	7.9	17	330	11.5	4.3	8.9	318	
Nematoscelis megalops											
Whole euphausiid			4.9			102	8.8			78	
Molts[a]			13.4			845	15.7			224	
			(27)			(83)	(18)			(29)	
Eggs	2.22	0.52	6.3		28	200	17.4		11.1	389	
Sergestes arcticus											
Whole shrimp		1.03	1.1		30	45	5.8			55	
Molts[a]			11.9			283	19.1			127	
			(76)			(44)	(23)			(16)	
Fecal pellets	16.2	13.9	52		145	6600	120			813	
Pasiphaea sivado											
Whole shrimp		1.28	1.0		53	17	2.1			50	
Eggs	2.42	0.14	4.3	0.1	253	27	2.8	7.3	9.0	66	
Mixed copepods		0.9			10	129	5.5			71	
Fecal pellets		1.8			950	15400	277			915	
Microplankton[c]	0.67	2.1	0.87	4.9	39	570	17.9	8.1	11	483	0.30

Sample	Cs	Eu	Hg	Sb	Sc	Se	Sr	^{210}Po[b]	$^{239,240}Pu$[b]	PCB
Meganyctiphanes norvegica										
Whole euphausiid	0.062	0.0023	0.35	0.071	0.009	4.4	117	1100	0.41	0.62
Molts[a]	0.019	0.0077	0.17	0.80	0.030	1.9	350	360	4.8	1.4
	(2)	(26)	(4)	(87)	(26)	(3)	(23)	(2.5)	(90)	(17)
Fecal pellets	6.0	0.66	0.34	71	2.8	6.6	78	17500	98	16.0
Eggs			0.11					800		
Mixed copepods										0.18
Fecal pellets										1.3
Microplankton[c]	0.080	0.013	0.05	0.22	0.13	2.7	520	7500	4.0	4.5

[a] Numbers in parentheses represent percent of element or compound body burden contained in the molt.
[b] pCi kg⁻¹ dry.
[c] Principally copepods, phytoplankton, chaetognaths, and detritus which serve as food for larger pelagic crustaceans.

situ filtration system, Bishop et al.[228] found that 99% of the vertical mass flux through the top 388 m in the north equatorial Atlantic was by fecal pellets and aggregated fecal matter which accounted for only 4% of the suspended mass. Particle collectors set at depths ranging from 1000 to 5000 m in different oceanic areas have measured fecal pellet fluxes ranging from 10^3 to 10^5 pellets m^{-2} d^{-1}.[229-231] The occurrence of molts and carcasses in collectors is much more rare, and where detailed characterization of trapped particulates has been made,[229-231] most organic particulates are fecal pellets. Thus it appears that passively sinking fecal material is the principal biological pathway by which pollutants reach deep waters.

Studies addressing this aspect of pollutant transport are relatively few. Small and Fowler[224] used data on the zinc content of euphausiids (*Meganyctiphanes norvegica*) and their particulate products[147] to examine the vertical transport of zinc in the Ligurian Sea. By combining sinking rates with biodetritas disintegration rates, they were able to assess the relative importance of the different particulates in zinc transport across two boundaries at 500 and 2500 m. They found that even though a large fraction of the zinc was lost during descent, fecal pellets still accounted for 81 to 98% of the zinc transport by euphausiids to both depths under various conditions, with molts and carcasses comprising the remainder. Under conditions of sufficient food availability,[147] sinking biogenic debris from *M. norvegica* transported 98 and 15% of the animal's body zinc concentration each day below 500 and 2500 m, respectively. With euphausiids feeding under marginal food conditions, the percentages dropped to 36 and 6% at the same depths. Based on the above data,[224] Davies[9] compared the fraction of seawater zinc leaving the top 50 m via sinking biodetritus and reaching 500 and 2500 m with the fraction of zinc sedimenting out by a purely geochemical means. The comparison indicated that even in waters of relatively low fertility like the Mediterranean, zinc transport out of the surface waters by euphausiid-produced biodetritus would exceed by two orders of magnitude that occurring by geochemical processes. However, Davies[9] points out that due to loss of zinc from sinking biodetritus, the zinc flux by both processes would approach similar values in deeper waters.

Although the most detailed information on vertical transport by biogenic detritus pertains to zinc, the downward transport of some other contaminants appears to follow the same pathway. Recent geological studies[305] suggest that past and present-day accumulation rates of Ba, B, Cu, Ni and other elements in deep-sea sediments can be readily explained by sinking particulate matter of planktonic origin. Cherry et al.[232] demonstrated a high correlation between the concentration factors of 15 elements in fecal pellets and their oceanic residence times. Their theoretical model, based on the extreme assumption that zooplankton fecal pellets were the only route by which elements reached the sediments, agreed well with their observations, leading the authors to suggest that the oceanic residence times of these elements are controlled primarily by fecal pellet flux. The vertical transport of plutonium has also been linked to sinking biogenic particles.[26,27] Higgo et al.[27] calculated that plutonium could be removed from the upper mixed layer in 5 years by zooplankton metabolic activity alone. This value was similar to estimates of the total upper mixed layer removal time ($\simeq$ 1 yr) of plutonium, indicating that zooplankton defecation, molting, and soluble excretion could contribute substantially to the downward transport of $^{239,240}Pu$ from surface waters.

Fecal pellets are also implicated in the vertical transport of organic pollutants. Elder and Fowler[178] found high concentrations of PCBs in fecal pellets of the euphausiid *Meganyctiphanes norvegica* and, using known fecal pellet production rates for this species,[147] computed delivery rates of PCBs to Ligurian Sea sediments of 1.4 to 4.1 μg m^{-2} yr^{-1}. These values were one to two orders of magnitude lower than independent estimates of PCB delivery derived from sediment cores. However, the fact that *M. norvegica* comprises only 1 to 5% of the zooplankton biomass in the region suggested that the two estimates of PCB delivery would be more alike if total zooplankton defecation were considered. Oil particles have also been observed in zooplankton fecal pellets.[128,129] After the wreck of the tanker Arrow in Chedabucto Bay, Nova Scotia, Conover[129] found that copepod fecal pellets contained up to 7% bunker C oil by weight. Using estimates of oil-particle concentration, zooplankton biomass, grazing rates, and assimilation efficiencies, he concluded that 20% or more of the suspended oil in the bay was sedimented in zooplankton fecal pellets. In a similar fashion, Corner[88] calculated that for a dispersed oil droplet concentration of 1.5 mg ℓ, a single female *Calanus helgolandicus* in 1 ℓ of sea water could release a maximum of 30% of the oil particles in fecal pellets each day. Besides oil particles, water-soluble hydrocarbons may also be transported in fecal pellets; roughly 40% of the naphthalene in phytoplankton ingested by copepods is excreted with their feces.[88,97] Despite the evidence linking zooplankton fecal pellets to the vertical transport of oil, it has only been very recently that direct measurements were made which indicate the nature of the compounds being sedimented. Comparing sediment trap flux data with those derived from ^{210}Pb dated sediments in Dabob Bay, Washington, Prahl and Carpenter[314] found that fecal pellets quantitatively accounted for essentially 100% of the polycyclic aromatic hydrocarbon fluxes based on dated sediments. Their findings suggest that in this particular bay, virtually the total removal of fluoranthene, benzofluoranthene, perylene, pyrene, and benzo[*a*]pyrene to sediments is controlled by sinking zooplankton fecal pellets.

Sinking phytoplankton cells or fragments, planktonic tests, crustacean molts, and dead organisms also transport pollutants downward. All have been observed in particle collector samples taken at depth but, usually, they are in lesser abundance than fecal pellets. For example, sinking phytoplankton has been implicated in the removal of polycyclic aromatic hydrocarbons[85] and copper[233] added to a controlled ecosystem enclosure suspended in the marine environment. While cells or algal fragments might be instrumental in carrying down pollutants in shallow waters, this mechanism is of only limited importance in deeper areas because of the extremely slow sinking rates of these small particles.[234] Smayda[234] points out the need for a mechanism of accelerated phytoplankton sinking to account for the presence of phytoplankton in certain deep areas and suggests incorporation into fecal pellets as a probable one. Zooplankton molts, which are shed frequently,[164] contain relatively high concentrations of contaminants (Table 6), and sink rapidly for short periods of time[224] also contribute to vertical transport. However, even though fresh molts sink rapidly, they disintegrate much faster than fecal pellets[224] and, hence, are more likely involved in pollutant cycling in the upper layers of the water column. The same holds for carcasses; fast settling rates[224] are offset by both rapid decomposition times and predation. The general lack of molts and carcasses in particulate samples collected at depth[229-231] supports this contention. Sinking nonviable eggs from pelagic species also can transport pollutants;[147] however, the few data suggest that this route will be inconsequential vis à vis other biogenic particulates. For example, euphausiid eggs are released only seasonally, and even during the reproductive period their production rates are much lower than those for fecal pellets, molts, and carcasses.[147,224] Furthermore, most planktonic eggs have specific gravities near that of sea water which reduces their settling rate. Hence, these products most likely take part in pollutant cycling near where they are released.

Macrophytic detritus or macrofaunal carcasses are other possible carriers of pollutants into the deep sea. Large patches of turtle grass, *Thalassia testudinum,* and *Sargassum* blades have been noted at great depths and apparently serve as food for many benthic organisms.[235] On the other hand, large animal remains have never been observed on the deep sea floor,[235] hence it is difficult to evaluate their importance in transporting pollutants to the deep ocean. Most probably, large carcasses are consumed well up in the water column which implies a role in pollutant cycling in the upper water layers.

2. Remineralization, Scavenging, and Recycling

The degree to which sinking biogenic particulates transport their associated contaminants to depth is a function of the rate of remineralization of these substances from decomposing detritus. That this process occurs is inferred from the vertical distributions of inorganic contaminants[6,7] and radionuclides[3] in the water column which correlate well with the vertical distributions of nutrients. More direct evidence has recently been obtained from *in situ* particle traps. Biodetritus collected at depth has higher C/N ratios than that from upper water layers, suggesting that relatively more nitrogen is leached from the material and/or recycled than carbon as the particles sink through the water column.[231] Similarly, fecal material in particle trap samples collected at 100 m in the Ligurian Sea had significantly less PCBs than that of freshly produced zooplankton fecal pellets released in the overlying surface waters.[236] This observation suggests that PCBs are lost from the pellets during their descent.

Direct measurement of contaminant remineralization rates under natural conditions is difficult, if not impossible. On the other hand, some information has been derived from laboratory studies in which the loss of various radionuclides from fecal pellets, molts, and dead euphausiids has been followed over time. The results are summarized in Table 7. The half-times suggest that a significant fraction of certain elements is lost before the particles either disintegrate or reach bottom in deep areas. It is interesting to note that methyl mercury is remineralized from biodetritus much more rapidly than the inorganic form. This probably results from enhanced degradation of the organic compound by the bacteria population associated with decomposing detritus. In most cases the longest half-times are for loss from fecal pellets. Relatively strong retention, slow disintegration, and rapid sinking rates over long periods of time give fecal pellets great potential to transfer much of their pollutant content to depth. This will be particularly true once fecal pellets reach cooler, deep waters where microbial attachment to — and biodegradation of — the pellets is greatly reduced.[298] Molts and carcasses which are more friable tend to release their contaminant load fairly rapidly and disintegrate in a few days. Furthermore, as these materials disintegrate, their descent slows considerably.[224] All these factors suggest that molts and dead organisms produced near the surface are principally involved only in the recycling of contaminants in the upper layers of the oceans.

To what extent biologically active elements like zinc are rapidly remineralized from biodetritus and retained in the upper waters rather than carried to depth is still an open question. Pearcy et al.[237] studied ^{65}Zn specific activities (^{65}Zn/total Zn) in pelagic and benthic organisms from different depths off Oregon to estimate the vertical transport rate of zinc. Their results suggested that vertical transport was slower in upper waters, due to recycling of zinc within biological communities, and more rapid below 500 m. The time necessary to transport zinc from the surface to abyssobenthic animals was estimated to be about 2 years, which contrasts sharply with the rapid times (days)[224] suggested if fecal pellets are the major conveyors of zinc to depth. To explain their findings Pearcy et al.[237] hypothesize that zinc may be rapidly lost from fecal pellets and efficiently recycled through organisms in the upper waters. That fecal pellets lose zinc during descent seems clear from both radiotracer studies (Table 7) and data which indicate that fecal pellets collected in deep traps (5000 m) contain 1 to 2 orders of magnitude less zinc[238] than in freshly released zooplankton fecal pellets.[28,147] However, Small and Fowler[224] estimated that despite a more than 90% zinc loss from euphausiid fecal pellets sinking to 2500 m, these particulates could still account for the majority of the zinc transport to this depth by euphausiids, principally because of the initial high zinc concentration and rapid sinking rates of euphausiid fecal pellets. Although the specific activity approach used by Pearcy et al.[237] operates under several assumptions which, if not valid, tend to overestimate transport times, their observations and those of others[215] on the vertical distribution of ^{65}Zn in marine species need

Table 7
HALF-TIME ESTIMATES FOR THE RELEASE OF ELEMENTS AND RADIONUCLIDES FROM BIOGENIC PARTICULATES PRODUCED BY THE EUPHAUSIID *MEGANYCTIPHANES NORVEGICA* — CALCULATIONS OF LOSS RATE FOR EACH ELEMENT ARE BASED ON A SINGLE EXPONENTIAL MODEL

		Half-time (days)			
Element or radionuclide	Tracer	Fecal Pellets	Molts	Carcasses	Ref.
Zn	$^{65}ZnCl_2$	2.0	1.7	7.8	224
Hg	$^{203}HgCl_2$	14.1	2.3	3.1	19
	$CH_3\,^{203}HgCl$	6.0	1.4	2.6	19
Se	$Na_2\,^{75}SeO_3$	3.9	—	—	21
$^{239,240}Pu$	$^{237}Pu\,(+6)$	9.7	4.7	5.3	20
	$^{237}Pu\,(+4)$	22.9	—	—	20
$^{141,144}Ce$	$^{141}CeCl_3$	7.5	10.6	17.3	8
^{210}Po	natural	3.5	—	—	226

explaining. Hopefully, more detailed study on the fecal transport and remineralization mechanisms operating with other zooplankton species will help clarify the relative importance of particulates in conveying biologically essential elements like zinc to depth.

At present, no data are available on the release of organic pollutants from sinking particulates. Nevertheless, remineralization rates for both chlorinated and petroleum hydrocarbons in biodetritus should be rapid due to the microbial biodegration expected in the environment[323]. Thus, easily degradable petroleum hydrocarbon fractions associated with biodetritus may not reach great depth, whereas more refractory compounds (e.g., DDE and PCB) which degrade slowly, should reach bottom sediments in deeper areas.

While the high concentration of contaminants in biogenic particulates creates a gradient favorable for loss of contaminants to water, it is conceivable that certain highly reactive substances in water might be scavenged by the detrital particles sinking through the water column. The removal from sea water of higher weight aromatic hydrocarbons such as chrysenes, benzathracenes, and benzpyrenes is thought to occur by this mechanism and thus leads to short residence times of these compounds in the water column.[85,323] Nevertheless, direct measurements to test this hypothesis are few,[323] and therefore it is difficult to evaluate adsorption to biodetritus as a pathway for pollutant removal. Clearly, more data are needed on the adsorption-desorption kinetics of contaminants associated with biogenic particles.

The above examples of pollutant transport by biogenic particulates pertain primarily to open waters. There is some evidence that in shallow bays and estuaries the impact of sinking fecal pellets and other biodetritus on pollutant transport to the sediments may be much greater than in deeper areas. Seki et al.[239] found that fecal pellets and phytodetritus brought about by intense copepod grazing formed the major fraction of the organic debris reaching the bottom of Tokyo Bay. A similar study in a semienclosed Scottish loch showed that about 27% of the annual primary production reached the bottom mostly as zooplankton fecal pellets.[240] It seems clear that in these shallower areas, shorter transit times with lesser decomposition and remineralization results in more of the incorporated pollutant being transported to sediments. In fact, one study[241] concludes that defecation of unassimilated trace metals may be a major biological process in the cycling of these elements through highly productive coastal-plain estuaries.

3. General Consequences

The consequences of biodetritus on pollutant transport in general are several. Sinking biogenic particulates offer an attractive mechanism for explaining the rapid transport of surface-introduced pollutants to a depth beyond that which pelagic species normally encounter during vertical migration. As the particles sink, they can either release contaminants back to water in the deeper layers, scavenge certain materials from the water column, or they can be ingested by mesopelagic and bathypelagic biota. Biogenic particulates such as fecal material and carcasses serve as food sources for a number of shallow and deep-water organisms[242-246,299] and in this way constitute another direct source of pollutants to the consumer. Those particles which are neither ingested nor decompose during descent will reach the sediments with some fraction of their original pollutant concentration. What remains of the contaminant in particles reaching the sediments depends on the nature of the biogenic particle, sinking rates, remineralization rates, and depth of the water column.

C. Benthic Boundary Layer Processes

Generally, many anthropogenic and natural contaminants associated with inorganic and organic particulates eventually become incorporated in the sediments. This is particularly true in estuaries and coastal or relatively shallow oceanic areas.[3,247] In polluted zones, sediments usually contain the major fraction of the total contaminant in a given ecosystem as well as the highest concentrations of the pollutant.[2-4,208,247-249] Because sediments might be the ultimate pollutant reservoir, it is not surprising that more attention is being given to pathways by which contaminants are recycled through benthic ecosystems. Evidence of biological processes that remobilize, transfer, and transport sediment-associated pollutants in both shallow and deep-sea areas are discussed in this section.

1. Microbial Action

The primary step in biological remobilization of many pollutants at the benthic boundary layer may be through the action of microorganisms in the sediments. Sediment associated microorganisms are considered to be responsible for the methylation of mercury,[250] although its rate of formation appears to be low since methylmercury typically represents less than 0.1% of the total mercury in sediments.[247,250,251] There is evidence[251] that biological methylation occurs primarily in the upper oxidized layers of the sediment and is considerably reduced under anoxic conditions. On the other hand, recent work of Olson and Cooper[250] has shown that the net production of methyl mercury in bay sediments was greater under anaerobic conditions than aerobic conditions. Nevertheless, the authors caution that this difference may be more apparent than real, particularly if demethylating microorganisms are more active under aerobic conditions. Because the rate of methyl mercury formation is low, it appears that biological methylation processes do not play a major role in transferring mercury out of near-shore sediments.[247,251] While information on the biological methylation of mercury is accruing, little is yet known about other metals and metalloids which may undergo methylation by similar processes in the marine environment.[252,253]

Besides direct transformation reactions, microorganisms affect the distribution and transport of elements associated with sediments in other ways. In areas of high productivity, particularly those in shallow, quiet waters, large populations of oxidizing bacteria often produce reducing environments containing hydrogen sulfide. If certain metals and radionuclides contact the hydrogen sulfide, they precipitate as sulfides and become fixed in bottom sediments.[3] Microbial epiphyton also associate strongly with particulate material and may cause the high element concentration factors characteristically found in sediments.[3] Microbiota on sediment particles could thus act as pollutant carriers into food chains whose organisms feed on fine particles.

Microorganisms are also implicated in the metabolism of persistent chlorinated hydrocarbons. These processes should be particularly important in marine sediments where many of these compounds are found in relatively high concentrations. In general, chlorinated hydrocarbons will be degraded only slowly in the marine environment.[254] The major degradative reaction with DDT is reductive dechlorination to DDD. This reaction occurs aerobically with either o,p′- or p,p′-DDT, although there is evidence that it also proceeds under anaerobic conditions. In addition to DDD, varying amounts of polar compounds are always produced. Compounds such as aldrin, dieldrin, and endrin are also degraded aerobically by microorganisms. Little information exists on the microbial degradation of polychlorinated biphenyls; however, recent laboratory experiments of Carey and Harvey[255] demonstrated that mixed cultures of near-shore marine bacteria can metabolize PCBs to an acid lactone metabolite. The same reaction did not occur under anaerobic conditions, suggesting that anoxic sediments would serve as long-term sinks for PCBs. Thus, microbial remobilization of chlorinated hydrocarbons in sediments probably occurs only in the upper oxygenated layers with metabolites from these reactions becoming available for uptake by other benthic biota.

The introduction of petroleum hydrocarbons to marine sediments greatly increases the hydrocarbon degrading microbial population.[256,257] Biodegradation occurs aerobically as evidenced by unaltered oil in the anoxic layers of sediments.[256,257,259] Besides oxygen other factors such as temperature, pH, nutrient concentration, and hydrocarbon substrate affect the rate of microbial degradation.[208] In areas contaminated by oil, n-alkanes are degraded most rapidly followed by slower attack on branched alkanes, cycloalkanes, and aromatic hydrocarbons.[258,260] One study[311] has shown that microbial degradation rates in sediments vary with polycyclic aromatic hydrocarbon structure and are reduced for the high molecular weight compounds, particularly benzo[*a*]pyrene. The polycyclic aromatic compounds are further broken down to cis-diol intermediates with carbon dioxide as an end product.[256] Oil biodegrades slowly despite the fact that populations of oil-degrading bacteria, yeasts, and fungi are found in all marine environments.[258] However, obtaining accurate rates at which hydrocarbons are degraded by microorganisms is difficult, both because of the many different hydrocarbons in the sea and the multiplicity of ways these compounds are measured and the data evaluated.[208]

Pollutant transformation and degradation by microbial processes in the deep sea have not yet been examined. Nevertheless, present information on metabolic rates[261] indicates that oxygen consumption in abyssobenthic microorganisms may be as much as 100 times lower than for microbes inhabiting the continental shelf. It seems probable that rates of pollutant biotransformation and biodegradation will be correspondingly low.

2. Transfer from Sediments

Correlations between pollutant concentrations in marine species and contaminated sediments in their surroundings indicate that sediments are a source of many pollutants to the benthos.[25,61,62,101,209,247-249,262] Despite this, relatively little work has been done to elucidate the pathways of pollutant transfer from sediments to organisms. There are several ways in which sediment-associated pollutants may be accumulated by benthic biota. Mechanical action of waves or currents and biological activity resuspend sedimented pollutants, which can then be filtered from the water by suspension feeding organisms[182,263] (See Section II.B). If these particles have bacterial populations associated with them, pollutant uptake into filter feeders may be enhanced.[264] In addition, pollutants which diffuse out of the sediments in soluble or colloidal form can be adsorbed or absorbed by organisms in proximity to the sediments.[265,266] Epifaunal detritivores and predators can also accumulate pollutants in their search for food in or near the surface sediments. The presence of this pathway is supported by observations of tar balls in deep-sea starfish,[157] radionuclides,[217] insecticides,[218] and petroleum hydrocarbons[219] in bathy-demersal fish and the uptake of chlorinated hydrocarbons from biodetritus and sediments by benthic crabs[267] and shrimp.[101,248]

Perhaps most important in uptake are the sediment ingestion activities of deposit feeders, a functional group including members of several phyla. For example, deep-sea holothurians have been found to contain elevated levels of radionuclides,[215] heavy metals,[220] and petroleum hydrocarbons.[219] Bowen et al.[25] reported that tissues of various invertebrate infauna may contain plutonium concentration 5 to 10 times those in the sediments, which suggests availability of sediment-associated transuranics to biota. Although these field studies underline the importance of sediments in transferring pollutants to benthic organisms, the exact mechanism for uptake is still unclear and may vary with the different pollutants. Beasley and Fowler[153] exposed the polychaete *Nereis diversicolor* to sediments contaminated with plutonium and americium and found that concentration factors based on sediment were very low (CF $\simeq 10^{-3}$). These concentration factors plus those based on direct uptake from water (CF $\simeq$ 200) applied to natural conditions led to the conclusion that water, possibly sediment pore water, was the predominant pathway for plutonium uptake by worms. Studies on the uptake of other radionuclides and heavy metals by polychaetes have led to similar conclusions.[61,62,265,266] However, leaching of sorbed inorganic contaminants from ingested sediments and subsequent incorporation into tissue can occur in some deposit feeders. For example, Bryan and Uysal[268] found that usually more than 75% of Cd, Co, Cr, Ni, Pb, and Zn in the soft parts of the deposit-feeding bivalve *Scrobicularia plana* was located in the digestive gland, suggesting that ingested sediment was the principal source of these metals. Furthermore, the fraction of sediment-sorbed element taken up by the animal depends on the element involved,[268] its physical-chemical form,[269] and the type of sediment.[269]

Corresponding studies with polychlorinated biphenyls indicate that polychaetes readily accumulate these compounds from contaminated sediments.[103,270,271] During long-term exposure to contaminated sediments, uptake of PCB by *Nereis diversicolor* was shown to be dose dependent (Figure 10). Furthermore, PCB thus accumulated was strongly retained with depuration half-times of several weeks. Equilibrium concentration factors following uptake from sediment were roughly 3 to 4 compared to 800 for direct absorption from water.[103] Fowler et al.[103] applied these concentration factors to environmental PCB concentrations in water and sediments from coastal and open ocean areas in the Mediterranean and showed that under these conditions 85% or more of the uptake could be attributed to the sediments. Similar calculations[103] based on known PCB distribution coefficients suggested that uptake from interstitial waters would be insignificant. Thus, if only sediments and water are concerned, ingested sediments would be the primary pathway by which infaunal worms accumulate PCB.

Results for PCBs differ significantly from similar studies which show that benthic infauna take up little, if any, petroleum hydrocarbons associated with sediments.[136,137,188,272,273,316,322,325] Polychaetes,[136] sipunculid worms,[188] detritivorous clams,[272] and benthic amphipods[322] failed to accumulate significant amounts of naphthalenes in their tissues following long-term exposure to contaminated sediments. Roesijadi et al.[272] exposing the clam *Macoma inquinata* to sediments containing higher molecular weight aromatic compounds, found that phenanthrene and chrysene reached equilibrium in 2 and 14 days, respectively. Tissue hydrocarbon concentrations then declined, reflecting new equilibrium concentrations in response to the declining levels in sediment (Figure 11). However, despite the loss of benzo[*a*]pyrene from sediments, concentrations of this compound continued to increase in the clam throughout expo-

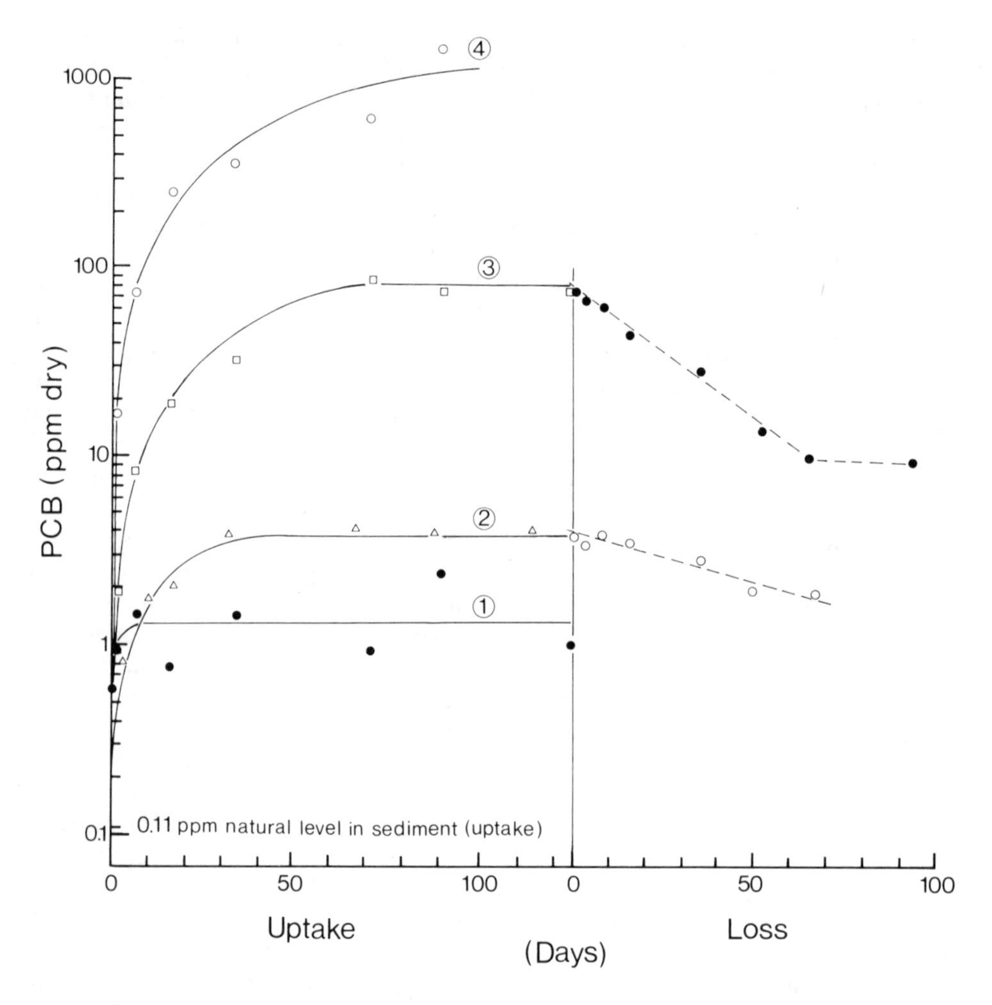

FIGURE 10. *Nereis diversicolor.* Accumulation of PCBs from sediments spiked with 0.65 ppm (curve 2), 9.3 ppm (curve 3), and 80 ppm dry (curve 4). Curve 1 represents PCB levels in worms living in unspiked sediments containing 0.11 ppm PCB. Depuration in unspiked sediments for two treatments is also shown.[103,271]

sure, indicating a relatively strong persistence of this high molecular weight hydrocarbon in detritivorous clams. Nevertheless, uptake of higher molecular weight compounds from sediments by this species is relatively low; tissue-sediment concentration factors computed after 7 days were $\leqslant 0.2$ for phenanthrene, chrysene, dimethylbenz[*a*]anthracene and benzo[*a*]pyrene. On the other hand, corresponding tissue to sea water concentration factors were high ($\gg 1$), suggesting strong bioavailability of these compounds when dispersed in the aqueous phase. These observations and the fact that the compounds were rapidly leached from sediments led the authors to conclude that the small amounts of petroleum hydrocarbons in the clam resulted from direct absorption of the fraction released to water. Their results and those of others[136,188,273,316,322] suggest that petroleum hydrocarbons present in interstitial water are the main source of contamination for infaunal species.

Experimentally derived tissue-sediment concentration factors for several pollutants in benthic infauna are given in Table 8. Despite the fact that quite different experimental designs were used, it seems clear that of the elements and compounds tested, only PCBs and alkylnaphthalenes are taken up from sediments to any significant degree. Further work should show if this behavior is typical for higher molecular weight chlorinated hydrocarbons.

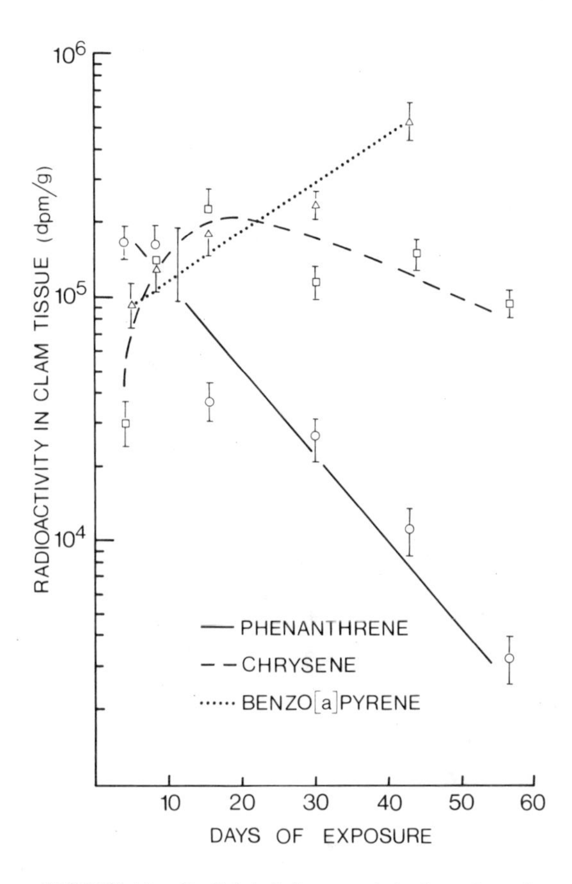

FIGURE 11. Radiolabeled aromatic hydrocarbons in tissue of the deposit-feeding clam *Macoma inquinata* during long-term exposure to contaminated sediments. (From Roesijadi, G., Anderson J. W., and Blaylock, J. W., *J. Fish. Res. Board Can.*, 35, 608, 1978. With permission.)

3. Redistribution and Bioturbation

Sediment reworking activities such as ingestion and egestion of sediments are probably the primary processes by which benthic organisms redistribute pollutants. Biodeposition is known to be important to biogeochemical cycles, especially in intertidal areas.[274] Kraeuter[274] recently summarized several studies on rates of sediment reworking and showed that many species in coastal areas are capable of biodepositing several kilograms of feces or pseudofeces m^{-2} y^{-1}. Biodeposition has been proposed as an important mechanism in transporting pollutants in the benthos[3] and several recent studies support this contention. For example, clams filtering particulate radionuclides from sea water release feces containing concentrations several orders of magnitude over those in water.[263] Courtney and Denton[182] found more higher-chlorinated PCB isomers in clam feces and mud samples than in clam tissue and suggested that defecation exerted both a qualitative and quantitative effect on PCB distribution in the ecosystem. The mussel *Mytilus galloprovincialis* can filter and release oil as pseudofeces consisting of oil connected by mucous strands.[275] Recent work[308] with mussels has shown that the oil fractions in this material as well as true feces differ both qualitatively and quantitatively from the original oil ingested. These examples indicate that suspension feeding activities play a major role in consolidating pollutants and incorporating them in the sediments.

Table 8
CONCENTRATION FACTORS FOR THE UPTAKE OF CONTAMINANTS FROM SEDIMENTS BY BENTHIC INFAUNA — NUMBERS IN PARENTHESES ARE CONCENTRATION FACTORS BASED ON UPTAKE FROM WATER

Species	Days of exposure	$^{239,240}Pu$	^{241}Am	^{55}Fe	^{95}Zr—^{95}Nb	^{137}Cs	^{106}Ru	^{60}Co	Cd	Phenan-threne	Chry-sene	Di-methyl-benz[a] anthra-cene	Benzo [a] pyr-ene	Naph-thalene	Alkyl-naph-thalenes	PCB (DP-5)	Ref.
Polychaete																	
Nereis diversicolor	40 (15)	0.0014 (200)	0.0005														153
Nereis diversicolor	88			0.019													330
Nereis diversicolor	125 (14)															3—4 (800)	103,271
Nereis japonica	8 (12)								0.12 (22)								265
Nereis japonica	11 (11)				0.01 (4)	0.2 (6)	0.006 (6)	0.06 (6)									266
Arenicola marina	0.2 (0.4)													0.1—4 (20—300)			316
Mollusk																	
Macoma inquinata	7 (7)									0.2 (10)	0.04 (694)	0.06 (1349)	0.09 (861)				272
Crustacean																	
Anonyx laticoxae	4 (1—27)														2—4 (10—1000)		322

Defecation processes in other benthic feeding groups are also instrumental in pollutant cycling. Crabs ingesting algae have been found to release fecal pellets rich in heavy metals.[173] This process is particularly significant, because certain metal concentrations in feces are increased leading to a reprocessed, metal-rich food source for detritivores. Fecal material has also been shown to be the major pathway for the removal of petroleum hydrocarbons and metabolites from crabs.[111] Several studies suggest that deposit feeding infauna living in contaminated sediments also produce contaminated feces. For example, feces production is thought to be the main route of elimination of PCBs in nereid worms.[185] Ernst et al.[185] noted that certain PCB isomers were leached from worm feces, thus suggesting another pathway by which PCBs are recycled back into the water column. The low degree of assimilation of ingested petroleum hydrocarbons in deposit-feeding worms indicates that fecal casts will be instrumenta l in redistributing this group of compounds in the sediments. The above examples highlight the importance of defecation by benthic species in the transfer and transport of pollutants within sediments. Coprophagy is thought to be an important mode of nutrition in the benthos;[242,276] thus, ingestion of feces facilitates reentry of pollutants into benthic food chains. Equally important is the fact that fecal material contains large populations of microbes which enhance the degradation and release of many organic pollutants.

Bioturbation as it affects pollutant redistribution has had little quantitative study but there are indications[25,137,300,311] that it exerts a strong influence on pollutant recycling through the benthic boundary layer, especially in highly productive areas. While bioturbative processes are generally thought to affect only the upper few centimeters of sediment, there is evidence[25,300] that some fast-moving species of worms like *Arenicola* or *Nereis* penetrate sediments to depths of several tens of centimeters. Recent data[25,300] suggest that the filtering activities of infaunal worms draw down radionuclides from the overlying waters and fix them at some depth within the sediment. In the case of $^{239,240}Pu$, ^{241}Am, and ^{55}Fe fixed in sediment by this mechanism, Livingston and Bowen[300] have hypothesized that these radionuclides may be resolubilized in the presence of anionic organic complexes released deep in the sediments by the metabolic activities of worms. Once in soluble form the radionuclides are free to migrate slowly upwards through the sediments via interstitial waters or rapidly through the worm burrows themselves, eventually being released to the overlying waters. While this specific hypothesis neatly explains certain features of radionuclide profiles in sediments from some areas, there are no comparable data which would allow assessing its general applicability to other types of pollutants. Nevertheless, Gordon et al.[137] have observed that bioturbative activities of *Arenicola* can substantially accelerate the weathering of sediment-bound oil. They calculated that an average population of this species in coastal areas could remove all the oil in 1 m^2 of sediment in approximately 2 to 4 years. Studies such as these also illustrate the importance of worm burrowing activities in remobilizing pollutants from anaerobic to aerobic conditions within the sediments.

Despite an abundance of observations on the qualitative aspects of bioturbation, little effort has been put into quantifying these processes.[301] Biological mixing rates are extremely difficult to measure directly but they have been estimated to range between roughly 1 and 10^3 cm^2 Kyr^{-1} at abyssal depths and to increase to 10^3—10^6 cm^2 Kyr^{-1} in more densely populated nearshore sediments.[277,302,321] (see Chapter 3).

Several other routes by which the benthos recycle pollutants are worth noting. Contaminants accumulated by benthic species may be excreted in a form different from that taken up. Contaminated shrimp have been reported to excrete different physicochemical forms of zinc,[278] and several marine species are known to eliminate organic metabolites of the parent compound originally incorporated (see Section II.D.2). More important in pollutant recycling in the benthic boundary layer is the release or secretion of biogenic particulates in the form of molts, mucus, byssal threads, etc. Crustacean exoskeletons often contain elevated levels of inorganic contaminants; hence molting has been cited as an important facet of benthic biogeochemical cycles.[24,35,65,71] However, in deeper, colder waters this process may be much less important since deep sea crustaceans tend to molt rarely.[165] Mucus is known to sequester certain contaminants[30,32,103] and the shedding of mucus sheaths and their subsequent rapid disintegration[279] suggest enhanced turnover of pollutants associated with these secretions. Byssal thread formation by mussels has been proposed as a mechanism for the excretion of certain elements.[34,72,280] The accumulation of large mats of byssal threads thus might offer a concentrated source of pollutant to detritivores feeding on this material. Bivalve mollusks also secrete phosphorite concretions which contain extremely high concentrations of heavy metals.[281] The concretion and its associated pollutant load eventually end up in the sediments either by excretion or when the animal dies. Reproductive processes of benthic biota may represent an additional vector for pollutant redistribution. Release of contaminated eggs[186] with subsequent retention of petroleum hydrocarbons in developing larvae[282] serves as a mechanism for conserving these contaminants in benthic populations.

4. Transport from Sediments to Upper Waters

Current interest in using deep-sea sediments for containing radioactive wastes has focused attention on the possible biological pathways by which these toxic materials might be transported upward to the surface layers.[283] Given the fact that our present knowledge of deep-sea biological processes is so scanty, any quantitative predictions can only be considered as speculative. Nevertheless, in a qualitative sense the large-scale processes of bioturbation and vertical and horizontal movements of animals will tend to disperse sediment-associated pollutants. The recent discovery of large populations of highly mobile deep-sea amphipods[284,285] suggests that this component of the abyssal biomass may play an important role in pollutant redistribution. These scavengers along with other abyssal megafauna probably cover large areas in their search for food and certainly have the potential to transport incorporated pollutants horizontally by defecation, molting, or being eaten by other predators. Some upward movement of pollutants might also be expected, since there is evidence that certain abyssopelagic fish species undertake vertical migration.[286] Pollutants would enter mesopelagic food chains when these animals, or their fecal pellets, are consumed during forays into the upper waters; however, the magnitude of this pathway is expected to be small.[283] Benthic larvae may be the one biological agent which will disperse pollutants over the greatest vertical distance. Benthic fish[287] and mollusks[288] living at great depth have larvae which have been caught in surface waters; thus, in this way, contaminants in the benthos could be directly transferred to epipelagic food chains. Another biological mechanism effecting the upward vertical transport of pollutants has been linked to the "rising particle hypothesis".[303] This model envisages the rapid ascent of lipid-rich particles with their associated pollutants following their release at depth through the intense feeding activities of deep-sea amphipods and the decay of dead organisms. Stokes' Law calculations based on the size and density of lipid particles recently isolated from deep-sea amphipods show that particles released at 5000 m could reach the surface within 1 week to a year. These mechanisms are, of course, only speculative, as there are no quantitative data on which to judge their relative effectiveness in dispersing pollutants in the deep sea. However, based on present knowledge it seems safe to say that biological processes are probably extremely unimportant on a macroscale when compared to advective transport such as upwelling.

One final point worth mentioning is the possible future development of deep-sea fisheries which could bring man in direct contact with pollutants in habitats distant from his own. One such fishery might be squid. Considering the extremely large numbers of these highly mobile organisms[289] and the large distance over which they can migrate vertically,[332] some effort would be well spent in examining their potential for accumulating and transporting pollutants.

IV. SUMMARY

Data on pollutant transfer and transport by marine species are accruing at a rapid rate and as a result, a general picture is now emerging about the role played by these organisms in pollutant biogeochemical cycles in the sea. However, there are several aspects relating to pollutant transfer mechanisms in general which remain to be clarified. Definitely, one key need for future research is to determine the relative importance of the food and water pathways in pollutant uptake. There is considerable confusion on this point in the literature, particularly in the case of chlorinated and petroleum hydrocarbons and at low trophic levels in general. This problem is difficult to address and novel experimental approaches will be needed. This information, while important in elucidating transfer pathways, is essential in order to assess pollutant biomagnification in marine food chains, a second point of major controversy.

Of the several biological transport mechanisms highlighted in the previous sections, the most striking is the release and rapid settling of biogenic particulates like fecal pellets and molts enriched with various contaminants. These materials have been cited as prime agents in the redistribution of contaminants in the upper waters as well as conveyors of pollutants to depth, yet quantitative information on remineralization rates and vertical fluxes is still generally lacking. This situation will hopefully improve by continued use of *in situ* particle collectors in conjunction with measurements on freshly produced biogenic particulates.

Sediments are likely to be the major recipient of anthropogenic pollutants, both through natural sedimentation processes and by intentional dumping. Pollutants have been measured in many species living in deep-sea sediments, but because of our general ignorance about deep-sea biological processes, we have no knowledge about pollutant flux through populations at depth. It is generally considered that biomass and metabolic rates in the deep sea are low; however, recent technological advances have led to discoveries which may radically change our ideas about the biomass of at least certain deep-sea species. More quantitative biological studies in the deep ocean are needed before we can accurately estimate the degree to which deep-sea biota can remobilize and return pollutants towards the surface.

ACKNOWLEDGMENTS

I wish to thank Mr. J. La Rosa for his invaluable help in compiling the references and also preparing the illustrations. I am also grateful to Drs. C. L. Osterberg, R. J. Pentreath, and A. G. Davies for editorial and critical comments and to Mmes. E. Illes and C. Sunderland for their efforts in typing the manuscript.

GLOSSARY

Adipose	Fatty tissues
Anadromous	Relating to marine species which return up rivers to spawn
Bile	Digestive fluid secreted by liver
Bioturbation	Sediment mixing caused by animal burrowing activities
Chitin	Horny polysaccharide substance forming the outer integument of crustaceans and other invertebrates

Copepod	Small planktonic crustacean
Coprophagy	Ingestion of feces
Elasomobranch	Cartilaginous fishes, e.g., sharks, rays, etc.
Epidermis	Outer epithelial layer of the external integument
Epifauna	Animals living on the surface of sediments
Epineuston	Organisms strictly associated with the surface film of the sea; also called pleuston
Epiphyton	Algae or plants attached to surfaces
Epithelium	Membranous cellular tissue covering free surfaces of organs and outer integuments
Euphausiid	Shrimplike macroplanktonic crustacea
Exoskeleton	Generally hard, external supportive covering of many invertebrates; periodically shed by crustacea during growth
Exudates	Generally organic matter exuded or excreted by certain organisms
Feral	Of or relating to wild animals in the natural environment
Gut wall	Internal epithelial surface of the digestive tract
Hepatopancreas	Digestive gland of many invertebrates; analogous function to that of vertebrate liver and pancreas
Heterotroph	An organism requiring a supply of organic material (food) from its environment
Infauna	Organisms living most of the time within the sediments, e.g., worms, molluscs, microbes, etc.
Integument	Outer covering of organisms, e.g., skin or exoskeleton
Limpet	Gastropod mollusk with low conical shell that browses over rocks, etc. in the littoral zone
Lipophilic	Having an affinity for fats or other lipids
Macruran	Relating to a suborder of decapod crustaceans characterized by a well-developed, elongated and extended abdomen, e.g., shrimps, prawns, and lobsters
Molting	Act of shedding the molt or exoskeleton of crustaceans during growth
Mucus	Viscous proteinaceous secretion exuded by many organisms to protect their outer surface
Mysid	Small, usually pelagic, crustacea
Phyla	Major taxonomic groups used in classifying animals
Pseudofeces	Undigested food material rejected by many mollusks when filtering dense suspensions of particulates
Specific activity	Ratio of the radionuclide concentration to that of the corresponding stable element(s) in a substance
Spleen	Mass of lymphoid tissue near stomach or intestine of most vertebrates concerned with blood storage and lymphocyte production
$Tb_{1/2}$	Biological half-life or time necessary to lose one-half of an element or compound from some pool within an organism
Teleost	Pertaining to fishes with bony skeletons
Transdermal	Movement across the skin or epidermis
Whelk	Large marine snail

REFERENCES

1. **Polikarpov, G. G.,** *Radioecology of Aquatic Organisms,* Reinhold, New York, 1966, 314.
2. **Bowen, V. T., Olsen, J. S., Osterberg, C. L., and Ravera, J.,** Ecological interactions of marine radioactivity, in *Radioactivity in the Marine Environment,* National Academy of Sciences, Washington, D.C., 1971, chap. 8.
3. **Lowman, F. G., Rice, T. R., and Richards, F. A.,** Accumulation and redistribution of radionuclides by marine organisms, in *Radioactivity in the Marine Environment,* National Academy of Sciences, Washington, D.C., 1971, chap. 7.
4. **Bryan, G. W.,** Heavy metal contamination in the sea, in *Marine Pollution,* Johnston, R., Ed., Academic Press, London 1976, 185.
5. **Pentreath, R. J.,** Radionuclides in marine fish, *Oceanogr. Mar. Biol. Annu. Rev.,* 15, 365, 1977.
6. **Bruland, K. W., Knauer, G. A., and Martin, J. H.,** Cadmium in northeast Pacific waters, *Limnol. Oceanogr.,* 23, 618, 1978.
7. **Bruland, K. W., Knauer, G. A., and Martin, J. H.,** Zinc in north-east pacific water, *Nature (London),* 271, 741, 1978.
8. **Fowler, S. W., Heyraud, M., Small, L. F., and Benayoun, G.,** Flux of ^{141}Ce through a euphausiid crustacean, *Mar. Biol.,* 21, 317, 1973.
9. **Davies, A. G.,** Pollution studies with marine plankton. II. Heavy metals, in *Adv. Mar. Biol.,* 15, 381, 1978.
10. **Schulz-Baldes, M. and Lewin, R. A.,** Lead uptake in two marine phytoplankton organisms, *Biol. Bull.,* 150, 118, 1976.
11. **Glooschenko, W. A.,** Accumulation of ^{203}Hg by the marine diatom *Chaetoceros costatum, J. Phycol.,* 5, 224, 1969.
12. **Sick, L. V. and Windom, H. L.,** Effects of environmental levels of mercury and cadmium on rates of metal uptake and growth physiology of selected genera of marine phytoplankton, in *Mineral Cycling in Southeastern Ecosystems,* Howell, F. G., Gentry, J. B., and Smith, M. H., Eds., U.S. Energy and Research Development Administration, Washington, D.C., 1975, 239.
13. **Davies, A. G.,** An assessment of the basis of mercury tolerance in *Dunaliella tertiolecta, J. Mar. Biol. Assoc. U.K.,* 56, 39, 1976.
14. **Davies, A. G.,** The kinetics of and a preliminary model for the uptake of radiozinc by *Phaeodactylum tricornutum* in culture, in *Radioactive Contamination of the Marine Environment,* International Atomic Energy Agency, 1973, 403.
15. **Rice, T. R.,** Review of zinc in ecology, in *Proc. 1st Natl. Symp. Radioecology,* Schultz, V. and Klement, A. W., Jr., Eds., Reinhold, New York, 1963, 619.
16. **Sick, L. V. and Baptist, G. J.,** Cadmium incorporation by the marine copepod *Pseudodiaptomus coronatus, Limnol. Oceanogr.,* 24, 453, 1979.
17. **Fowler, S. W., Small, L. F., and Dean, J. M.,** Metabolism of zinc-65 in euphausiids, in *Symp. on Radioecology,* Proc. 2nd Natl. Symp., Conf 670503, USAEC TID-4500, Nelson, D. J. and Evans, F. C., Eds., Ann Arbor, Michigan, 1969, 399.
18. **Benayoun, G., Fowler, S. W., and Oregioni, B.,** Flux of cadmium through euphausiids, *Mar. Biol.,* 27, 205, 1974.
19. **Fowler, S. W., Heyraud, M., and La Rosa, J.,** Mercury kinetics in marine zooplankton, in *Activities of the International Laboratory of Marine Radioactivity, 1976 Report,* International Atomic Energy Agency, Vienna, 1976, 20.
20. **Fowler, S. W., Heyraud, M., and Cherry, R. D.,** Accumulation and retention of plutonium by marine zooplankton, in *Activities of the International Laboratory of Marine Radioactivity, 1976 Report,* International Atomic Energy Agency, Vienna, 1976, 42.
21. **Fowler, S. W. and Benayoun, G.,** Selenium kinetics in marine zooplankton, *Mar. Sci. Commun.,* 2, 43, 1976.
22. **Phillips, D. J. H.,** The use of biological indicator organisms to monitor trace metal pollution in marine and estuarine environments — a review, *Environ. Pollut.,* 13, 281, 1977.
23. **Pentreath, R. J. and Fowler, S. W.,** Irradiation of aquatic animals by radionuclide incorporation, in *Methodology For Assessing Impacts of Radioactivity on Aquatic Ecosystems,* Tech. Rept. Ser. No. 190, International Atomic Energy Agency, Vienna, 1979, 129.
24. **Bertine, K. K. and Goldberg, E. D.,** Trace elements in clams, mussels, and shrimp, *Limnol. Oceanogr.,* 17, 877, 1972.
25. **Bowen, V. T., Livingston, H. D., and Burke, J. C.,** Distribution of transuranium nuclides in sediment and biota of the North Atlantic Ocean, in *Transuranium Nuclides in the Environment,* International Atomic Energy Agency, Vienna, 1976, 107.
26. **Higgo, J. J. W., Cherry, R. D., Heyraud, M., and Fowler, S. W.,** Rapid removal of plutonium from the oceanic surface layer by zooplankton faecal pellets, *Nature (London),* 266, 623, 1977.

27. **Higgo, J. J. W., Cherry, R. D., Heyraud, M., Fowler, S. W., and Beasley, T. M.,** The vertical oceanic transport of alpha radioactive nuclides by zooplankton fecal pellets, in *Natural Radiation Environment III,* Vol. I, Gesell, T. F. and Lowder, W. N., Eds., NTIS CONF-780422, 1980, 502.
28. **Fowler, S. W.,** Trace elements in zooplankton particulate products, *Nature,* 269, 51, 1977.
29. **Guary, J. C., Fowler, S. W., and Beasley, T. M.,** MS in preparation, 1981.
30. **Coombs, T. L., Fletcher, T. C., and White, A.,** Interaction of metal ions with mucus from the plaice (*Pleuronectes platessa* L.), *Biochem. J.,* 128, 128, 1972.
31. **Chow, T. J., Patterson, C. C., and Settle, D.,** Occurrence of lead in tuna, *Nature (London),* 251, 159, 1974.
32. **Fowler, S. W., Heyraud, M., and Beasley, T. M.,** Experimental studies on plutonium kinetics in marine biota, in *Impacts of Nuclear Releases into the Aquatic Environment,* International Atomic Energy Agency, Vienna, 1975, 157.
33. **Somero, G. N., Chow, T. J., Yancey, P. H., and Snyder, C. B.,** Lead accumulation rates in tissues of the estuarine teleost fish, *Gillichthys mirabilis*: salinity and temperature effects, *Arch. Environ. Contam. Toxicol.,* 6, 337, 1977.
34. **Pentreath, R. J.,** The accumulation from water of ^{65}Zn, ^{54}Mn, ^{58}Co and ^{59}Fe by the mussel, *Mytilus edulis, J. Mar. Biol. Assoc. U.K.,* 53, 127, 1973.
35. **Fowler, S. W. and Benayoun, G.,** Accumulation and distribution of selenium in mussel and shrimp tissues, *Bull. Environ. Contam. Toxicol.,* 16, 339, 1976.
36. **Pentreath, R. J.,** The accumulation of inorganic mercury from sea water by the plaice, *Pleuronectes platessa* L., *J. Exp. Mar. Biol. Ecol.,* 24, 103, 1976.
37. **Nimmo, D. R., Lightner, D. V., and Bahner, L. H.,** Effects of cadmium on the shrimp, *Penaeus duorarum, Palaemonetes pugio,* and *Palaemonetes vulgaris,* in *Physiological Responses of Marine Biota to Pollutants,* Vernberg, F. J., Calabrese, A., Thurberg, F. P. and Vernberg, W. B., Eds., Academic Press, New York, 1977, 131.
38. **George, S. G., Pirie, B. J. S., and Coombs, T. L.,** Absorption, accumulation, and excretion of iron-protein complexes by *Mytilus edulis* (L), in *Int. Conf. Heavy Metals in the Environment, Proc. 2,* Hutchinson, T. C., Ed., University of Toronto, Toronto, 1977, 887.
39. **Pentreath, R. J.,** The accumulation of organic mercury from sea water by the plaice, *Pleuronectes platessa* L., *J. Exp. Mar. Biol. Ecol.,* 24, 121, 1976.
40. **Fowler, S. W., Heyraud, M., and La Rosa, J.,** Factors affecting methyl and inorganic mercury dynamics in mussels and shrimp, *Mar. Biol.,* 46, 267, 1978.
41. **Pentreath, R. J.,** The accumulation of arsenic by the plaice and thornback ray: some preliminary observations, C. M., 1977/E:17, Fisheries Improvement Comm., ICES, Copenhagen, 1977.
42. **Fowler, S. W. and Ünlü, M. Y.,** Factors affecting bioaccumulation and elimination of arsenic in the shrimp *Lysmata seticaudata, Chemosphere,* 7, 711, 1978.
43. **Cross, F. A., Hardy, L. H., Jones, N. V., and Barber, R. T.,** Relation between total body weight and concentrations of manganese, iron, copper, zinc, and mercury in white muscle of bluefish (*Pomatomus saltatrix*) and a bathy-demersal fish *Antimona rostrata, J. Fish. Res. Board Can.,* 30, 1287, 1973.
44. **Fowler, S. W.,** The effect of organism size on the content of certain trace metals in marine zooplankton, *Rapp. Comm. Int. Mer. Médit.,* 22, 145, 1974.
45. **Boyden, C. R.,** Effect of size upon metal content of shellfish, *J. Mar. Biol. Assoc. U. K.,* 57, 675, 1977 .
46. **Cunningham, P. A. and Tripp, M. R.,** Factors affecting the accumulation and removal of mercury from tissues of the American oyster *Crassostrea virginica, Mar. Biol.,* 31, 311, 1975.
47. **Barber, R. T., Vijayakumar, A., and Cross, F. A.,** Mercury concentrations in recent and ninety-year-old benthopelagic fish, *Science,* 178, 636, 1972.
48. **Davies, A. G.,** Studies of the accumulation of radio-iron by a marine diatom, in *Radioecological Concentration Processes,* Aberg, B. and Hungate, F. P., Eds., Pergamon Press, Oxford, 1967, 983.
49. **Bernhard, M. and Zattera, A.,** A comparison between the uptake of radioactive and stable zinc by a marine unicellular alga, in *Symposium on Radioecology,* Proc. 2nd Natl. Symp., CONF 670503, USAEC TID-4500, Nelson, D. J. and Evans, F. C., Eds., Ann Arbor, Michigan, 1969, 389.
50. **Myers, V. B., Iverson, R. L., and Harriss, R. C.,** The effect of salinity and dissolved organic matter on surface charge characteristics of some euryhaline phytoplankton, *J. Exp. Mar. Biol. Ecol.,* 17, 59, 1975.
51. **Sunda, W., and Guillard, R. R. L.,** The relationship between cupric ion activity and the toxicity of copper to phytoplankton, *J. Mar. Res.,* 34, 511, 1976.
52. **Cossa, D.,** Sorption du cadmium par une population de la diatomée *Phaeodactylum tricornutum* en culture, *Mar. Biol.,* 34, 163, 1976.
53. **Lowman, F. G. and Ting, R. Y.,** The state of cobalt in seawater and its uptake by marine organisms and sediments, in *Radioactive Contamination of the Marine Environment,* International Atomic Energy Agency, Vienna, 1973, 369.

54. **Small, L. F.**, Experimental studies on the transfer of ^{65}Zn in high concentration by euphausiids, *J. Exp. Mar. Biol. Ecol.*, 3, 106, 1969.
55. **Kečkeš, S., Ozretić, B., and Krajnović, M.**, Metabolism of Zn^{65} in mussels (*Mytilus galloprovincialis* Lam.). Uptake of Zn^{65}, *Rapp. Comm. Int. Mer Medit.*, 19, 949, 1969.
56. **George, S. G. and Coombs, T. L.**, The effects of chelating agents on the uptake and accumulation of cadmium by *Mytilus edulis, Mar. Biol.*, 39, 261, 1977.
57. **George, S. G. and Coombs, T. L.**, Effects of high stability iron complexes on the kinetics of iron accumulation in *Mytilus edulis* (L), *J. Exp. Mar. Biol. Ecol.*, 28, 133, 1977.
58. **Corner, E. D. S. and Rigler, F. H.**, The modes of action of toxic agents. III. Mercury chloride and n-amylmercuric chloride on crustaceans, *J. Mar. Biol. Assoc. U.K.* 37, 85, 1958.
59. **Mandelli, E. F.**, The inhibitory effects of copper on marine phytoplankton, *Contrib. Mar. Sci.*, 14, 47, 1969.
60. **Fowler, S. W. and Small, L. F.**, Procedures involved in radioecological studies with marine zooplankton, in *Design of Radiotracer Experiments in Marine Biological Systems*, Tech. Rept. Ser. No. 167, International Atomic Energy Agency, Vienna, 1975, 63.
61. **Bryan, G. W. and Hummerstone, L. G.**, Adaption of the polychaete *Nereis diversicolor* to estuarine sediments containing high concentrations of heavy metals. I. General observations and adaption to copper, *J. Mar. Biol. Assoc. U.K.*, 51, 845, 1971.
62. **Bryan, G. W. and Hummerstone, L. G.**, Adaption of the polychaete *Nereis diversicolor* to estuarine sediments containing high concentrations of zinc and cadmium, *J. Mar. Biol. Assoc., U.K.*, 53, 839, 1973.
63. **Chipman, W. A.** Uptake and accumulation of chromium-51 by the clam, *Tapes decussatus*, in relation to physical and chemical form, in *Disposal of Radioactive Wastes into Seas, Oceans and Surface Waters*, International Atomic Energy Agency, Vienna, 1966, 571.
64. **Schulz-Baldes, M.**, Lead uptake from sea water and food, and lead loss in the common mussel, *Mytilus edulis, Mar. Biol.*, 25, 177, 1974.
65. **Fowler, S. W. and Benayoun, G.**, Experimental studies on cadmium flux through marine biota, in *Comparative Studies of Food and Environmental Contamination*, International Atomic Energy Agency, Vienna, 1974, 159.
66. **Fowler, S. W. and Benayoun, G.**, Influence of environmental factors on selenium flux in two marine invertebrates, *Mar. Biol.*, 37, 59, 1976.
67. **Bryan, G. W.**, The metabolism of Zn and ^{65}Zn in crabs, lobsters, and fresh water crayfish, in *Radioecological Concentration Processes*, Aberg, B. and Hungate, F. P., Eds., Pergamon Press, 1967, 1005.
68. **Kečkeš, S., Pucar, Z., and Marazovic, L.**, Accumulation of electrodialytically separated physico-chemical forms of Ru-106 by mussels, *Int. J. Oceanol. Limnol.*, 1, 246, 1967.
69. **Kečkeš, S., Fowler, S. W., and Small, L. F.**, Flux of different forms of ^{106}Ru through a marine zooplankter, *Mar. Biol.*, 13, 94, 1972.
70. **Cross, F. A., Dean, J. M., and Osterberg, C. L.**, The effect of temperature, sediment and feeding on the behavior of four radionuclides in a marine benthic amphipod, in *Symposium on Radioecology*, Proc. 2nd Natl. Symp., CONF 670503, USAEC TID-4500, Nelson, D. J. and Evans, F. C., Eds., Ann Arbor, Michigan, 1969, 450.
71. **Van Weers, A. W.**, The effect of temperature on the uptake and retention of ^{60}Co and ^{65}Zn by the common shrimp *Crangon crangon* (L), in *Combined Effects of Radioactive, Chemical and Thermal Releases to the Environment*, International Atomic Energy Agency, Vienna, 1975, 35.
72. **Ünlü, M. Y. and Fowler, S. W.**, Factors affecting the flux of arsenic through the mussel *Mytilus galloprovincialis, Mar. Biol.*, 51, 209, 1979.
73. **Pentreath, R. J.**, Radiobiological studies with marine fish, in *Design of Radiotracer Experiments in Marine Biological Systems*, Tech. Rept. Ser. No. 167, International Atomic Energy Agency, Vienna, 1975, 137.
74. **O'Hara, J.**, Cadmium uptake by fiddler crabs exposed to temperature and salinity stress, *J. Fish. Res. Board Can.*, 30, 846, 1973.
75. **Phillips, D. J. H.**, The common mussel *Mytilus edulis* as an indicator of pollution by zinc, cadmium, lead and copper. I. Effect of environmental variables on uptake of metals, *Mar. Biol.*, 38, 59, 1976.
76. **Jackim, E., Morrison, G., and Steele, R.**, Effects of environmental factors on radiocadmium uptake by four species of marine bivalves, *Mar. Biol.*, 40, 303, 1977.
77. **Wright, D. A.**, The effect of salinity on cadmium uptake by the tissues of the shore crab *Carcinus maenas, J. Exp. Biol.*, 67, 137, 1977.
78. **George, S. G., Carpene, E., and Coombs, T. L.**, The effect of salinity on the uptake of cadmium by the common mussel, *Mytilus edulis* (L.), in *Physiology and Behaviour of Marine Organisms*, McLusky, D. S. and Berry, A. J., Eds., Pergamon Press, Oxford, 1978.
79. **Gutknecht, J.**, Zn^{65} uptake by benthic marine algae, *Limnol. Oceanogr.*, 8, 31, 1963.

80. **Grimes, D. J. and Morrison, S. M.,** Bacterial bioconcentration of chlorinated hydrocarbon insecticides from aqueous systems, *Microb. Ecol.*, 2, 43, 1975.
81. **Cox, J. L.,** Low ambient level uptake of ^{14}C- D.D.T. by three species of marine phytoplankton, *Bull. Environ. Contam. Toxicol.*, 5, 218, 1970.
82. **Harding, L. W., Jr. and Phillips, J. H., Jr.,** Polychlorinated biphenyl (PCB) uptake by marine phytoplankton, *Mar. Biol.*, 49, 103, 1978.
83. **Harding, L. W., Jr. and Phillips, J. H., Jr.,** Polychlorinated biphenyls: transfer from microparticulates to marine phytoplankton and the effects on photosynthesis, *Science*, 202, 1189, 1978.
84. **Addison, R. F.,** Organochlorine compounds in aquatic organisms: their distribution, transport and physiological significance, in *Effects of Pollutants on Aquatic Organisms*, Lockwood, A. P. M., Ed., Cambridge University Press, Cambridge, 1976, 127.
85. **Lee, R. F., Gardner, W. S., Anderson, J. W., Blaylock, J. W., and Barwell-Clarke, J.,** Fate of polycyclic aromatic hydrocarbons in controlled ecosystem enclosures, *Environ. Sci. Technol.*, 12, 832, 1978.
86. **Kauss, P., Hutchinson, T. C., Soto, C., Hellebust, J., and Griffiths, M.,** The toxicity of crude oil and its components to freshwater algae, in *Proc. 1973 Conf. Prevention and Control of Oil Spills*, American Petroleum Institute, Washington, D.C., 1973, 703.
87. **Soto, C., Hellebust, J. A., and Hutchinson, T. C.,** Effect of naphthalene and aqueous crude oil extracts on the green flagellate *Chlamydomonas angulosa*. II. Photosynthesis and the uptake and release of naphthalene, *Can. J. Bot.*, 53, 118, 1975.
88. **Corner, E. D. S.,** Pollution studies with marine plankton. I. Petroleum hydrocarbons and related compounds, in *Adv. Mar. Biol.*, 15, 289, 1978.
89. **Ware, D. M. and Addison, R. F.,** PCB residues in plankton from the Gulf of St. Lawrence, *Nature (London)*, 246, 519, 1973.
90. **Harvey, G. R., Miklas, H. P., Bowen, V. T., and Steinhauer, W. G.,** Observations on the distribution of chlorinated hydrocarbons in Atlantic ocean organisms, *J. Mar. Res.*, 32, 103, 1974.
91. **Fowler, S. W. and Elder, D. L.,** PCB and DDT residues in a Mediterranean pelagic food chain, *Bull. Environ. Contam. Toxicol.*, 19, 244, 1978.
92. **Cox, J. L.,** Uptake, assimilation and loss of DDT residues by *Euphausia pacifica*, a euphausiid shrimp, *Fish. Bull.*, 69, 627, 1971.
93. **Harding, G. C. H. and Vass, W. P.,** Uptake from sea water and clearance of ^{14}C-p,p′-DDT by the marine copepod *Calanus finmarchicus*, *J. Fish. Res.* Board Can., 34, 177, 1977.
94. **Clayton, J. R., Jr., Pavlou, S. P., and Breitner, N. F.,** Polychlorinated biphenyls in coastal marine zooplankton: bioaccumulation by equilibrium partitioning, *Environ. Sci. Technol.*, 11, 676, 1977.
95. **Darrow, D. C. and Harding, G. C.,** Accumulation and apparent absence of DDT metabolism by marine copepods, *Calanus* spp., in culture, *J. Fish. Res. Board Can.*, 32, 1845, 1975.
96. **Lee, R. F.,** Fate of petroleum hydrocarbons in marine zooplankton, in *Proc. 1975 Conf. Petroleum and Control of Oil Pollution*, American Petroleum Institute, Washington, D.C., 1975, 549.
97. **Harris, R. P., Berdugo, V., Corner, E. D. S., Kilvington, C. C., and O'Hara, S. C. M.,** Factors affecting the retention of a petroleum hydrocarbon by marine planktonic copepods, in *Fate and Effects of Petroleum Hydrocarbons in Marine Organisms and Ecosystems*, Wolfe, D. A., Ed., Pergamon Press, Oxford, 1977, chap. 30.
98. **Lowe, J. I., Parrish, P. R., Patrick, J. M., Jr., and Forester, J.,** Effects of the polychlorinated bipheynl arochlor 1254 on the American oyster *Crassostrea virginica*, *Mar. Biol.*, 17, 209, 1972.
99. **Sanders, H. O. and Chandler, J. H.,** Biological magnification of a polychlorinated bipheynl (Aroclor® 1254) from water by aquatic invertebrates, *Bull. Environ. Contam. Toxicol.*, 7, 257, 1972.
100. **Vreeland, V.,** Uptake of chlorobiphenyls by oysters, *Environ. Pollut.*, 6, 135, 1974.
101. **Nimmo, D. R., Forester J., Heitmuller, P. T., and Cook, G. H.,** Accumulation of Aroclor® 1254 in grass shrimp (*Palaemonetes pugio*) in laboratory and field exposures, *Bull. Environ. Contam. Toxicol.*, 11, 303, 1974.
102. **Nimmo, D. R., Hansen, D. J., Couch, J. A., Cooley N. R., Parrish, P. H., and Lowe, J. I.,** Toxicity of Aroclor® 1254 and its physiological activity in several estuarine organisms, *Arch. Environ. Contam. Toxicol.*, 3, 221, 1975.
103. **Fowler, S. W., Polikarpov, G. G., Edler, D. L., Parsi, P., and Villeneuve, J. P.,** Polychlorinated biphenyls: accumulation from contaminated sediments and water by the polychaete *Nereis diversicolor*, *Mar. Biol.*, 48, 303, 1978.
104. **Sheridan, P. F.,** Uptake, metabolism and distribution of DDT in organs of the blue crab, *Callinectes sapidus*, *Chesapeake Sci.*, 16, 20, 1975.
105. **Lee, R. F., Sauerheber, R., and Benson, A. A.,** Petroleum hydrocarbons: uptake and discharge by the marine mussel *Mytilus edulis*, *Science*, 177, 344, 1972.
106. **Stegeman, J. J. and Teal, J. M.,** Accumulation, release and retention of petroleum hydrocarbons by the oyster *Crassostrea virginica*, *Mar. Biol.*, 22, 37, 1973.

107. **Neff, J. M. and Anderson, J. W.,** Accumulation, release and distribution of benzo[*a*]pyrene-C^{14} in the clam *Rangia cuneata*, in *Proc. 1975 Conf. Control of Oil Pollution*, American Petroleum Institute, Washington, D.C., 1975, 469.
108. **Neff, J. M., Cox, B. A., Dixit, D., and Anderson, J. W.,** Accumulation and release of petroleum-derived aromatic hydrocarbons by four species of marine animals, *Mar. Biol.*, 38, 279, 1976.
109. **Lee, R. F.,** Accumulation and turnover of petroleum hydrocarbons in marine organisms, in *Fate and Effects of Petroleum Hydrocarbons in Marine Organisms and Ecosystems*, Wolfe, D. A., Ed., Pergamon Press, Oxford, 1977, chap. 6.
110. **Varanasi, U. and Malins, D. C.,** Metabolism of petroleum hydrocarbons: accumulation and biotransformation in marine organisms, in *Effects of Petroleum on Arctic and Subarctic Marine Environments and Organisms*, Vol. II, Malins, D. C., Ed., Academic Press, New York, 1977, chap. 3.
111. **Lee, R. F., Ryan, C., and Neuhauser, M. L.,** Fate of petroleum hydrocarbons taken up from food and water by the blue crab *Callinectes sapidus*, *Mar. Biol.*, 37, 363, 1976.
112. **Fucik, K. W. and Neff, J. M.,** Effects of temperature and salinity of naphthalenes uptake in the temperature clam, *Rangia cuneata*, and the boreal clam, *Protothaca staminea*, in *Fate and Effects of Petroleum Hydrocarbons in Marine Organisms and Ecosystems*, Wolf, D. A., Ed., Pergamon Press, Oxford, 1977, chap. 31.
113. **Lee, R. F., Sauerheber, R., and Dobbs, G. H.,** Uptake, metabolism, and discharge of polycyclic aromatic hydrocarbons by marine fish, *Mar. Biol.*, 17, 201, 1972.
114. **Fowler, S. W., Small, L. F., and Dean, J. M.,** Distribution of ingested zinc-65 in the tissues of some marine crustaceans, *J. Fish. Res. Board Can.*, 27, 1051, 1970.
115. **Bryan, G. W.,** Zinc regulation in the lobster *Homarus vulgaris*, *J. Mar. Biol. Assoc. U.K.*, 44, 549, 1964.
116. **Bryan, G. W. and Ward, E.,** The absorption and loss of radioactive and non-radioactive manganese by the lobster, *Homarus vulgaris*, *J. Mar. Assoc. U.K.*, 45, 65, 1965.
117. **Amiard-Triquet, C. et Amiard, J. C.,** L'organotrophisme du ^{60}Co chez *Scrobicularia plana* et *Carcinus maenas* en fonction du vecteur de contamination, *Oikos*, 27, 122, 1976.
118. **Pentreath, R. J.,** The accumulation of mercury from food by the plaice, *Pleuronectes platessa* L., *J. Exp. Mar. Biol. Ecol.*, 25, 51, 1976.
119. **Pentreath, R. J.,** The accumulation of cadmium by the plaice, *Pleuronectes platessa* L. and the thornback ray, *Raja clavata* L., *J. Exp. Mar. Biol. Ecol.*, 30, 223, 1977.
120. **Pentreath, R. J.,** The accumulation of ^{110m}Ag by the plaice *Pleuronectes platessa* L. and the thornback ray, *Raja clavata* L., *J. Exp. Mar. Biol. Ecol.*, 29, 315, 1977.
121. **Pentreath, R. J.,** ^{237}Pu experiments with plaice *Pleuronectes platessa*, *Mar. Biol.*, 48, 327, 1978.
122. **Pentreath, R. J.,** ^{237}Pu experiments with the thornback ray *Raja clavata*, *Mar. Biol.*, 48, 327, 1978.
123. **Hoss, E. D.,** Accumulation of zinc-65 by flounder of the genus *Paralichtys*, *Trans. Am. Fish. Soc.*, 93, 364, 1964.
124. **Wieser, W.,** Conquering terra firma: the copper problem from the isopod's point of view, *Helgol. Wiss. Meersunters.*, 15, 282, 1967.
125. **Amiard-Triquet, C. and Amiard, J. C.,** Etude expérimental du transfert du cobalt-60 dans une chaine trophique marine benthique, *Helgol. Wiss. Meeresunters.*, 27, 283, 1975.
126. **Fowler, S. W. and Guary, J. C.,** High absorption efficiency for ingested plutonium in crabs, *Nature (London)*, 266, 827, 1977.
127. **Jennings, J. R. and Rainbow, P. S.,** Studies on the uptake of cadmium by the crab *Carcinus maenas* in the laboratory. I. Accumulation from seawater and a food source, *Mar. Biol.*, 50, 131, 1979.
128. **Freegarde, M., Hatchard, C. G., and Parker, C. A.,** Oil spilt at sea: its identification, determination, and ultimate fate, *Lab. Pract.*, 20, 35, 1971.
129. **Conover, R. J.,** Some relations between zooplankton and Bunker C oil in Chedabucto bay following the wreck of the tanker "Arrow", *J. Fish. Res. Board Can.*, 28, 1327, 1971.
130. **Petrocelli, S. R., Anderson, J. W., and Hanks, A. R.,** Controlled food-chain transfer of dieldrin residues from phytoplankters to clams, *Mar. Biol.*, 31, 215, 1975.
131. **Langston, W. J.,** Accumulation of polychlorinated biphenyls in the cockle *Cerastoderma edule* and the tellin *Macoma balthica*, *Mar. Biol.*, 45, 265, 1978.
132. **Goerke, H. and Ernst, W.,** Fate of ^{14}C-labelled di-, tri- and pentachlorobiphenyl in the marine annelid *Nereis virens*. I. Accumulation and elimination after oral administration, *Chemosphere*, 6, 551, 1977.
133. **Blackman R. A. A.,** Effects of sunken crude oil on the feeding and survival of the brown shrimp - *Crangon crangon*, C.M. 1972/ K:13, *Fisheries Improvement Comm., ICES*, Copenhagen, 1972.
134. **Fossato, V. U. and Canzonier, W. J.,** Hydrocarbon uptake and loss by the mussel *Mytilus edulis*, *Mar. Biol.*, 36, 243, 1976.
135. **Dunn, B. P. and Stich, H. F.,** Release of the carcinogen benzo[*a*]pyrene from environmentally contaminated mussels, *Bull. Environ. Contam. Toxicol.*, 15, 398, 1976.

136. **Rossi, S. S.,** Bioavailability of petroleum hydrocarbons from water, sediments and detritus to the marine annelid, *Neanthes arenaceodentata,* in *Proc. Joint Conf. Prevention and Control of Oil Spills,* American Petroleum Institute, Washington, D.C., 1977, 621.
137. **Gordon, D. C., Jr., Dale, J., and Keizer, P. D.,** Importance of sediment working by the deposit-feeding polychaete *Arenicola marina* on the weathering rate of sediment-bound oil, *J. Fish. Res. Board Can.,* 35, 591, 1978.
138. **Zitko, V.,** Uptake of chlorinated paraffins and PCB from suspended solids and food by juvenile Atlantic salmon, *Bull. Environ. Contam. Toxicol.,* 12, 406, 1974.
139. **Gruger, E. H., Jr., Karrick, N. L., Davidson, A. I., and Hruby, T.,** Accumulation of 3,4,3′,4′-tetrachlorobiphenyl and 2,4,5,2′,4′,5′- and 2,4,6,2′,4′,6′-hexachlorobiphenyl in juvenile coho salmon, *Environ. Sci. Technol.,* 9, 121, 1975.
140. **Horn, M. H., Teal, J. M., and Backus, R. H.,** Petroleum lumps on the surface of the sea, *Science,* 168, 245, 1970.
141. **Hardy, R., Mackie, P. R., Whittle, K. J., and McIntyre, A. D.,** Discrimination in the assimilation of n-alkanes in fish, *Nature (London),* 252, 577, 1974.
142. **Blackman, R. A. A. and Mackie, P.,** Preliminary results of an experiment to measure the uptake of n-alkane hydrocarbons by fish, C. M. 1973/E:23, *Fisheries Improvement Comm., ICES,* Copenhagen, 1973.
143. **Corner, E. D. S., Harris, R. P., Whittle, K. J., and Mackie, P. R.,** Hydrocarbons in marine zooplankton and fish, in *Effects of Pollutants on Aquatic Organisms,* Lockwood, A. P. M., Ed., Cambridge University Press, Cambridge, 1976, 71.
144. **Roubal, W. T., Collier, T. K., and Malins, D. C.,** Accumulation and metabolism of carbon-14 labelled benzene, naphthalene and anthracene by young coho salmon (*Oncorhynchus kisutch*), *Arch. Environ. Contam. Toxicol.,* 5, 513, 1977.
145. **Fowler, S. W. and Small, L. F.** Moulting of *Euphausia pacifica* as a possible mechanism for vertical transport of zinc-65 in the sea. *Int. J. Oceanol. Limnol.,* 1, 237, 1967.
146. **Nassogne, A.,** First heterotrophic level of the food chain, in *Studies on the Radioactive Contamination of the Sea, Annual Report 1972,* Bernhard, M., Ed., CNEN Report No. RT/B10(74)35, CNEN, Fiascherino, La Spezia, Italy, 1974, 30.
147. **Small, L. F., Fowler, S. W., and Keckes, S.,** Flux of zinc through a macroplanktonic crustacean, in *Radioactive Contamination of the Marine Environment,* International Atomic Energy Agency, Vienna, 1973, 437.
148. **Preston, E. M.,** The importance of ingestion in chromium-51 accumulation by *Crassostrea virginica* (Gmelin), *J. Exp. Mar. Biol. Ecol.,* 6, 47, 1971.
149. **Baptist, J. P. and Lewis, C. W.,** Transfer of ^{65}Zn and ^{51}Cr through an estuarine food chain, in *Symposium on Radioecology,* Proc. 2nd Nat. Symp., CONF 670503, USAEC TID-4500, Nelson, D. J. and Evans, F. C., Eds., Ann Arbor, Michigan, 1969, 420.
150. **Young, M. L.,** The transfer of ^{65}Zn and ^{59}Fe along a *Fucus serratus* (L) *Littorina obtusata* (L) food chain, *J. Mar. Biol. Assoc. U.K.,* 55, 583, 1975.
151. **Renfro, W. C., Fowler, S. W., Heyraud, M., and La Rosa, J.,** The relative importance of food and water in long term zinc-65 accumulation by marine biota, *J. Fish. Res. Board Can.,* 32, 1339, 1975.
152. **Pentreath, R. J.,** The roles of food and water in the accumulation of radionuclides by marine teleost and elasmobranch fish, in *Radioactive Contamination of the Marine Environment,* International Atomic Energy Agency, Vienna, 1973, 421.
153. **Beasley, T. M. and Fowler, S. W.,** Plutonium and americium: uptake from contaminated sediments by the polychaete *Nereis diversicolor., Mar. Biol.,* 38, 95, 1976.
154. **Epifanio, C. E.,** Dieldrin uptake by larvae of the crab *Leptodius floridanus, Mar. Biol.,* 19, 320, 1973.
155. **Scura, E. D. and Theilacker, G. H.,** Transfer of the chlorinated hydrocarbon PCB in a laboratory marine food chain, *Mar. Biol.,* 40, 317, 1977.
156. **Bahner, L. H., Wilson, A. J., Jr., Sheppard, J. M., Patrick, J. M., Jr., Goodman, L. R., and Walsh, G. E.,** Kepone® bioconcentration, accumulation, loss and transfer through estuarine food chains, *Chesapeake Sci.,* 18, 299, 1977.
157. **Teal, J. M.,** Food chain transfer of hydrocarbons, in *Fate and Effects of Petroleum Hydrocarbons in Marine Organisms and Ecosystems,* Wolfe, D. A., Ed., Pergamon Press, Oxford, 1977, chap. 7.
158. **Burns, K. A. and Teal, J. M.,** Hydrocarbons in the pelagic *Sargassum* community, *Deep Sea Res.,* 20, 207, 1973.
159. **Morris, B. F., Cadwallader, J., Geiselman, J., and Butler, J. N.,** Transfer of petroleum and biogenic hydrocarbons in the *Sargassum* community, in *Marine Pollutant Transfer,* Windom, H. L. and Duce, R. A., Eds., Lexington Books, Lexington, Mass., 1976, chap. 11.
160. **Rice, T. R. and Willis, V. M.,** Uptake, accumulation and loss of radioactive cerium-144 by marine planktonic algae, *Limnol. Oceanogr.,* 4, 277, 1959.

161. **Kuenzler, E. J.,** Elimination of iodine, cobalt, iron and zinc by marine zooplankton, in *Symposium on Radioecology,* Proc. 2nd Natl. Symp., CONF 670503, USAEC TID-4500, Nelson, D. J. and Evans, F. C., Eds., Ann Arbor, Michigan, 1969, 462.
162. **Kuenzler, E. J.,** Elimination and transport of cobalt by marine zooplankton, in *Symposium on Radioecology,* Proc. 2nd Nat. Symp., CONF 670503, USAEC TID-4500, Nelson, D. J. and Evans, F. C., Eds., Ann Arbor, Michigan, 1969, 483.
163. **Fowler, S. W., Small, L. F., and Dean, J. M.,** Experimental studies on elimination of zinc-65, cesium-137 and cerium-144 by euphausiids, *Mar. Biol.,* 8, 224, 1971.
164. **Fowler, S. W., Small, L. F., and Kečkeš, S.,** Effects of temperature and size on molting of euphausiid crustaceans, *Mar. Biol.,* 11, 45, 1971.
165. **Childress, J. J. and Price, M. H.,** Growth rate of the bathypelagic crustacean *Gnathophausia ingens* (Mysidacea: Lophogastridae). I. Dimentional growth and population structure, *Mar. Biol.,* 50, 47, 1978.
166. **Zattera, A., Bernhard, M., and Galli, C.,** Radiotracer experiments with benthic marine algae, in *Design of Radiotracer Experiments in Marine Biological Systems,* Tech. Rept. Ser. No. 167, International Atomic Energy Agency, Vienna, 1975, 85.
167. **Luoma, S. N.,** The dynamics of biologically available mercury in a small estuary, *Estuar. Coast. Mar. Sci.,* 5, 643, 1977.
168. **Amiard-Triquet, C.,** Etude de la décontamination d'*Arenicola marina* (annelide, polychete) après contamination expérimentale par le caesium-137 ou le cobalt-60, *Mar. Biol.,* 26, 161, 1974.
169. **Young, D. R. and Folsom, T. R.,** Loss of ^{65}Zn from the California sea mussel *Mytilus californianus, Biol. Bull.,* 133, 438, 1967.
170. **Van Weers, A. W.,** Uptake and loss of ^{65}Zn and ^{60}Co by the mussel *Mytilus edulis* L., in *Radioactive Contamination of the Marine Environment,* International Atomic Energy Agency, Vienna, 1973, 385.
171. **Guary, J. C. and Fowler, S. W.,** Elimination et répartition du ^{241}Am et du ^{237}Pu chez la moule *Mytilus galloprovincialis* dans son environnement naturel, *Rapp. Comm. Int. Mer Médit.,* 25/26, 53, 1979.
172. **Guary, J. C. and Fowler, S. W.,** Biokinetics of neptunium-237 in mussels and shrimp, *Mar. Sci. Commun.,* 3, 211, 1977.
173. **Boothe, P. N. and Knauer, G. A.,** The possible importance of fecal material in the biological amplification of trace and heavy metals, *Limnol. Oceanogr.,* 17, 270, 1972.
174. **Fowler, S. W., La Rosa, J., Heyraud, M., and Renfro, W. C.,** Effect of different radiotracer labelling techniques on radionuclide excretion from marine organisms, *Mar. Biol.,* 30, 297, 1975.
175. **Cunningham, P. A. and Tripp, M. R.,** Accumulation, tissue distribution and elimination of $^{203}HgCl_2$ and $CH_3{}^{203}HgCl$ in the tissues of the American oyster *Crassostrea virginica, Mar. Biol.,* 31, 321, 1975.
176. **Dillon, T. M. and Neff, J. M.,** Mercury and the estuarine marsh clam, *Rangia cuneata* Gray. II. Uptake, tissue distribution and depuration, *Mar. Environ. Res.,* 1, 67, 1978.
177. **Baptist, J. P., Hoss, D. E., and Lewis, C. W.,** Retention of ^{51}Cr, ^{59}Fe, ^{60}Co, ^{65}Zn, ^{85}Sr, ^{95}Nb, ^{141m}In and ^{131}I by the Atlantic croaker (*Micropogon undulatus*), *Health Phys.,* 18, 141, 1970.
178. **Elder, D. L. and Fowler, S. W.** Polychlorinated biphenyls: penetration into the deep ocean by zooplankton fecal pellet transport, *Science,* 197, 459, 1977.
179. **Mason, J. W. and Rowe, D. R.,** The accumulation and loss of dieldrin and endrin in the eastern oyster, *Arch. Environ. Contam. Toxicol.,* 4, 349, 1976.
180. **Butler, P. A.,** Residues in fish, wildlife and estuaries, *Pestic. Monit. J.,* 6, 238, 1973.
181. **Nimmo, D. R., Blackman, R. R., Wilson, A. J., Jr., and Forester, J.,** Toxicity and distribution of Aroclor® 1254 in the pink shrimp *Penaeus duorarum, Mar. Biol.,* 11, 191, 1971.
182. **Courtney, W. A. M. and Denton, G. R. W.,** Persistence of polychlorinated biphenyls in the hard clam (*Mercenaria mercenaria*) and the effect upon the distribution of these pollutants in the estuarine environment, *Environ. Pollut.,* 10, 55, 1976.
183. **Bend, J. R., Bend, S. G., Guarino, A. M., Rall, D. P., and Fouts, J. R.,** Distribution of ^{14}C-2,4,5,2′5′-pentachlorobiphenyl in the lobster *Homarus americanus* at various times after a single injection into the pericardial sinus, *Bull. Mt. Desert Isl. Biol. Lab.,* 13, 1, 1973.
184. **Langston, W. J.,** Persistence of polychlorinated biphenyls in marine bivalves, *Mar. Biol.,* 46, 35, 1978.
185. **Ernst, W., Goerke, H., and Weber, K.,** Fate of ^{14}C-labelled di-, tri- and pentachlorobiphenyl in the marine annelid *Nereis virens.* II. Degradation and faecal elimination, *Chemosphere,* 6, 559, 1977.
186. **DiSalvo, L. H., Guard, H. E., and Hunter, L.,** Tissue hydrocarbon burden of mussels as a potential monitor of environmental hydrocarbon insult, *Environ. Sci. Technol.,* 9, 247, 1975.
187. **Anderson, J. W.,** An assessment of knowledge concerning the fate and effects of petroleum hydrocarbons in the marine environment, in *Marine Pollution: Functional Responses,* Vernberg, W. B., Calabrese, A., Thurberg, F. P., Vernberg, F. J., Eds., Academic Press, New York, 1979, 3.

188. **Anderson, J. W., Moore, L. J., Blaylock, J. W., Woodruff, D. L., and Kiesser, S. L.,** Bioavailability of sediment-sorbed naphthalenes to the sipunculid worm, *Phascolosoma agassizii,* in *Fate and Effects of Petroleum Hydrocarbons in Marine Organisms and Ecosystems,* Wolfe, D. A., Ed., Pergamon Press, Oxford, 1977, chap. 29.
189. **Zinck, M. E. and Addison, R. F.** The fate of 2-, 3-, and 4-chlorobiphenyl following intravenous administration to the thorny skate (*Raja radiata*) and the winter skate *(Raja ocellata), Arch. Environ. Contam. Toxicol.,* 2, 52, 1974.
190. **Khan, M. A. Q.,** Elimination of pesticides by aquatic animals, in *Pesticides in Aquatic Environments,* Khan, M. A. Q., Ed., Plenum Press, New York, 1977, 107.
191. **Ogata, M. and Miyake, Y.,** Disappearance of aromatic hydrocarbons and organic sulfur compounds from fish flesh reared in crude oil suspension, *Water Res.,* 12, 1041, 1978.
192. **Connell, D. W. and Bycroft, B. M.,** Maximum biological half lives of n-alkanes (C_9—C_{13}) in the sea mullet *(Mugil cephalus), Chemosphere,* 7, 779, 1978.
193. **Peden, J. D., Crothers, J. H., Waterfall, C. E., and Beasley, J.,** Heavy metals in Somerset marine organisms, *Mar. Pollut. Bull.,* 4, 7, 1973.
194. **Noshkin, V. E., Bowen, V. T., Wong, K. M., and Burke, J. C.,** Plutonium in north Atlantic ocean organisms; ecological relationships, in *Radionuclides in Ecosystems,* Proc. 3rd Natl. Symp. Radioecology, USAEC-CONF 710501, Oak Ridge, Tenn., 1971, 681.
195. **Fowler, S. W., Guary, J. C., Heyraud, M., and La Rosa, J.,** Assimilation of plutonium in selected marine invertebrates, *Rapp. Comm. Int. Mer Médit.,* 24, 3, 1977.
196. **Jernelov, A.** Heavy metals, metalloids, and synthetic organics, in *The Sea,* Vol. 5, Goldberg, E. D., Ed., John Wiley & Sons, New York, 1974, 799.
197. **Gardner, W. S., Windom, H. L., Stephens, J. A., Taylor, F. E., and Stickney, R. R.,** Concentrations of total mercury and methyl mercury in fish and other coastal organisms: implications to mercury cycling, in *Mineral Cycling in Southeastern Ecosystems,* Howell, F. G., Gentry, J. B., and Smith, M. H., Eds., U.S. Energy Research and Development Administration, Washington, D.C., 1975, 268.
198. **Windom, H. L., Gardner, W. S., Dunstan, W. M., and Paffenhöfer, G. A.,** Cadmium and mercury transfer in a coastal marine ecosystem, in *Marine Pollutant Transfer,* Windom, W. L. and Duce, R. A., Eds., Lexington Books, Lexington, Mass., 1976, chap. 6.
199. **Ratkowsky, D. A., Dix, T. G., and Wilson, K. C.,** Mercury in fish in the Derwent estuary, Tasmania, and its relation to the position of the fish in the food chain, *Aust. J. Mar. Freshwat. Res.,* 26, 223, 1975.
200. **Knauer, G. A. and Martin, J. H.,** Mercury in a marine pelagic food chain, *Limnol. Oceanogr.,* 17, 868, 1972.
201. **Williams, P. M. and Weiss, H. V.,** Mercury in the marine environment: concentration in sea water and in a pelagic food chain, *J. Fish. Res. Board Can.,* 30, 293, 1973.
202. **Parrish, K. M. and Carr, R. A.,** Transport of mercury through a laboratory two-level marine food chain, *Mar. Pollut. Bull.,* 7, 90, 1976.
203. **Woodwell, G. M., Wurster, C. F., Jr., and Isaacson, P. A.,** DDT residues in an east coast estuary: a case of biological concentration of a persistent insecticide, *Science,* 156, 821, 1967.
204. **Jensen, S., Johnels, A. G., Olsson, M., and Otterlind, G.,** DDT and PCB in marine animals from Swedish waters, *Nature (London),* 224, 247, 1969.
205. **Portman, J. E.,** The bioaccumulation and effects of organochlorine pesticides in marine animals, *Proc. R. Soc. London,* 189, 291, 1975.
206. **Schaefer, R. G., Ernst, W., Goerke, H., and Eder, G.,** Residues of chlorinated hydrocarbons in North Sea animals in relation to biological parameters, *Ber. Dt. Wiss. Kommn. Meeresforsch.,* 24, 225, 1976.
207. **Baird, R. C., Thompson, N. P., Hopkins, T. L., and Weiss, W. R.,** Chlorinated hydrocarbons in mesopelagic fishes of the eastern Gulf of Mexico, *Bull. Mar. Sci.,* 25, 473, 1975.
208. NAS, *Petroleum in the Marine Environment,* National Academy of Sciences, Washington, D.C., 1975.
209. **Aarkrog, A.,** Environmental behavior of plutonium accidentally released at Thule, Greenland, *Health Phys.,* 32, 271, 1977.
210. **Vinogradov, M. E.,** Feeding of the deep-sea zooplankton, *Rapp. P. V. Réun. Cons. Int. Explor. Mer,* 153, 114, 1962.
211. **Longhurst, A. R.,** Vertical migration, in *The Ecology of the Seas,* Cushing, D. H. and Walsh, J. J., Eds., Blackwell Scientific Publications, Oxford, 1976, chap. 6.
212. **Ketchum, B. H. and Bowen, V. T.,** Biological factors determining the distribution of radioisotopes in the sea, in *Proc. 2nd Intern. Conf. Peaceful Uses Atomic Energy,* Geneva, 1958, 429.
213. **Pearcy, W. G. and Osterberg, C. L.,** Depth, diel, seasonal, and geographic variations in zinc-65 of midwater animals off Oregon, *Int. J. Oceanol. Limnol.,* 1, 103, 1967.
214. **Osterberg, C. L., Pattulo, J., and Pearcy, W.,** Zinc-65 in euphausiids as related to Columbia River water off the Oregon coast, *Limnol. Oceanogr.,* 9, 249, 1964.

215. **Osterberg, C. L., Carey, A. G., and Curl, H.,** Acceleration of sinking rates of radionuclides in the ocean, *Nature (London)*, 200, 1276, 1963.
216. **Williams, P. M., McGowan, J. A., and Stuiver, M.,** Bomb carbon-14 in deep sea organisms, *Nature (London)*, 227, 375, 1970.
217. **Pearcy, W. G. and Vanderploeg, H. A.,** Radioecology of benthic fishes off Oregon, in *Radioactive Contamination of the Marine Environment*, International Atomic Energy Agency, Vienna, 1973, 245.
218. **Meith-Avcin, N., Warlen, S. M., and Barber, R. T.,** Organochlorine insecticide residues in a bathyl-demersal fish from 2,500 meters, *Environ. Lett.*, 5, 215, 1973.
219. **Teal, J. M.,** Hydrocarbon uptake by deep sea benthos, in *Sources, Effects and Sinks of Hydrocarbons in the Aquatic Environment*, American Institute of Biological Sciences, Washington, D.C., 1976, 359.
220. **Arima, S., Marchand, M., and Martin, J. L.,** Pollutants in deep-sea organisms and sediments, in *The Deep Sea, Ecology and Exploitation, Ambio* Spec. Rept. No. 6, 97, 1979.
221. **Fowler, S. W. and Small, L. F.,** Sinking rates of euphausiid fecal pellets, *Limnol. Oceanogr.*, 17, 293, 1972.
222. **Small, L. F., Fowler, S. W., and Ünlü, M. Y.,** Sinking rates of natural copepod fecal pellets, *Mar. Biol.*, 51, 233, 1979.
223. **Martin, J. H.,** The possible transport of trace metals via moulted copepod exoskeletons, *Limnol. Oceanogr.*, 15, 756, 1970.
224. **Small, L. F. and Fowler, S. W.,** Turnover and vertical transport of zinc by the euphausiid *Meganyctiphanes norvegica* in the Ligurian Sea, *Mar. Biol.*, 18, 284, 1973.
225. **Fowler, S. W. and Oregioni, B.** Elemental concentrations of zooplankton and their particulate products, in *Activities of the International Laboratory of Marine Radioactivity, 1974 Report*, International Atomic Energy Agency, Vienna, 1974, 55.
226. **Heyraud, M., Fowler, S. W., Beasley, T. M., and Cherry, R. D.,** Polonium-210 in euphausiids: a detailed study, *Mar. Biol.*, 34, 127, 1976.
227. **McCave, I. N.,** Vertical flux of particles in the ocean, *Deep Sea Res.*, 22, 491, 1975.
228. **Bishop, J. K. B., Edmond, J. M., Ketten, D. R., Bacon, M. P., and Silker, W. B.,** The chemistry, biology, and vertical flux of particulate matter from the upper 400 m of the equatorial Atlantic Ocean, *Deep Sea Res.*, 24, 511, 1977.
229. **Wiebe, P. H., Boyd, S. H., and Winget, C.,** Particulate matter sinking to the deep sea floor at 2000 m in the Tongue of the Ocean, Bahamas, with a description of a new sedimentation trap, *J. Mar. Res.*, 34, 341, 1976.
230. **Honjo, S.,** Sedimentation of materials in the Sargasso Sea at a 5,367 m deep station, *J. Mar. Res.*, 36, 469, 1978.
231. **Knauer, G. A., Martin, J. H., and Bruland, K. W.,** Fluxes of particulate carbon, nitrogen, and phosphorus in the upper water column of the northeast Pacific, *Deep Sea Res.*, 26, 97, 1979.
232. **Cherry, R. D., Higgo, J. J. W., and Fowler, S. W.,** Zooplankton faecal pellets and element residence times in the ocean, *Nature (London)*, 274, 246, 1978.
233. **Topping, G. and Windom, H. L.,** Biological transport of copper at Loch Ewe and Saanich Inlet; controlled ecosystem pollution experiment, *Bull. Mar. Sci.*, 27, 135, 1977.
234. **Smayda, T. J.,** Normal and accelerated sinking of phytoplankton in the sea, *Mar. Geol.*, 11, 105, 1971.
235. **Rowe, G. T. and Staresinic, N.,** Sources of organic matter to the deep-sea benthos, in *The Deep Sea, Ecology and Exploitation, Ambio* Spec. Rep. No. 6, 19, 1979.
236. **Fowler, S. W., Small, L. F., Elder, D. L., Ünlü, M. Y., and La Rosa, J.,** The role of zooplankton fecal pellets in transporting PCBs from the upper mixed layer to the benthos, in *IVes Journees d'Etudes, Pollutions*, C.I.E.S.M., Monaco, 1979, 289.
237. **Pearcy, W. G., Krygier, E. E., and Cutshall, N. H.,** Biological transport of zinc-65 into the deep sea, *Limnol. Oceanogr.*, 22, 846, 1977.
238. **Spencer, D. W., Brewer, P. G., Fleer, A., Honjo, S.,** Krishnaswami, S., and Nozaki, Y., Chemical fluxes from a sediment trap experiment in the deep Sargasso sea, *J. Mar. Res.*, 36, 493, 1978.
239. **Seki, H., Tsuji, T., and Hattori, A.,** Effect of zooplankton grazing on the formation of the anoxic layer in Tokyo bay, *Estuar. Coast. Mar. Sci.*, 2, 145, 1974.
240. **Davies, J. M.,** Energy flow through the benthos in a Scottish sea loch, *Mar. Biol.*, 31, 353, 1975.
241. **Cross, F. A., Willis, J. N., Hardy, L. H., Jones, N. Y., and Lewis, J. M.,** Role of juvenile fish in cycling of Mn, Fe, Cu, and Zn in a coastal plain estuary, in *Estuarine Research*, Vol. 1, Academic Press, New York, 1975, 45.
242. **Johannes, R. E. and Satomi, M.,** Composition and nutritive value of fecal pellets of a marine crustacean, *Limnol. Oceanogr.*, 11, 191, 1966.
243. **Wheeler, E. H., Jr.,** Copepod detritus in the deep sea, *Limnol. Oceanogr.*, 12, 697, 1967.

244. **Paffenhöfer, G. A. and Strickland, J. D.,** A note on the feeding of *Calanus helgolandicus* on detritus, *Mar. Biol.*, 5, 97, 1970.
245. **Harding, G. C. H.,** The food of deep sea copepods, *J. Mar. Biol. Assoc. U.K.*, 54, 141, 1974.
246. **Sedberry, G. R. and Musick, J. A.,** Feeding strategies of some demersal fishes of the continental slope and rise off the mid-Atlantic coast of the U.S.A., *Mar. Biol.*, 44, 357, 1978.
247. **Windom, H., Gardner, W., Stephens, J., and Taylor, F.,** The role of methylmercury production in the transfer of mercury in a salt marsh ecosystem, *Estuar. Coast. Mar. Sci.*, 4, 579, 1976.
248. **Nimmo, D. R., Wilson, P. D., Blackman, R. R., and Wilson, A. J., Jr.,** Polychlorinated biphenyl absorbed from sediments by fiddler crabs and pink shrimp, *Nature (London)*, 231, 50, 1971.
249. **Young, D. R., McDermott-Ehrlich, D., and Heesen, T. C.** Sediments as sources of DDT and PCB, *Mar. Pollut. Bull.*, 8, 254, 1977.
250. **Olson, B. H. and Cooper, R. C.,** Comparison of aerobic and anaerobic methylation of mercuric chloride by San Francisco bay sediments, *Water Res.*, 10, 113, 1976.
251. **Andren, A. W. and Harris, R. C.,** Methylmercury in estuarine sediments, *Nature (London)*, 245, 256, 1973.
252. **Braman, R. S. and Foreback, C. C.,** Methylated forms of arsenic in the environment, *Science*, 182, 1247, 1973.
253. **Ridley, W. P., Dizikes, L. J., and Wood, J. M.,** Biomethylation of toxic elements in the environment, *Science*, 197, 329, 1977.
254. **Fries, G. F.,** Degradation of chlorinated hydrocarbons under anaerobic conditions, in *Fate of Organic Pesticides in the Aquatic Environment*, Advances in Chemistry Series 111, American Chemical Society, Washington, D.C., 1972, Chap. 13.
255. **Carey, A. E. and Harvey, G. R.,** Metabolism of polychlorinated biphenyls by marine bacteria, *Bull. Environ. Contam. Toxicol.*, 20, 527, 1978.
256. **Lee, R. F.,** Metabolism of petroleum hydrocarbons in marine sediments, in *Sources, Effects and Sinks of Hydrocarbons in the Aquatic Environment, Proc. Symp. American University*, AIBS, Washington, D.C., 1976, 334.
257. **Atlas, R. M., Horowitz, A., and Busdosk, M.,** Prudhoe crude oil in Arctic marine ice, water, and sediment ecosystems: degradation and interactions with microbial and benthic communities, *J. Fish. Res. Board Can.*, 35, 585, 1978.
258. **Butler, J. N. and Levy, E. M.,** Long-term fate of petroleum hydrocarbons after spills — compositional changes and microbial degradation, *J. Fish. Res. Board Can.*, 35, 604, 1978.
259. **Anderson, J. W. and Malins, D. C.,** Physiological stresses and response in chronically oiled organisms, *J. Fish. Res. Board Can.*, 35, 679, 1978.
260. **Blumer, M., Ehrhardt, M., and Jones, J. H.,** The environmental fate of stranded crude oil, *Deep Sea Res.*, 20, 239, 1973.
261. **Jannasch, H. W. and Wirsen, C. O.,** Deep-sea microorganisms: *in situ* response to nutrient enrichment, *Science*, 180, 641, 1973.
262. **Teal, J. M. and Farrington, J. W.,** A comparison of hydrocarbons in animals and their benthic habitats, in *Petroleum Hydrocarbons in the Marine Environment*, McIntyre, A. D. and Whittle, K. J., Eds., *Rapp, P. V. Réun. Cons. Int. Explor. Mer*, 171, 1977, 79.
263. **Andrews, H. L., and Warren, S.,** Ion scavenging by the eastern clam and quahog, *Health Phys.*, 17, 807, 1969.
264. **Sayler, G. S., Nelson, J. D., Jr., and Colwell, R. R.,** Role of bacteria in bioaccumulation of mercury in the oyster *Crassostrea virginica*, *Applied Microbiol.*, 30, 91, 1975.
265. **Ueda, T., Nakamura, R., and Suzuki, Y.,** Comparison of ^{115m}Cd accumulation from sediments and sea water by polychaete worms, *Bull. Jpn. Soc. Sci. Fish.*, 42, 299, 1976.
266. **Ueda, T., Nakamura, R., and Suzuki, Y.,** Comparison of influences of sediments and sea water on accumulation of radionuclides by worms, *J. Radiat. Res.*, 18, 84, 1977.
267. **Odum, W. E., Woodwell, G. M., and Wurster, C. F.,** DDT residues absorbed from organic detritus by fiddler crabs, *Science*, 164, 576, 1969.
268. **Bryan, G. W., Uysal, H.,** Heavy metals in the burrowing bivalve *Scrobicularia plana* from the Tamar estuary in relation to environmental levels, *J. Mar. Biol. Assoc. U.K.*, 58, 89, 1978.
269. **Luoma, S. N. and Jenne, E. A.,** Estimating bioavailability of sediment-bound trace metals with chemical extractants, in *Trace Substances in Environmental Health-X*, Hemphill, D. D., Ed., University of Missouri Press, Columbia, 1976, 343.
270. **Courtney, W. A. M. and Langston, W. J.,** Uptake of polychlorinated biphenyl (Aroclor® 1254) from sediment and from seawater in two intertidal polychaetes, *Environ. Pollut.*, 15, 303, 1978.
271. **Elder, D. L., Fowler, S. W., and Polikarpov, G. G.,** Remobilization of sediment-associated PCBs by the worm *Nereis diversicolor*, *Bull. Environ. Contam. Toxicol.*, 21, 448, 1979.
272. **Roesijadi, G., Anderson, J. W., and Blaylock, J. W.,** Uptake of hydrocarbons from marine sediments contaminated with Prudhoe Bay crude oil: influence of feeding type of test species and availability of polycyclic aromatic hydrocarbons, *J. Fish. Res. Board Can.*, 35, 608, 1978.

273. **Roesijadi, G., Woodruff, D. L., and Anderson, J. W.,** Bioavailability of naphthalenes from marine sediments artificially contaminated with Prudhoe Bay crude oil, *Environ. Pollut.*, 15, 223, 1978.
274. **Kraeuter, J. N.,** Biodeposition by salt-marsh invertebrates, *Mar. Biol.*, 35, 215, 1975.
275. **Alyakrinskaya, I. O.,** Behavior and filtering ability of the Black Sea *Mytilus galloprovincialis* on oil polluted water, *Zool. Zh.*, 45, 998, 1966.
276. **Tsuchiya, M. and Kurihara, Y.,** The feeding habits and food sources of the deposit-feeding polychaete, *Neanthes japonica* (Izuka), *J. Exp. Mar. Biol. Ecol.*, 36, 79, 1979.
277. **Guinasso, N. L., Jr. and Schink, D. R.,** Quantitative estimates of biological mixing rates in abyssal sediments, *J. Geophys. Res.*, 80, 3032, 1975.
278. **Small, L. F., Kečkeš, S., and Fowler, S. W.,** Excretion of different forms of zinc by the prawn *Palaemon serratus* (Pennant), *Limnol. Oceanogr.*, 19, 789, 1974.
279. **Paul, A. Z., Thorndike, E. M., Sullivan, L. G., Heezen, B. C., and Gerard, R. D.,** Observations of the deep-sea floor from 202 days of time-lapse photography, *Nature (London)*, 272, 812, 1978.
280. **George, S. G., Pirie, B. J. S., and Coombs, T. L.,** The kinetics of accumulation and excretion of ferric hydroxide in *Mytilus edulis* (L.) and its distribution in the tissues, *J. Exp. Mar. Biol. Ecol.*, 23, 71, 1976.
281. **Doyle, L. J., Blake, N. J., Woo, C. C., and Yevich, P.,** Recent biogenic phosphorite: concretions in mollusk kidneys, *Science*, 199, 1431, 1978.
282. **Rossi, S. S. and Anderson, J. W.,** Accumulation and release of fuel-oil-derived diaromatic hydrocarbons by the polychaete *Neanthes arenaceodentata*, *Mar. Biol.*, 39, 51, 1977.
283. **Rice, A. L.,** Radioactive waste disposal and deep-sea biology, *Oceanol. Acta*, 1, 483, 1978.
284. **Hessler, R. R., Ingram, C. L., Yayanos, A. A., and Burnett, B. R.,** Scavenging amphipods from the floor of the Philippine Trench, *Deep Sea Res.*, 25, 1029, 1978.
285. **Thurston, M. H.,** Scavenging abyssal amphipods from the Northeast Atlantic ocean, *Mar. Biol.*, 51, 55, 1979.
286. **Pearcy, W. G.,** Pelagic capture of abyssopelagic macrourid fishes, *Deep Sea Res.*, 23, 1065, 1976.
287. **Merrett, N. R.,** On the identity and pelagic occurrence of larvae and juvenile stages of rattail fishes (Family Macrouridae) from 60°N, 20°W, and 53°N, 20°W, *Deep Sea Res.*, 25, 147, 1978.
288. **Bouchet, P.,** Mise en évidence d'une migration des larves véligères entre l'étage abyssal et la surface, *C.R. Acad. Sci.*, Paris, 283, 821, 1976.
289. **Clarke, M. R.,** A review of the systematics and ecology of oceanic squids, in *Adv. Mar. Biol.*, 4, 91, 1966.
290. **Harding, G. C. H. and Vass, W. P.,** Uptake from seawater and clearance of p,p′-DDT by marine planktonic crustacea, *J. Fish. Res. Board Can.*, 36, 247, 1979.
291. **Nunes, P. and Benville, P. E., Jr.,** Uptake and depuration of petroleum hydrocarbons in the Manila clam, *Tapes semidecussata* Reeve, *Bull. Environ. Contam. Toxicol.*, 21, 719, 1979.
292. **Ernst, W. and Goerke, H.,** Anreicherung, verteilung, umwandlung und ausscheidung von DDT-^{14}C bei *Solea solea* (Pisces: Soleidae), *Mar. Biol.*, 24, 287, 1974.
293. **Pentreath, R. J.,** The accumulation of mercury by the thornback ray, *Raja clavata* L., *J. Exp. Mar. Biol. Ecol.*, 25, 131, 1976.
294. **Jefferies, D. F. and Hewett, C. J.,** The accumulation and excretion of radioactive caesium by the plaice (*Pleuronectes platessa*) and the thornback ray (*Raia clavata*), *J. Mar. Biol. Assoc. U.K.*, 51, 411, 1971.
295. **Pennacchioni, A., Marchetti, R., and Gaggino, G. F.,** Inability of fish to methylate mercuric chloride *in vivo*, *J. Environ. Qual.*, 5, 451, 1976.
296. **Goerke, H., Eder, G., Weber, K., and Ernst, W.,** Patterns of organochlorine residues in animals of different trophic levels from the Weser estuary, *Mar. Pollut. Bull.*, 10, 127, 1979.
297. **Leatherland, T. M., Burton, J. D., Culkin, F., McCartney, M. J., and Morris, R. J.,** Concentrations of some trace metals in pelagic organisms and of mercury in Northeast Atlantic ocean water, *Deep Sea Res.*, 20, 679, 1973.
298. **Turner, J. T.,** Microbial attachment to copepod fecal pellets and its possible ecological significance, *Trans. Am. Micros. Soc.*, 98, 131, 1979.
299. **Paffenhöfer, G. A. and Knowles, S. C.,** Ecological implications of fecal pellet size, production and consumption by copepods, *J. Mar. Res.*, 37, 35, 1979.
300. **Livingston, H. D. and Bowen, V. T.,** Pu and ^{137}Cs in coastal sediments, *Earth Planet. Sci. Lett.*, 43, 29, 1979.
301. **McCave, I. N., Ed.,** *The Benthic Boundary Layer*, Plenum Press, New York, 1976.
302. **Schink, D. R. and Guinasso, N. L., Jr.,** Effects of bioturbation on sediment-seawater interaction, *Mar. Geol.*, 23, 133, 1977.
303. **Yayanos, A. A., and Nevenzel, J. C.,** Rising particle hypothesis: a mechanism for the rapid ascent of matter from the deep ocean, *Naturwissenschaften*, 65, 225, 1978.

304. **Wiebe, P. H., Madin, L. P., Haury, L. R., Harbison, G. R., and Philbin, L. M.,** Diel vertical migration by *Salpa aspera* and its potential for large scale particulate organic matter transport to the deep-sea, *Mar. Biol.,* 53, 249, 1979.
305. **Boström, K., Moore, C., and Joensuu, O.,** Biological matter as a source for Cenozoic deep-sea sediments in the Equatorial Pacific, in *The Deep Sea, Ecology and Exploitation, Ambio* Spec. Rept. No. 6, 11, 1979.
306. **Hiraizumi, Y., Takahashi, M., and Nishumura, H.,** Adsorption of polychlorinated biphenyl onto sea bed sediment, marine plankton and other adsorbing agents, *Environ. Sci. Technol.,* 13, 580, 1979.
307. **Boehm, P. D. and Quinn, J. G.,** The effect of dissolved organic matter in sea water on the uptake of mixed individual hydrocarbons and number 2 fuel oil by a marine filter-feeding bivalve *(Mercenaria mercenaria), Estuar. Coast. Mar. Sci.,* 4, 93, 1976.
308. **Mironov, O. G. and Shchekaturina, T. L.,** Oil change in excretory products of mussels (*Mytilus galloprovincialis*), *Mar. Pollut. Bull.,* 10, 232, 1979.
309. **Wyman, K. D. and O'Connors, H. B., Jr.,** Implications of short term PCB uptake by small estuarine copepods (Genus *Acartia*) from PCB contaminated water, inorganic sediments and phytoplankton, *Estuar. Coast. Mar. Sci.,* 11, 121, 1980.
310. **Roubal, W. T., Stranahan, S. I., and Malins, D. C.,** The accumulation of low molecular weight aromatic hydrocarbons of crude oil by coho salmon (*Oncorhynchus kisutch*) and starry flouder (*Platichthys stellatus*), *Arch. Environ. Contam. Toxicol.,* 7, 237, 1978.
311. **Gardner, W. S., Lee, R. F., Tenore, K. R., and Smith, L. W.,** Degradation of selected polycyclic aromatic hydrocarbons in coastal sediments: importance of microbes and polychaete worms, *Water, Air, Soil Pollut.,* 11, 339, 1979.
312. **Alzieu, C.,** Etude de l'épuration "*in situ*" d'huîtres *Crassostrea gigas* contaminées par les diphénylpolychlorés, *Sci. Peche,* 208, 7, 1979.
313. **Miramand, P., Guary, J. C., and Fowler, S. W.,** Vanadium transfer in the mussel *Mytilus galloprovincialis, Mar. Biol.,* 56, 281, 1980.
314. **Prahl, F. G. and Carpenter, R.,** The role of zooplankton fecal pellets in the sedimentation of polycyclic aromatic hydrocarbons in Dabob Bay, Washington, *Geochim. Cosmochim. Acta,* 43, 1959, 1979.
315. **Hansen, N., Jensen, V. B., Appelquist, H., and Mørch, E.,** The uptake and release of petroleum hydrocarbons by the marine mussel *Mytilus edulis, Prog. Water Technol.,* 10, 351, 1978.
316. **Lyes, M. C.,** Bioavailability of a hydrocarbon from water and sediment to the marine worm, *Arenicola marina, Mar. Biol.,* 55, 121, 1979.
317. **Janssen, H. H. and Scholz, N.,** Uptake and cellular distribution of cadmium in *Mytilus edulis, Mar. Biol.,* 55, 133, 1979.
318. **Boehm, P. D. and Quinn, J. G.,** The persistence of chronically accumulated hydrocarbons in the hard shell clam *Mercenaria mercenaria, Mar. Biol.,* 44, 227, 1977.
319. **Young, M. L.,** The roles of food and direct uptake from water in the accumulation of zinc and iron in the tissues of the dog-whelk, *Nucella lapillus* (L.), *J. Exp. Mar. Biol. Ecol.,* 30, 315, 1977.
320. **Miramand, P., Guary, J.-C., and Fowler, S. W.,** Uptake, assimilation and excretion of vanadium in the shrimp *Lysmata seticaudata* (Risso) and the crab *Carcinus maenas* (L.) *J. Exp. Mar. Biol. Ecol.,* 49, 267, 1981.
321. **Dayal, R., Okubo, A., Duedall, I. W., and Ramamoorthy, A.,** Radionuclide redistribution mechanisms at the 2800-m Atlantic nuclear waste dump site, *Deep Sea Res.,* 26A, 1329, 1979.
322. **Anderson, J. W., Kiesser, S. L., and Blaylock, J. W.,** Comparative uptake of naphthalenes from water and oiled sediment by benthic amphipods, in *1979 Oil Spill Conference,* American Petroleum Institute, Washington, D.C., 1979, 579.
323. **Gearing, J. N., Gearing, P. J., Wade, T., Quinn, J. G., McCarty, H. B., Farrington, J. and Lee, R. F.,** The rates of transport and fates of petroleum hydrocarbons in a controlled marine ecosystem, and a note on analytical variability, in *1979 Oil Spill Conference,* American Petroleum Institute, Washington, D.C., 1979, 555.
324. **Fowler, S. W. and Heyraud, M.,** Biologically-transformed zinc and its availability for bioaccumulation by marine organisms, in *Management of Environment,* Patel, B., Ed., Wiley Eastern Ltd., New Delhi, 1980, 389.
325. **Neff, J. M.,** *Polycyclic Aromatic Hydrocarbons in the Aquatic Environment; Sources, Fates and Biological Effects,* Applied Science, London, 1979, 262.
326. **Lee, R. F., Singer, S. C., Tenore, K. R., Gardner, W. S., and Philpot, R. M.,** Detoxification system in polychaete worms: importance in the degradation of sediment hydrocarbons, in *Marine Pollution; Functional Responses,* Vernberg, W. B., Thurberg, F. P., Calabrese, A., and Vernberg, F. J., Eds., Academic Press, New York, 1979, 23.
327. **Cheng, L., Alexander, G. V., and Franco, P. J.,** Cadmium and other heavy metals in sea-skaters (Gerridae: *Halobates,* Rheumatobates), *Water, Air, Soil Pollut.,* 6, 33, 1976.

328. **Moore, M. N.,** Cellular responses to polycyclic aromatic hydrocarbons and phenobarbital in *Mytilus edulis, Mar. Environ. Res.,* 2, 255, 1979.
329. **Fisher, N. S., Olson, B. L., and Bowen, V. T.,** Plutonium uptake by marine phytoplankton in culture, *Limnol. Oceanogr.,* 25, 823, 1980.
330. **Jennings, C. D. and Fowler, S. W.,** Uptake of ^{55}Fe from contaminated sediments by the polychaete *Nereis diversicolor, Mar. Biol.,* 56, 277, 1980.
331. **Fowler, S. W., Benayoen, G., Parsi, P., Essa, M. W. A., and Schulte, E. H.,** Experimental studies on the bioavailability of technetium in selected marine organisms, in *Impacts of Radionuclide Releases into the Marine Environment,* International Atomic Energy Agency, Vienna, in press.
332. **Roper, C. F. E. and Young, R. E.,** Vertical distribution of pelagic cephalopods, *Smithson. Contrib. Zool.,* 209, 51, 1975.
333. **Dobroski, C. J. and Epifanio, C. E.,** Accumulation of benzo[a]pyrene in a larval bivalve via trophic transfer, *Can. J. Fish. Aquat. Sci.,* 37, 2318, 1980.
334. **Thomann, R. V.,** Equilibrium model of fate of microcontaminants in diverse aquatic food chains, *Can. J. Fish. Aquat. Sci.,* 38, 280, 1981.

Chapter 2

THE CONCENTRATION, MINERALOGY, AND CHEMISTRY OF TOTAL SUSPENDED MATTER IN SEA WATER

R. Chester

TABLE OF CONTENTS

I. INTRODUCTION

Particulate material in the oceans plays a very significant role in the marine geochemical cycles of many elements, and the fluxes of particulates through the water column form a vital stage in the processes which control the fate of both dissolved and solid phases introduced into the marine environment. In a recent paper Turekian[2] has highlighted the importance of particulates in the oceanic water column and has concluded that they dominate the behavior of the dissolved species in sea water; a phenomenon he has termed "the great particle conspiracy".

This volume is primarily concerned with the transfer and transport of pollutants in the oceans. However, a knowledge of the distribution and behavior of natural particulate phases in the sea is necessary in order to understand the processes which control the fate of anthropogenic material introduced into this environment. To this end, the present chapter is an attempt to summarize our knowledge of the distribution, mineralogy, and chemistry of the particulates in sea water, and in doing so draws heavily on data relating to the natural components of these solids. The aim is to give at least some small understanding of Turekian's "great particle conspiracy".

II. THE CONCENTRATION OF THE TOTAL SUSPENDED MATTER IN SEA WATER

There are a number of techniques which have been used to measure the concentration of total suspended matter (TSM) in sea water. These include: methods by which the concentrations of the particulates are measured *in situ* but samples are not collected, e.g., those based on optical phenomena such as light absorption (transmissometry) and light scattering (nephetometry); methods by which particulates are collected and subsequently weighed, e.g., filtration and centrifugation; and chemical methods, e.g., the use of aluminum to estimate the aluminosilicate fraction of TSM. These various techniques have been reviewed by Sackett.[2]

One of the first major studies of the distribution of TSM in the World Ocean was made by Jerlov[3], using light-scattering measurements. Although the instrumentation employed was much less sophisticated than that used at present, Jerlov was able to establish a number of overall trends in the distribution of oceanic TMS, including the identification of what were later to be called nepheloid layers. The work was an important forerunner of subsequent studies. The next principal phase in the investigation of TSM in the oceans was dominated by Russian scientists who, during the 1950s, made ~18,000 TSM collections using either membrane filtration or centrifugation; much of this work has been summarized by Lisitzin.[4] The next principal development occurred during the 1960s and came from a group at the Lamont-Doherty Geological Observatory, who employed both optical and centrifugation techniques to determine the concentrations of TSM in ocean waters. Some of the values reported by this group were lower than those found by the Russian scientists. For example, TSM concentrations reported for Indian Ocean waters by Jacobs and Ewing[5] were an order of magnitude lower than those given by Lisitzin.[4] This discrepancy may be a function of the centrifugation technique used by Jacobs and Ewing,[5] which, according to Bassin et al.,[6] might not collect the TSM quantitatively. However, Chester and Stoner,[7] who used a filtration technique, similarly found TSM concentrations in "open-ocean" Indian Ocean surface waters which were about an order of magnitude lower than those given by Lisitzin. Furthermore, the results from subsequent studies, in which improved techniques were employed, also found TSM concentrations of the same order of magnitude as those given by the Lamont-Doherty scientists. For example, Sackett[2] has summarized the data given by Jacobs and Ewing[5] and has reported that they found a range of 0.5 to 247 μgl^{-1} for the Atlantic water column; this may be compared to a range of 5 to 300 μ kg^{-1} found in the same ocean during the GEOSECS program.[19]

One of the most important findings to emerge from the work initiated in the 1960s was the identification of nepheloid layers (i.e., layers of relatively turbid water) above the sea bed in many parts of the World Ocean (see, e.g., Ewing and Thorndike,[8] Hunkins et al.[9] Eittriem et al.,[10-12] Ewing and Connary,[13] Ewing et al.,[14] Eittreim and Ewing,[11] Connary and Ewing[15]). Plank et al.[119] have investigated the size distribution of TSM in nepheloid layers and have concluded that it is little different from that of material in the clearer waters above. These authors concluded that this supports the hypothesis that material in the nepheloid layer has settled from above and has increased in concentration near the bottom as a result of turbulent diffusion. Other important investigations of TSM using optical methods were carried out by workers at the University of Washington (see e.g., Baker et al.[16]) and at Texas A and M University (see, e.g., Bassin,[17] Feely[18]). The most recent ocean-wide studies on the quantitative distribution of TSM in the oceans have been those of the GEOSECS program (see e.g., Brewer et al.[19]).

From these various investigations an overall picture is beginning to emerge from which it is apparent that the spatial and temporal distributions of TMS in sea water are controlled by the classical oceanographic parameters such as water circulation, primary production, and topography. One of the most important recent studies which has attempted to explain the factors controlling the ocean-wide distribution of TSM was that made by Biscaye and Eittreim[20] for the Atlantic Ocean. These authors converted nephelometer data into units of absolute TMS concentrations and were able to construct a generalized profile of the distribution of TSM in the oceanic water column. This profile is illustrated in Figure 1, from which it can be seen that, when simplified, the distribution of TSM can be explained in terms of a three layer model; the three elements in it being surface waters, the clear-water minimum (ranging from $\sim$1000 m to $\sim$3000 m in depth), and deep waters.

A. Surface Waters

In surface waters the concentrations of TSM are higher in coastal and estuarine areas than in "open-ocean" regions. For example, Chester and Stoner[7] found surface water TSM concentrations ranging up to $\sim$2600 $\mu g\ l^{-1}$ in the English Channel and Irish Sea, compared to average values in the "open-ocean" surface water of $\sim$100 $\mu g\ l^{-1}$. The higher concentrations of TSM in coastal waters are the result of factors such as river run-off and primary production. The effect of river run-off on the concentration of TSM in coastal waters has been dramatically illustrated by Gibbs,[21] who found that in periods of high river discharge, TSM transported by the Amazon River resulted in a band extending seaward as far as $\sim$100 km in some places in which TSM concentrations were $> 5000\ \mu g\ l^{-1}$. Other workers who have found relatively high concentrations of TSM in coastal waters include: Manheim et al.,[22] Meade et al.,[23] and Spencer and Sachs.[24] High primary production occurs around the periphery of the oceans, particularly in the regions of upwelling off the western margins of the continents, e.g., off Peru and South West Africa (Namibia). High concentrations of TSM in these regions have been reported by various authors (see e.g., Chester and Stoner,[7] Brewer et al.[19]). According to Biscaye and Eittreim.[20] The distribution of TSM at the clear water minimum (see below), which is similar to that at the surface, parallels the distribution of surface water primary production which is controlled by processes such as the addition of nutrients at the ocean margins from river runoff, upwelling, and the seasonal mixing of more productive shallow coastal waters.

In the surface waters, areas of relatively low TSM concentrations are found in "open-ocean" areas of low primary production, which occurs in the central regions of the main surface circulation gyres; in these areas TSM concentrations are usually $\leqslant$ 100 $\mu g\ l^{-1}$ (see, e.g., Chester and Stoner,[7] Brewer et al.[19]). Investigations carried out

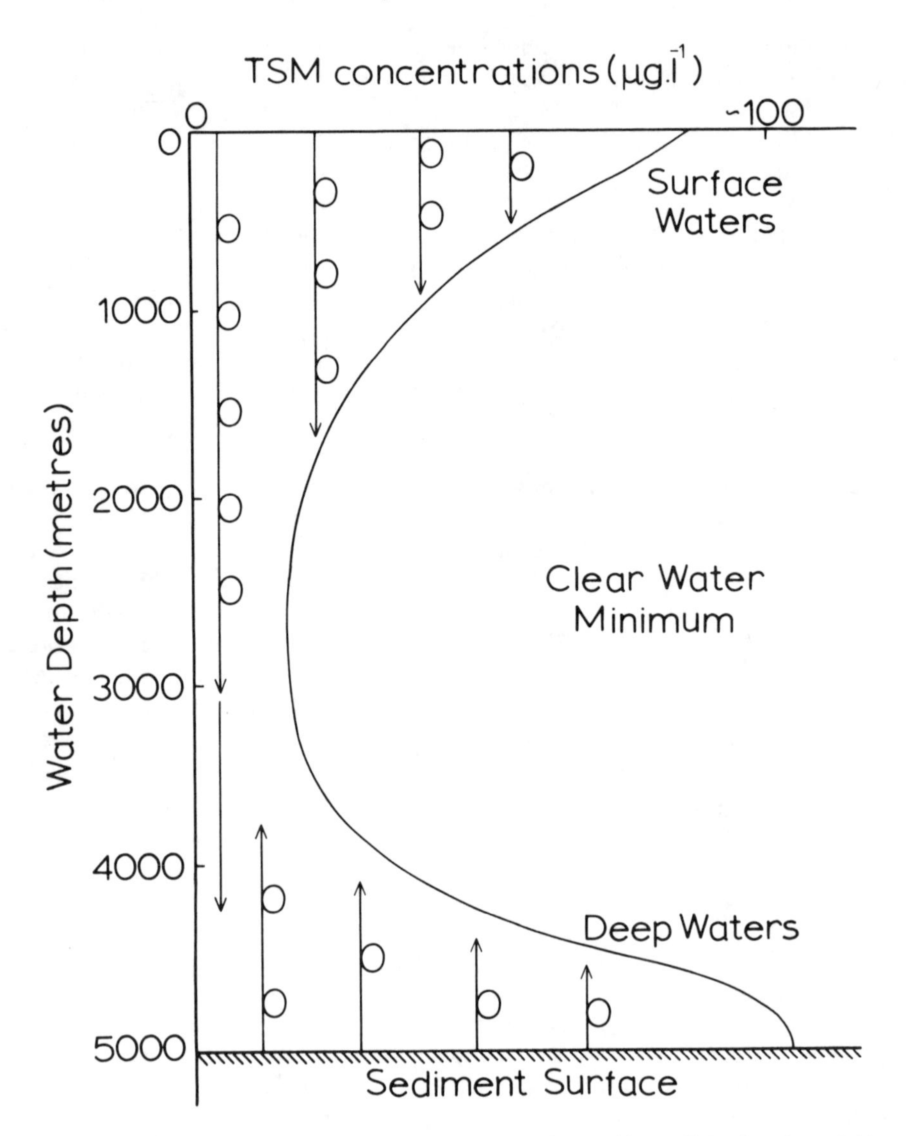

FIGURE 1. Typical oceanic TSM profile for a water column with a well-developed nepeloid layer. The curlicue arrows indicate that downward particle settling may be affected by horizontal advective processes. (Modified after Biscaye, P. E. and Eittriem, S. L., *Mar. Geol.*, 23, 155, 1977.)

on the distribution of TSM by the GEOSECS, and other programs, have been recently summarized by Lal.[25] This author concluded that in Atlantic surface waters, the distributions of minerals and biological particles show a pronounced latitudinal effect, with an overall minimum in equatorial regions and highest concentrations generally occurring at latitudes above ~40°N and ~40°S; the latter being due, in part, to primary production. In general, TSM concentratins in Atlantic surface waters are greater than those in the Pacific. Some data on the concentrations of TSM in various surface water populations is given in Table 1.

B. The Clear Water Minimum

According to the model proposed by Biscaye and Eittriem[20] the minimum in the concentrations of TSM in the average oceanic water column profile is termed the clear water minimum. The depth at which it is found varies in different parts of the World

Table 1
THE CONCENTRATIONS OF TOTAL SUSPENDED MATTER IN SURFACE SEA WATERS[a]

Oceanic area	No. of samples	Concentration; average (μg ⁻¹)	Time of year
Eastern margins of North Atlantic	25	467	April, July
Open-ocean South Atlantic	23	150	April, July
Open-ocean Indian Ocean	30	66	April, June
China Sea	22	127	May, June
Open-ocean; average	75	110	—
Near-shore localities; average	37	1050	—

[a] Data from Chester and Stoner.[7]

Ocean, and it results from a decrease in the concentration of TSM sinking from the surface layers. This decrease is caused by processes such as particle dissolution and decomposition, and advection, i.e., it represents the contribution of the surface particulates to this depth. The concentrations of TSM at the clear water minimum are considerably lower than those at the surface, although the spatial distribution is similar, i.e., the principle features are that the lowest concentrations are found underlying the zones of minimum primary production in the central regions of the main surface circulation gyres, and the highest concentrations are at the periphery of the ocean and under zones of high primary production, Biscaye and Eittriem[20] have concluded that the fact that the distribution of TSM at the clear water minimum parallels that at the surface reflects a balance between primary production, variable rates of particle setting, and dissolution and decomposition of the particles as they sink and are advected through the water column.

C. The Deep Waters

In the model postulated by Biscaye and Eittriem[20], there is an increase in the concentration of TSM below the clear water minimum towards the sea bed. This is the result of the resuspension of bottom sediments into the turbid nepheloid layer, in which the concentration of particles decreases upwards. The upper limit of this layer is defined as the clear water minimum. In order to determine the spatial distribution of TSM in this layer, Biscaye and Eittriem[20] distinguished between two particle populations: (1) the gross particulate standing crop, defined as the total particulate matter below the clear water minimum and (2) the net particulate standing crop, which is defined as the particulate matter below clear water which is in excess of the clear water concentration itself. The principal features in the distribution of the gross particulate standing crop are relatively low concentrations in the central portion of the North and South Atlantic, and the increases in concentration, sometimes up to an order of magnitude, in the western relative to the eastern basins. In these western basins the maximum particulate concentrations are coincident with the axis of the western boundary current. The distribution of the gross particulate standing crop, therefore, reflects the effect of two different kinds of processes: those which control the particulate distribution in the surface waters (see above), and those resulting from the superimposition of abyssal circulation, which are evidenced in high particulate concentrations in the western boundary currents. The effects of the abyssal circulation on the distribution of the particulates can be seen more clearly by considering the deep water net particulate standing crop, i.e., when the "noise" from surface-derived particulates is excluded. According to Biscaye and Eittriem[20] the deep water net particulate standing crop can be considered to reflect the effect of processes which raise particles into near-bottom waters and maintain them in suspension; these processes include the injection of particulates into the lower water column at the edges of the continents by turbidity currents, and the direct resuspension of material by bottom currents. The most outstanding features in the distribution of the deep water net particulate standing crop in the Atlantic are the high burden of resuspended particulates coincident with the western boundary currents, and the absence of the high concentrations in areas underlying primary production, i.e., the effect of abyssal circulation on the net crop is emphasized.

To summarize, therefore, it may be concluded that the quantitative distribution of TSM in the World Ocean is largely controlled by the classical oceanic features such as circulation of the major water masses, transport of material from the continents, and primary production. It is convenient to consider the distribution of the particulates in terms of a three layer model: (1) In the upper portion of the water column the distribution and concentration of TSM are controlled by surface processes; the highest concentration being in areas which receive large inputs of continentally derived solids (e.g., from river run-off or eolian transport, or in which primary production is at a maximum). (2) The surface water particulates fall down the water column and undergo dissolution and decomposition; together with other processes, such as incorporation into the food chain, this results in a decrease in TSM concentrations with lowest values found at the clear water minimum. (3) Below this minimum the concentrations of TSM rise towards a maximum in the nepheloid layer which is formed mainly by the resuspension of bottom sediments. Distributions of particulates in this layer are controlled largely by abyssal circulation, the highest concentrations being coincident with the errosive western boundary currents. The implications of this generalized model for the description of the distribution of TSM in the oceans are important when the effect of the particulates on marine geochemistry is considered.

III. THE ORIGINS OF THE TOTAL SUSPENDED MATTER IN SEA WATER

Total suspended particulate material in sea water consists of various components (see Section IV), some of which are brought to the oceans as solids and some of which are formed from material dissolved in sea water itself.

Processes by which solids are brought to the oceans include river, eolian, and ice transport, of which river run-off is by far the most important. Data on the discharge of suspended sediments by rivers have been provided by Holeman[26] and is summarized in Table 2. It is apparent from this table that sediment discharge varies greatly from one river to another, and according to Holeman[26] Asia produces by far the highest annual river-transported sediment yield with $\sim 600 \times 10^3$ kg for every km^2 of land surface, with Africa, Europe, and Australia each producing less than one tenth of this. The total river discharge of sediment to the oceans has been estimated to be $\sim 20 \times 10^{12}$ kg year^{-1}, of which $\sim 80\%$ originates in Asia which constitutes only $\sim 25\%$ of the total land area draining into the World Ocean.[26] However, only a small proportion of this riverine material reaches deep-sea areas by transport in surface waters, and much of the river-transported sediment load is initially deposited on the shelf and slope regions, which act as intermediate sediment traps from which material may be transported by bottom processes such as turbidity currents. In most oceanic regions the

Table 2
RIVER DISCHARGE OF SUSPENDED SEDIMENT INTO THE OCEANS[a]

River	Location	Drainage area (10^3 km)	Annual suspended sediment discharge ($g \times 10^{13}$)
Yellow (Hwang Ho)	China	666	208
Ganges	India	945	160
Brahmaputra	Bangladesh	658	80
Yangtze	China	1920	55
Indris	Pakistan	957	48
Amazon	Brazil	5710	40
Mississippi	U.S.	3180	34
Irrawaddy	Burma	425	33
Mekong	Thailand	786	19
Red	N. Vietnam	117	41
Nile	Egypt	2944	12
Congo (Zaire)	Africa	3968	7
Niger	Nigeria	1100	0.50
St. Lawrence	Canada	1275	0.40

[a] Data from Holeman.[26]

input of land-derived material by wind (i.e., eolian) transport is very small relative to that from rivers. The processes controlling the quantity and kind of eolian dust over marine areas include:

1. The strengths and circulation patterns of the major wind systems
2. The nature of the soils in the catchment region
3. The injection of anthropogenic materials into the atmosphere
4. The local effectiveness of the mechanisms by which the dust is removed from the atmosphere (i.e., dry deposition and precipitation scavenging)

In general soil-derived (i.e., "natural") atmospheric particulates over marine areas are $\gtrsim 1\ \mu m$ in size, whereas those produced in industrial processes involving combustion are usually $\lesssim 1\ \mu m$.[27] The non-sea salt component of the marine atmospheric particulate population has both a natural and an anthropogenic fraction; the natural fraction includes some non-soil-derived material, e.g., from volcanic activity, which can have a particle size $\lesssim 1\ \mu m$. Various workers (see, e.g., Delany et al.,[28] Aston et al.,[29] Chester et al.[30]) have used mesh collection techniques to sample various particulates from the marine atmosphere. These meshes do not retain particulates which are $\lesssim 0.5\ \mu m$ in size and therefore collect a size-biased sample. However, they may be considered to give an estimate of the soil-sized fraction of the particulate population. Data on the soil-sized dust-loadings obtained in this manner are given in Table 3A. Prospero[31] has summarized data on the distribution of total atmospheric particulates. i.e., collected by a filtration system, over the World Ocean and the average dust-loadings are given in Table 3B. It may be concluded from these various investigations that the concentration of particulates in the marine "dust veil" is extremely variable both spatially and temporally. The highest particulate loadings are found over marine areas adjacent to the major desert and arid lands, which act as reservoirs of loose easily lifted surface soils.

The supply of solids brought to the oceans by ice transport has its immediate effects in the polar regions. One of the most important glacial weathering products is "rock flour", which is composed of debris ground to a fine, usually clay-sized, powder. Ice movement can also transport material ranging in size from silt to boulders.

Table 3A
ATMOSPHERIC SOIL-SIZED DUST-LOADINGS IN THE LOWER MARINE TROPHOSPHERE

Oceanic area	Principal wind system	Dust-loading ($\mu g\ m^{-3}$ of air)
North Atlantic[a]	Westerlies	$10^3 — 10^\circ$
	Northeast trades	$10^\circ — 10^2$
South Atlantic[a]	Southeast trades	$10^{-2} — 10^{-1}$
Northern Indian[a]	Northeast monsoon	$10^{-1} — 10^1$
	Southwest Monsoon	
Southern Indian[a]	Northwest monsoon	10^{-2}
	Southeast trades	
China Sea[a]	Southwest monsoon	10^{-1}
Eastern Mediterranean[b]	Variables	10^1

Table 3B
TOTAL ATMOSPHERIC MINERAL PARTICULATE-LOADINGS IN THE LOWER MARINE TROPOSPHERE[c]

Oceanic area	Arithmetic mean loading ($\mu g\ m^{-3}$ of air)
Central and northern North Atlantic	1.3
Tropical and equatorial North Atlantic	36.6
Tropical and central South Atlantic	1.35
Pacific (28°N — 40°S)	0.58
Mediterranean	4.57
Indian Ocean (15°S — 7°N)	7.20
South China Sea region	1.51

[a] Data from Aston et al.[29]
[b] Data from Chester et al.[30]
[c] Data from Prospero.[31]

Goldberg[32] has made an estimate of the overall fluxes of particulate material brought to the ocean annually. These fluxes are given in Table 4, from which it is evident that ~90% of the total solids brought to the oceans in the sedimentary cycle result from river run-off.

Particulates are formed within the oceans by both organic and inorganic processes. The fertility of the sea varies considerably from one region to another, and a number of geographical trends in the distribution of primary productivity can be recognized. (1) Productivity, and also the standing crop, tend to be higher in near-shore areas and over submarine banks than in the "open ocean". (2) Regions of very high productivity are found on the continental shelf off the western margins of the continents as the result of upwelling of nutrient-rich intermediate water. Examples of these regions include the shelves off Peru, South West Africa (Namibia) and north west Africa. (3) The Antarctic Ocean is one of the most fertile regions in the World Ocean. (4) Areas of moderately high productivity are found in the areas of equatorial upwelling in the Atlantic and Pacific; however, production in these areas is considerably less than in the polar regions, a feature which contributes to the overall minimum concentrations of TSM in the equatorial Atlantic (see above). Little is known of the geographical

Table 4
FLUXES OF SOME PARTICULATE MATERIAL TRANSPORTED TO THE OCEANS[a]

Particulate material	Transporting agency	Flux to oceans (10^{14} g year^{-1})
Suspended river solids	River run-off	180
Continental rock and soil particles	Atmosphere	1—5
Volcanic debris	Stratosphere	0.036
Volcanic debris	Atmosphere	< 1.5
Glacial debris from Antarctic	Oceans	20

[a] Data from Goldberg.[32]

distribution of inorganic particulates, e.g., iron and manganese hydrous oxides, produced in the water column. However, their geochemical importance in the TSM budget will be discussed below.

IV. THE COMPONENTS OF THE TOTAL SUSPENDED MATTER IN SEA WATER

The components of the TSM in sea water may be conveniently divided into a number of broad classes generally based on Goldberg's[33] classification of deep-sea sediments. Each of these classes is briefly discussed below.

A. Lithogenous Solids

According to Goldberg,[33] lithogenous material is defined as being produced by the weathering of the earth's crust, by submarine volcanoes, or from underwater weathering in which the solid phase undergoes no major change during its residence in sea water. As far as particulates in the general water column itself, as opposed to deposited sediments, are concerned the crustal weathering products are by far the most important class of lithogenous material. However, the situation is different in the nepheloid layer which contains a high proportion of resuspended sediment, including the products of submarine volcanism. The principal lithogenous minerals in marine TSM are the clay minerals and feldspars, which are aluminosilicates, and quartz (SiO_2), together with smaller amounts of minerals such as micas, amphiboles, pyroxenes, rutile, anatase, haematite, goethite, calcite, dolomite, and iron and managanese oxides.

A number of workers have reported analyses on the clay mineralogy of TSM; these include Lisitzin,[4] Jacobs and Ewing,[5] and Feely.[18] In general, it appears that the mutual proportions of the clay minerals in the suspended material are the same as those in the underlying sediments of the same region; an exception is montmorillonite in certain locations (see below).

From the overall three layer model of the distribution of TSM in the oceanic water column (see Section II), it would be expected that the clay mineralogy of particulates in the lower, or nepheloid layer, would be similar to that of the underlying sediments from which they are resuspended. For example, Spencer et al.[35] have concluded from sediment trap data that in the deep Sargasso Sea ~95% of the total clay flux to the bottom originates from resuspended sediments, and Honjo[36] has demonstrated a close similarity between the mineralogy of the upper portion of the sediment and the bulk of the material in the overlying trap. Feely[18] has shown that there is a significant increase in the aluminosilicate fraction of TSM in the nepheloid layer compared to the overlying water in the Gulf of Mexico. He has interpreted this as being the result of the resuspension of bottom sediment — a conclusion which is supported by the similarity in mineralogy between the nepheloid layer TSM and the underlying sediments.

The distribution of the principal clay mineral species in surface water TSM has been investigated by Behairy et al.[37] These authors collected TSM by a continuous centrifugation technique from surface waters along a transect running from ~53°N to ~35°S in the eastern margins of the Atlantic Ocean. The overall average concentrations of the four clays (expressed in terms of 100% clay sample) in the surface water TSM were: illite ~42%, kaolinite ~25%, chlorite ~19%, and montmorillonite ~14%. There were specific trends in the latitudinal distribution of illite and kaolinite in the TSM — illite decreasing in concentration and kaolinite increasing towards low latitudes. These trends are similar to those found in eolian particulates and deep-sea sediments from the same region, and these clays, together with chlorite, were considered to have a detrital, i.e., continental, origin. The concentrations of montmorillonite in the TSM were similar to those of the underlying deep-sea sediments in the North Atlantic. However, in the South Atlantic the sediments contained about twice as much montmorillonite as did the TSM. The authors concluded that whereas the montmorillonite in North Atlantic deep-sea sediments is largely detrital, that in South Atlantic sediments has both a detrital and an authigenic component (detrital components are those which have been transported throgh the sedimentary cycle, and brought to the site of deposition, as solids; authigenic components are those which have been formed, at least in part, from material in solution, although not necessarily at their present location). In TSM from the surface waters of the Indian Ocean there is an even bigger shortfall in montmorillonite compared to the concentrations in the underlying deep-sea sediments.[38] It may be concluded, therefore, that on an ocean-wide basis the major latitudinal trends in the distribution of illite, kaolinite and chlorite found in deep-sea sediments are also reflected in surface layer TSM. In contrast, montmorillonite does not appear to have been transported through the water column in those regions (e.g., the South Atlantic, Indian and Pacific Oceans) in which it is concentrated in the underlying sediments.

The distribution of quartz in surface water TSM from the Atlantic Ocean has been investigated by Krishnaswami et al.[39] However, they found that it could only be satisfactorily measured in samples collected in the equatorial region; these contained an average of ~10% quartz, mainly from wind transport, whereas in the waters of high latitude quartz was below the detection limit (i.e., <2%) of the XRD technique employed, probably due to dilution by organic detritus.

B. Hydrogenous Solids

These are defined as resulting from the formation of solid matter in the sea by inorganic reactions, i.e., nonbiological processes. Chester and Hughes[40] divided hydrogenous components into two groups, primary material produced directly from components dissolved in water, and secondary material which is formed from the submarine alteration of pre-existing minerals. This simple twofold classification has subsequently been considerably enlarged, and the types of solids more rigorously defined, by Elderfield.[41] Hydrogenous components in the TSM of sea water include, among others, iron and manganese hydrous oxides, barite, francolite, and some carbonate minerals.

Many hydrogenous solids, e.g., ferro-manganese nodules, are formed at the sediment-water interface. Others are either transported from the continental or estuarine regions or are precipitated in the water column, and so form part of the total suspended matter population. However, there is very little data on the concentrations and mineralogies of hydrogenous solids in sea water. Some workers have actually identified specific hydrogenous minerals in oceanic TSM. For example, Chesselet et al.[42] have reported the presence of barite in Atlantic waters. To a depth of ~3 km almost all the particulate barium occurred as discrete crystals, but below this level it appeared to be bound in complex aggregates. Some of the barite was found to be concentrated in waters over the crest of the Mid-Atlantic Ridge in TSM which was free of biogenic debris; this barite is probably hydrogenous in origin. However, in other regions the distribution of the barite correlated with surface water primary production, and probably has a biogenous origin. Other workers have deduced the presence of hydrogenous phases by indirect means. In this context, Feely[18] divided the components of TSM from the Gulf of Mexico into particulate organic matter, biogenic silica, particulate carbonate, aluminosilicate, and "other material". TSM from the deep water of the Gulf of Mexico contained ~16% "other material", which the author concluded was composed of free iron and manganese oxides and hydroxides, together with other nonaluminosilicate material. Krishnaswami and Sarin[43] have conlcuded that ~30% of the particulate iron in Atlantic surface water TSM exists in a form which is not associated with either aluminosilicate or organic carbon phases; some of this iron may be present as goethite or hematite particles. Hydrogenous precipitates, such as those mentioned above, are extremely important in the marine geochemical cycle of many trace elements, and this topic is discussed in Section VI.

C. Biogenous Components

These are defined as being produced in the biosphere and include: organic matter, e.g., phytoplankton, zooplankton, bacteria, yeasts, fungi, fecal pellets, and detritus (dead organisms) and inorganic shell material which consists principally of calcareous and siliceous tests. The distribution of organic carbon is discussed in Section V, and the present description of biogenous componets is limited to shell material.

The principal organisms which secrete calcium carbonate tests are forminifera, coccolithophorids, and pteropods. Calcium carbonate shell material is a common constituent of the TSM in sea water and can make up $\gtrsim$50% of the total in some surface waters; however, the contribution is usually much less than this. According to Lisitzin[4] the highest concentrations of calcium carbonate in surface water TSM are found in a broad band lying between ~50°N and ~50°S with a maximum in the equatorial zones. However, the carbonate is diluted by other components such as land-derived aluminosilicate and organic carbon. In order to overcome these effects Lisitzin[4] expressed his calcium carbonate concentrations in $\mu g\ l^{-1}$ of sea water. In this manner, the concentrations had a much more regular distribution with the highest values near the convergence in the Southern Hemisphere, near the polar front in the Northern Hemisphere, along the peripheries of the temperate zones, and in the equatorial regions. Similar distribution patterns have been found by Krishnaswami et al.[39] for particulate calcium carbonate in Atlantic Ocean surface waters. For example, when their results were expressed as $\mu g\ l^{-1}$ of sea water, the highest concentrations were found near the Antarctic convergence in the Southern Hemisphere, near the polar front in the Northern Hemisphere, and in the equatorial region. Calcium carbonate undergoes dissolution in the water column and within the sediment complex, with the result that there is a so-called "carbonate compensation depth" below which solution apparently exceeds supply; for a review of this subject see Berger.[44] The average concentration of C_aCO_3 in Atlantic surface waters is ~10 $\mu g\ l^{-1}$ compared to $\lesssim$ 2.5 $\mu g\ l^{-1}$ in water from depths $\gtrsim$500m.[25] However, the extent of C_aCO_3 dissolution varies from one region to another. For example, Krishnaswami et al.[39] estimated that in the Argentine and Cape basins $\gtrsim$90% of the surface water C_aCO_3 dissolves prior to reaching the sediment/water interface. In contrast, $\gtrsim$50% of surface water C_aCO_3 escapes dissolution in the Equatorial Atlantic.

The most abundant opal-secreting organisms in the oceans are diatoms, with radiolarians as the second major producer. According to Lisitzin[4] the concentrations of amorphous particulate silica have a latitudinally controlled distribution, the dominant feature being a belt of high concentration in the water around Antarctica, with pronounced, but less distinct, bands in the northern and equatorial regions of the World Ocean surface waters. In surface waters of the Atlantic Krishnaswami et al.[39] reported a maximum concentration of particulate opal in surface waters from latitudes south of ∼40°S, with minor maxima in near-equatorial regions. These authors also found that opaline silica constituted between <10% and ∼80% of the ashed TSM samples. The distribution of silicous skeletal material appears to follow less obvious patterns than does that of the carbonates. There is some evidence that opaline skeletons are more easily dissolved in the upper parts of the water column, and according to data summarized by Berger[44] there may be a silica compensation level above which the rate of dissolution exceeds the rate of supply. According to Lisitzin[4] the dissolution of opaline tests varies from species to species; for example, diatom frustules are destroyed most rapidly in the upper ∼100 m of the water column and the extent of their dissolution exceeds that of radiolarian skeletons. Berger[44] has established a dissolution scale for siliceous shells, the decreasing order of dissolution being silicoflagellates > diatoms > radiolarians > sponge specules. Krishnaswami et al.[39] have made tentative flux calculations and have concluded that in the Pacific-Antarctic region ⋧90% of the silica produced in surface waters is regenerated within the water column.

D. Cosmogeneous Components

These are derived from extraterrestrial sources. Various types of cosmogenous material have been identified in deep-sea sediments. These include cosmic spherules, tecktites, and a number of cosmogenic nuclides. However, cosmogenous material makes up only a trivial portion of both the sediments and suspended particulates in the oceans.

E. Anthropogenic Components

These are man-made materials which reach the oceans in a particulate form, or which subsequently become associated with particulates, include sewage products, nuclear components, various petroleum-associated hydrocarbons, metals, synthetic organic chemicals, and tars.

According to Goldberg,[32] nuclear explosions and the discharges from nuclear reactors have measurably increased the levels of radioactivity in the marine environment. Most nuclear bomb material which undergoes long-range transport consists of particles < 1 μm in size. Of the more important fission products, the radioactive nuclides of strontium and caesium are in forms which are almost wholly soluble in sea water, but those of cerium, yttrium, zirconium, ruthenium, and tellurium are mainly associated with particulate and colloidal material (see e.g., Freiling and Ballow[45], Joseph et al.[46]). In this context, Bowen and Sugihara[47] found that the relatively insoluble fission products, such as ^{144}Ce and ^{144}Pm, were transported vertically in the water column at a greater rate than the more soluble species (e.g., ^{90}Sr and ^{137}Cs) as a result of their association with other sinking particles; for a discussion of the sinking rates of radionuclides, see Sackett.[2] Artificially produced radionuclides brought to the oceans become involved in the major cycles which control the particulate and the dissolved chemistry of sea water, i.e., they are found in solution in association with both inorganic and organic particulates, in sediments and in interstitial waters, and for this reason have sometimes been used as indicators in pollutant transport studies (see, e.g., Osterberg et al.[49]). The involvement of radionuclides in the oceanic cycles was illustrated by Bowen and Sugihara,[47] who concluded that the distribution of ^{90}Sr in the

Atlantic Ocean bore no resemblance to fall-out patterns on land, but that hydrographic processes were responsible for the observed concentrations. This ^{90}Sr entered the oceans via atmospheric transport. There are, however, instances in coastal waters where the transport imprint is locally preserved following the input of radionuclides via rivers or outfalls (e.g., a plume of high ^{51}Cr activity has been found around the outflow of the Columbia River[48]). The involvement of radionuclides in the biological cycle has been studied by many authors. For example, in one of the classic studies, Osterberg et al.[49] suggested that vertical transport by the sinking of fecal pellets was responsible for the incorporation of some short half-life nuclides into deep-sea sediments; for a detailed review of this subject, see Fowler.[50] The general fate of radionuclides in the ocean has been covered in the treatment given by Burton.[51]

Petroleum-associated hydrocarbons (including tars) in the marine environment have been involved in some of the most recent dramatic incidents of oceanic pollution, e.g., the fouling of beaches, the production of slicks, and the killing off of fish and bird populations. According to Morris et al.,[52] $\sim 10^{12}$ to 10^{13} g of petroleum are introduced annually into the seas as floating oil slicks and tar lumps, particles (e.g., oil droplets and hydrocarbons adsorbed onto other particles) suspended in the water column, and as dissolved or colloidal material. Morris et al.[52] summarized the distribution of petroleum hydrocarbons in the waters of the Sargasso Sea. They concluded that at the sea surface tar lumps average ~ 10 mg m^{-2}, and that within the water column total hydrocarbons (i.e., particulate and dispersed) range from ~ 10 μg l^{-1} near the surface to <1 μg l^{-1} at depths $>$ 100 m; of these total hydrocarbons ~ 0.5 μg l^{-1} is in the form of tar particles with the remainder present either as colloidal micelles or adsorbed onto seston, with only the lightweight and highly polar hydrocarbons being in true solution. The fates of the various forms of petroleum in the sea have been the subject of many investigations, and the processes which the hydrocarbons undergo include evaporation, dissolution, emulsification, photochemical oxidation, and microbial degradation.

Pelagic tar (petroleum lumps, or tar balls) has been identified in the sea by Horn et al.,[53] and a detailed review of its occurrence around Bermuda and in the Sargasso Sea has been given by Butler et al.[54] These tar balls are black, or brown, irregularly shaped lumps having particle sizes ranging from ~ 1 mm to ~ 10 cm. According to Horn et al.[53] they contain the low boiling fraction of crude oils, suggesting they are relatively young.[32] The residence time of pelagic tar has been estimated to be ~ 1 year,[54,55] and its transfer from the ocean surface to the water column has been investigated by Morris et al.[52] The latter authors concluded that the highest concentrations of pelagic tar are found in the Sargasso Sea (~ 9 mg m^{-2} of ocean surface in 1970 to 1972) and in the Mediterranean Sea (~ 10 mg m^{-2} of ocean surface in 1974 to 1975). When they considered the fate of this tar the authors pointed out that in the open-ocean, and particularly in the central oceanic gyres such as the Sargasso Sea, stranding on beaches cannot be a major removal mechanism, and that evaporation is limited to the more volatile fractions (i.e., below C_{17}). Further, they reported that in the upper 100 m of the Sargasso Sea the estimated concentration of tar particles is ~ 40 μg beneath each m^{-2} of ocean surface. This is higher than the average surface concentration of ~ 10 mg m^{-2} in the Sargasso Sea. The authors suggested that tar particles in the water column, although derived from surface tar, may in fact exceed the surface concentrations at any given time by being more resistant to degradation and so having a longer residence time in the water column than do the parent tars on the surface. The residence time of surface tars was estimated to be about several months to 1 year and that of those suspended in the water column to be in the order of years before ultimately being oxidized or incorporated into sediments. Hydrocarbons may be incorporated with marine organisms via the adsorption of dissolved compounds and by ingestion (see, e.g., Burns and Teal,[56] Lee et al.,[57] Conover[58]); the effects of petroleum hydrocarbons on the biosphere have been reviewed by Goldberg,[32] Cowell,[59] and Morris et al[52].

The principal high molecular weight synthetic organic chemicals introduced into the marine environment are the DDTs and the PCBs (polychlorinated biphenyls); however, according to Risebrough et al.[60] the use of the terms DDT and PCB to describe diverse and frequently very different halogenated hydrocarbon compounds is no longer acceptable. The aliphatic halogenated hydrocarbons are important examples of low molecular weight synthetic chemicals reaching the seas. Synthetic organic chemicals are brought to the oceans via waste water outfalls, river run-off, and by atmospheric transport; the latter is particularly important for DDTs, PCBs, and their residues. Because of their low solubilities in water a substantial fraction of the higher molecular weight chlorinated hydrocarbons in the oceans will probably be associated with particulates, including plankton and other biota.[60] According to Bidleman et al.[61] chlorinated hydrocarbons are probably removed from the water column by adsorption onto sinking particles; the latter are expected to include inorganic particles, plankton, organic detritus, and fecal pellets.[60] Ultimately, some chlorinated hydrocarbons are incorporated into sediments. For example, Horne et al.,[62] have estimated that PCB and DDT residues have accumulated in the sediments of the Santa Barbara Basin at rates of $\sim 1.2 \times 10^{-4}$ g m^{-2} year^{-1} and $\sim 1.9 \times 10^{-4}$ g m^{-2}year^{-1}, respectively. In the open-ocean the rates appear to be lower. Harvey and Steinhaven[63] identified μg/g levels of PCBs in the top 2 cm of cores from the Hudson Canyon and the Hatteras Abyssal Plain, and using sediment trap data, they estimated an open-ocean PCB deposition rate of $\sim 5 \times 10^{-6}$ g m^{-2}year^{-1}.

Domestic and trade wastes have been introduced into the marine environment for many years. A detailed discussion of those types of wastes is clearly beyond the scope of the present chapter, and for a review of their chemical and physical characteristics the reader is referred to Painter[64] and Topping.[65]

It is now recognized that some metals are being mobilized to the environment by man at rates which are comparable to, and indeed sometimes exceed, those from national processes (see, e.g., Goldberg[32]). These metals are transported to the oceans mainly by river run-off and via the atmosphere, and those of particular environmental concern include Pb, Hg, Cd, Sn, Zn, Cu, Se, As, Sb, Ag, Fe, Sc, and Al. Those metals which reach the oceans via rivers undergo important reactions in the estuarine environment, and the histories of trace metal pollution have been recorded in sediments of the estuarine and coastal zones (see, e.g., Chow et al.,[66] Bruland et al.,[67] Perkins et al.,[68] Skei et al.,[69] Chester and Stoner[70]). The various chemical processes affecting the pollutant, and natural, trace metals are discussed in the following sections.

V. THE CHEMICAL COMPOSITION OF THE TOTAL SUSPENDED MATTER

A considerable amount of data is now available on the chemical composition of oceanic TSM from both the euphotic zone and deeper waters. It was shown above that the TSM in sea water consists of a number of components — principally organic matter, biogenous shells, hydrogenous solids, and lithogenous minerals. Some of these components have markedly different chemical compositions (e.g., planktonic matter has a very different trace element assemblage than that of lithogenous aluminosilicates) and the overall chemical composition of TSM will be controlled largely by the mutual proportions of these various components. Two of the most important of these are aluminosilicates and organic carbon, and it is of interest to examine their distributions in the waters of the World Ocean.

Chester et al.[71] have given data on the concentrations of aluminosilicates in particulates from the surface waters of a number of contrasting oceanic environments. These include: the eastern margins of the North Atlantic, a region which receives a relatively

Table 5
THE CONCENTRATIONS OF ALUMINOSILICATES IN SOME SURFACE SEA WATER PARTICULATES[a]

Oceanic region	Aluminosilicate as percentage of total particulate material
Eastern North Atlantic margins	9.6
Open-ocean South Atlantic	2.5
Open-ocean Indian Ocean	5.2
China Sea	62

[a] Data from Chester et al.[71]

large input of wind-transported solids; the China Sea, which has a large input of river-transported material; and the open-ocean areas of the South Atlantic and Indian Oceans. The authors were able to tentatively relate the concentrations of aluminosilicates to the supply of land-derived solids to the oceans, since most of the aluminosilicates in surface sea waters have a continental origin (see, e.g., Sackett and Arrhemus[72]). The average percentages of aluminosilicates in the particulates are given in Table 5, from which it can be seen that there are very large variations between the oceanic regions. Three important conclusions regarding the distribution of aluminosilicates can be drawn from this work: (1) TSM from the eastern margins of the North Atlantic contains ~10% aluminosilicate, compared to $\lesssim$2% in those from the eastern margins of the South Atlantic. This is probably a result of the input of eolian dust to the former region. (2) The China Sea lies adjacent to those regions of Asia from which the river-transported sediment yield is extremely high (see Table 2), and of the ~20 10^{12} kg year^{-1} total discharge of river solids ~80% has been estimated to originate in Asia, which constitutes only ~25% of the total land area draining into the World Ocean.[26] The bulk of river-transported sediment is deposited in near-shore environments (e.g., estuaries, deltas). Furthermore, very little sediment flows directly into open-ocean coastal areas, and the bulk of the discharge is into marginal seas. The China Sea is such a marginal sea, and the relatively large input of river-transported sediment is reflected in the composition of the particulates which contain an average of ~62% aluminosilicates. (3) Open-ocean areas of the South Atlantic and Indian Oceans have surface water TSM which contains, on the average, $\lesssim$5% of land-derived aluminosilicates. Similar conclusions have been reached by Wallace et al.[73] and Krishnaswami and Sarin.[43]

The quantitative distribution of particulate aluminium ($Al_{(p)}$) in oceanic waters has been investigated by various workers (see, e.g., Spencer and Sachs,[24] Feely et al.,[74] Sackett and Arrhenius,[72] Joyner,[75] Krishnaswami and Sarin,[43] Brewer et al.,[76] Chester et al[71]). A number of general conclusions on the overall distribution of $Al_{(p)}$ in surface sea waters can be drawn from these studies: (1) The concentrations of $Al_{(p)}$ vary by several orders of magnitude, ranging from $\lesssim$0.1 μg l^{-1} in some open-ocean areas ~100 μg l^{-1} in parts of the China Sea. (2) Relatively high concentrations of $Al_{(p)}$ are found in certain coastal, or marginal sea, surface waters, e.g., at river outflows (see Joyner[75]). (3) There is a generally small input of $Al_{(p)}$ to most open-ocean areas, which results in an average concentration of $\lesssim$0.15 μg $Al_{(p)}$ l^{-1}, such as those found in the South Atlantic Ocean. However values above this general background are found in some open-ocean areas (e.g., in waters underlying the north east trade wind system in the Atlantic

Table 6
THE CONCENTRATIONS OF PARTICULATE ORGANIC CARBON (POC) IN SOME SURFACE SEA WATERS[a]

Ocean Area	POC ($\mu g\ l^{-1}$)
Eastern North Atlantic margins	100
Open-ocean South Atlantic	25
Open-ocean Indian Ocean	9.7
China Sea	30

Data from Chester and Stoner.[78]

Ocean where the fall-out of eolian particulates results in $Al_{(p)}$ concentrations of up to $\gtrsim 1\ \mu g\ l^{-1}$, or in polar areas in which there is an influx of glacial material). These trends are apparent in the data given by Krishnaswami and Sarin[43] on the distribution of $Al_{(p)}$ in the surface waters of the Atlantic. Relatively high concentrations are found in the high latitudes and in the equatorial regions, and relatively low concentrations are found in the open-ocean South Atlantic. The analytical data generated by the GEOSECS program indicate that $Al_{(p)}$ concentrations in the water column do not vary by more than a factor of two at depths above the nepheloid layer (Spencer, in Krishnaswami and Sarin[43]). In the nepheloid layer itself $Al_{(p)}$ concentrations up to an order of magnitude higher than in surface waters are found in some regions; these result from the resuspension of bottom sediment which, if it is a clay, will contain ~10% Al.

To summarize, the distribution of $Al_{(p)}$ in surface sea waters appears to be controlled largely by the input of aluminosilicates from the continents by the principal transport paths, i.e., rivers, wind and ice, with high concentrations in some areas in which any of these agencies predominate. Further, the mineralogical compositions of the particulate aluminosilicates in general reflect those of the soils of the surrounding land masses. At depth in the water column the concentrations of $Al_{(p)}$ are increased due to sediment resuspension in the nepheloid layer.

The distribution of particulate organic carbon (POC) in sea water has recently been reviewed by several authors (see, e.g., Parsons[77]). The principal trends are summarized as follows: (1) In general the quantity of detrital POC in the oceans makes up a large fraction of the total POC, often exceeding the phytoplankton POC by a factor of 10, or more, an exception being during large phytoplankton blooms in near-surface waters. (2) Near-shore waters usually have higher concentration of POC than open-ocean waters. (3) Waters of the euphotic zone generally have more POC than those lying beneath it. (4) In open-ocean areas the highest surface concentrations of POC are found in productive waters, e.g., in areas of upwelling, and the lowest concentrations in the central oceanic regions, e.g., the Sargasso Sea and the Central Pacific.

Chester and Stoner[78] have given data on the concentrations of POC in surface waters from several regions of the World Ocean. Their results are summarized in Table 6, from which it is apparent that, on average, the concentration of POC in surface waters decreased in the following order: northeastern Atlantic > China Sea > South Atlantic > Indian Ocean. The POC is a quantitatively important constituent of the total suspended particulate material, and organic matter (calculated as organic carbon X2)[79] usually makes up ~25% to $\gtrsim$80% of the total solids in surface water TSM; however, at depth the percentage decreases (see, e.g., Krishnaswami and Sarin,[43] Feely et al.[74]). It is apparent, therefore, that the overall distribution of POC in sea water is controlled by the major oceanographic features, such as upwelling and surface circulation patterns, with a general decrease in concentration with depth in the water column as a result of both the decomposition of POC and of its incorporation in the marine food chain.

The mutual proportions of the various components of oceanic TSM will control its overall chemical composition, and it is of interest to examine the trace element assemblages of some of these individual components. The average trace element composition of marine organisms, fecal pellets, river-transported sediment, and atmospheric particulates are given in Table 7. It must be understood, however, that these average values are, at best, only tentative. For example, there are very large variations in the trace element contents of marine plankton (for a review of this subject, see Brewer[80]), and the average elemental concentrations given in Table 7 can only be regarded as giving order of magnitude estimates of planktonic compositions. Nevertheless, when enrichment factors are used, the average data do serve to illustrate that some elements (e.g., Zn, Pb) are much more concentrated in marine plankton, relative to crustal material, than are others (e.g., Cr, Mn).

The manner in which the mutual proportions of the various components control the trace element composition of oceanic TSM can be illustrated by considering the effect of crustal-derived solids on the concentrations of Cu, Pb, and Zn in surface water particulates from two of the oceanic environments for which Chester et al.[71] described the distribution of $Al_{(p)}$, i.e., the China Sea and the northeastern Atlantic. Data on the concentrations of Cu, Pb, and Zn, in these particulates have been provided by Chester et al.[81] and Chester and Stoner,[82] and are given in Table 8. To make a first order comparison between the particulates and other oceanic solids, Chester et al.[81] used an enrichment factor calculated in the following manner: $E.F._{(Al)}$ = (ppm element in solid/ppm Al in solid)/(ppm element in average crustal material/ppm Al in average crustal material). The average enrichment factors for Cu, Pb, and Zn, in the two groups of particulates are also given in Table 8. Chester et al.[81] used these enrichment factors to tentatively relate the compositions of the particulates to the type of oceanic environment in which they were found. For example, the China Sea receives most of its lithogenous solids from river run-off, and there is a close agreement between the enrichment factors of the surface water particulates from this region and local river-transported sediment (see Table 9). Lithogeneous solids are brought to the eastern margins of the North Atlantic by both river run-off and eolian transport, and the enrichment factors for the surface water particulates from this area, together with those of local river solids and soil-sized northeast trade wind dusts, are given in Table 10. From this table it is evident that the enrichment factors for the surface water particulates are considerably higher than those in either the river solids of the trade wind dusts.

Wallace et al.[73] have analyzed a series of surface water particulates from the northwest Atlantic, and by examining the ratios of trace metals to both aluminium and organic carbon they concluded that organic matter was probably the most important regulator of particulate trace metal abundance in open-ocean surface sea waters, and possibly also in continental shelf and slope waters. Krishnaswami and Sarin[43] have given data on the compositions of Atlantic Ocean surface water particulates and have concluded that the concentrations of Mn are controlled by both a particulate organic carbon and an iron phase.

The concentrations of particulate trace metals (PTM) in surface sea water have been investigated by many workers, and a compilation of some of these is given in Table 11. However, the distribution of the trace metals with depth in the water column is less well known, although one apparent feature of the depth distribution of TSM is the increase in the proportion of aluminosilicates, and the decrease in the proportion

Table 7
THE AVERAGE TRACE ELEMENT COMPOSITIONS OF SOME OF THE COMPONENTS OF THE TOTAL SUSPENDED MATTER IN SEA WATER AND OF SURFACE AND DEEP WATER TSM (ppm)

Element	Phytoplankton; Monterey Bay, California[b]	Zooplankton Monterey Bay, California[b]	Microplankton North Pacific[b]	Average; marine organisms[c]	Fecal pellets			River-transported solids[g]	Soil-sized atmospheric particulatesl[a] E.F.$_{(Al)}$[h]	Total atmospheric particulates,[a] E.F.$_{(Al)}$[i]	Oceanic TSM[i]	
					A[d]	B[e]	C[f]				Surface water	Deep water
Al	7—2850	<8—313	72—108	~160	—	20800	74900	94000	1.0	1.0	3000	7500
Fe	49—3120	54—1070	1030—4000	~900	24000	21600	43600	47300	0.8	0.9	8800	15000
Mn	2.1—30	2.2—12	3.4—32	~10	243	2110	768	1050	0.5	1.1	140	320
Cu	1.3—45	4.4—23	40—104	~30	226	650	308	100	0.9	7.0	145	200
Ni	<0.5—13	0.5—12	11—12	~20	20	—	—	90	0.9	4.0	70	130
Co	<1	<1	<1	≲1	3.5	15	10	20	—	1.6	5	16
Cr	<1—21	<1	<1—3.7	≲1	38	—	—	100	1.0	2.4	125	170
V	<3	<3	<3	≲3	—	76	114	170	—	2.0	22	—
Ba	5—500	4—257	51—70	~60	—	192	526	600	—	—	—	—
Sr	53—3934	83—810	6800—9650	~900	—	1432	0	150	0.6	—	—	—
Pb	<1—47	<1—12	17—39	~20	34	—	—	150	7.7	390	180	570
Zn	3—703	53—279	285—4190	~250	950	<20	<20	350	2.0	28	640	1000
Cd	0.4—6	0.8—10	1.0—2.2	~5	9.6	—	—	~1	—	—	—	—
Hg	0.10—0.59	0.07—0.16	0.11—0.53	~0.15	—	—	—	—	—	450	16	36

[a] Because of the wide range of element compositions of some atmospheric particulates E.F.$_{(Al)}$ are given here.
[b] Data from Martin and Knauer.[115]
[c] Data from Chester and Aston.[117]
[d] Data for "green" fecal pellets, from Spencer et al.[35]
[e] Data for "red" fecal pellets, from Spencer et al.[35]
[f] Data from Fowler.[92]
[g] Data from Martin and Maybeck.[114]
[h] Data from Chester et al.[11a]
[i] Data from Baut-Menard and Chesselet,[83] Atlantic Ocean.

Table 8
THE AVERAGE CONCENTRATIONS AND ENRICHMENT FACTORS FOR Cu, Pb, AND Zn IN SURFACE WATER PARTICULATES FROM THE CHINA SEA AND NORTHEASTERN ATLANTIC[a]

	China Sea		Northeastern Atlantic	
Element	Conc. (ppm)	$E.F._{(Al)}$	Conc. (ppm)	$E.F._{(Al)}$
Cu	100	2.3	116	16
Pb	64	6.9	52	34
Zn	201	3.8	231	27

[a] Data from Chester et al.[81]

Table 9
THE AVERAGE ENRICHMENT FACTORS FOR Cu, Pb, AND Zn IN SURFACE WATER PARTICULATES FROM THE CHINA SEA AND LOCAL RIVER-TRANSPORTED SOLIDS

Element	China Sea particulates[a] $E.F._{(Al)}$	River-transported solids (Mekong)[b] $E.F._{(Al)}$
Cu	2.3	1.4
Pb	6.9	6.7
Zn	3.8	3.1

[a] Data from Chester et al.[81]
[b] Data from Martin and Maybeck.[114]

Table 10
THE AVERAGE ENRICHMENT FACTORS FOR Cu, Pb, AND Zn IN SURFACE WATER PARTICULATES FROM THE NORTHEASTERN ATLANTIC, LOCAL RIVER, AND WIND-TRANSPORTED SOLIDS

Element	Northeastern Atlantic particulates[a] $E.F._{(Al)}$	River-transported solids (Zaire)[b] $E.F._{(Al)}$	Eolian dust (Atlantic northeast trades)[c] $E.F._{(Al)}$
Cu	16	0.55	1.7
Pb	34	26	18
Zn	27	4.0	5.4

[a] Data from Chester et al.[81]
[b] Data from Martin and Maybeck.[114]
[c] Data from Martin and Knauer.[115]

Table 11
THE CONCENTRATIONS OF SOME PARTICULATE TRACE ELEMENTS IN SURFACE SEA WATER (ng l^{-1})

Element	China Sea	North eastern margins; Atlantic Ocean[a]	Open-ocean; South Atlantic and Indian Oceans[c]	Atlantic Ocean[b]	Northwestern Atlantic[c]	Atlantic Ocean[d]	Atlantic Ocean[e]	Gulf of Mexico[f]		Gulf of Maine[g]
								Coastal	Open-ocean	
Mn	778	54	9	3.5	11	7.0	7.6	1110	1200	187
Cu	74	43	10	3.6	3.4	9.1	25	790	430	166
Pb	47	19	5	4.5	2.8	—	—	—	—	—
Zn	149	85	22	16	11	—	23	1200	550	135
Fe	—	—	—	222	121	223	341	—	—	—
Cr	—	—	—	3.1	3.1	—	9.6	—	—	—
Ni	—	—	—	1.8	<1.2	5.2	—	—	—	—
V	41	14	6	—	—	—	—	—	—	—
Al	45559	3755	298	78	160	164	118	—	—	—

[a] Data from Chester et al.[81]
[b] Data from Baut-Menard and Chesselet.[83]
[c] Data from Wallace et al.[73]
[d] Data from Krishnaswami and Sarin.[43]
[e] Data for GEOSECS, from Baut-Menard and Chesselet.[83]
[f] Data from Slowey and Hood.[116]
[g] Data from Spencer and Sachs.[24]

Table 12
PARTICULATE TRACE ELEMENT CONCENTRATIONS IN SURFACE AND INTERMEDIATE AND DEEP WATERS OF THE ATLANTIC OCEAN (ng l^{-1})[a]

Element	Surface water (4 Samples) Arithmetic mean	Range	Intermediate and deep water (35 samples); geometric mean
Al	78	46—98	110
Mn	3.5	1.2—8.7	4.8
Fe	222	128—296	200
Cr	3.1	1.3—4.6	2.6
Ni	1.8	1.0—4.0	2.0
Cu	3.6	1.0—6.6	3.0
Zn	16	5.1—26	15
Pb	4.5	2.6—6.1	8.6

[a] Data from Baut-Menard and Chesselet.[83]

of organic carbon, in particulates from the near-bottom nepheloid layer (see, e.g., Feely et al.,[74] Feely[18]). This is reflected in the depth profile of $Al_{(p)}$ in the Atlantic Ocean; near-surface water concentrations range from ~50 to ~400 μg kg^{-1} with those in near-bottom waters as high as ~1000 to ~2000 μg kg^{-1} (Spencer et al., quoted in Reference 2). There is more uncertainty on the depth distribution of trace metals in the oceans. However, Buat-Menard and Chesselet[83] did not find any significant trends in the concentrations of particulate trace metals from depths of ~300 to 5000 m in the Atlantic Ocean water column, and their average values for these intermediate and deep water particulates are compared to those from surface waters in Table 12. The depth distribution of PTM in the oceanic water column is intimately related to the processes which occur during the downward transport of TSM, and these are discussed in Section VI.

VI. THE SOURCES AND VERTICAL TRANSPORT OF PARTICULATE TRACE ELEMENTS IN THE OCEANS

There is a considerable body of evidence which indicates that in some regions there is a rapid transport of particulate matter out of the mixed layer to depth in the water column. For example, in some oceanic regions there are markedly well-developed contours in the distributions of clay minerals in deep-sea sediments. In this context, Griffin et al.[84] have shown that the concentrations of kaolinite in North Atlantic deep-sea sediments decrease systematically westwards from the coast of West Africa. They concluded that a major proportion of this mineral was brought to the region by eolian transport (see also Chester et al.[85]). This distribution of kaolinite is also reflected in surface water TSM, and because the patterns are preserved in the underlying sediments some mechanism must be operative which allows the particulates to reach the sea floor in a relatively short time. For example, Delany et al.[28] have shown that from Stokes' law consideration a 4-μm particle would take ~100 days to fall ~100 m, by which time it could have been transported ~ 4000 km by the North Equatorial Current. To overcome this difficulty, Delany et al.[28] suggested that the particulate may be aggregated within fecal pellets. Another line of evidence which suggests that the rate of sinking of TSM in the oceans is in excess of that predicted by Stokes' law was provided by Osterberg et al.[49] who found that certain radionuclides with a relatively short half-life, e.g., ^{95}Sr (half-life, 65 days), could be detected in bottom feeders at a depth of ~2800 m in Pacific Ocean waters. These authors also suggested sinking by fecal pellets as the downward transport mechanism.

The theoretical aspects of the downward transport of TSM in the oceans have recently been examined by Lerman et al.,[86] McCave,[87] and Brun-Cotton.[88] Lerman et al.[86] studied the settling rates of particles of various shapes and concluded that for particles having the same volume the settling velocities decrease in the order cylinders > spheroids > discs. According to Sackett,[2] this means that clay minerals, which most nearly approximate to discs in shape, have relatively low settling velocities. Baun-Cotton[88] has stressed that particle density is at least as important as particle size in determining settling rates and further showed that Stokes' law cannot be applied to particles with diameters smaller than 4 to 5 μm. He concluded that the population of particles $\lesssim$ 5 μm might be conservative and move horizontally in a given water mass, and that it is mainly the larger (~50 to ~100 μm) particles, or aggregates, in the upper water masses that fall to the sea bed. In contrast, the smaller particles, which make up the major part of the deep-sea sediments, reach the bottom by mechanisms and routes other than those determined for the fast-settling Stokes particles. McCave[87] has concluded that the principal feature in the vertical flux of particles down the water column is that relatively rare, rapidly sinking large particles contribute most of the flux, whereas smaller particles contribute most of the concentration. The larger particles are formed by aggregation in the upper water layers of the oceans; such aggregation is sufficient to remove material rapidly to the bottom and make the distribution of particle properties (e.g., clay mineralogy) similar in both surface and sea bed samples.

These various investigations lead to the very important conclusion that the particles in sea water may, in general, be considered to consist of two populations: (1) a small-sized population, $\lesssim$ 5 μm in size, which undergoes large-scale horizontal movement; and (2) a large-sized population of mainly aggregated particles, $\gtrsim$50 μm in size, which undergoes vertical settling and transports material down the water column to the sea bed. The application of this two-population particle model to the actual water column has been demonstrated by Bishop et al.[89] who showed that ~95% of the downward flux of particulate organic carbon in the upper ~400 m of the equatorial Atlantic is associated with large fecal particles ($\gtrsim$50 μm), although these only account for $\lesssim$5% of the total particulate organic carbon. Despite the importance of these larger particles in the downward transport of material from the sea surface, they are rare in the water column and tend to be missed by those TSM sampling techniques which process relatively small volumes of water.

The nature of these large aggregates has been considered by various authors, and it is apparent that fecal pellets and fecal material make up a high proportion of the large particle population. A good summary of the sinking rates of fecal pellets has been provided by Sackett;[2] the measured rates varying between ~15 m day^{-1} to ~860 m day. The role of fecal pellets in the vertical transport of trace elements has recently been investigated by Fowler and his co-workers (see, e.g., Elder and Fowler,[90] Higgo et al.,[91] Fowler,[92] Cherry et al.[93]). Fowler [92] has given data on the trace element contents of fecal pellets produced by a planktonic euphausiid, and these are given in Table 13. The identification of fecal pellets at depth in the water column has been reported by a number of workers (see, e.g., Small and Fowler,[94] Soutar et al.,[95] Bishop et al.[96]) and at the sediment/water, interface by Honjo[36] and Spencer et al.[35]

There have been several very important recent papers which have considered the sources of trace elements in oceanic TSM and the roles which the elements play in pelagic sedimentation. Wallace et al.[73] investigated the influence of atmospheric deposition on the trace metal concentrations in TSM from surface waters of the northwest Atlantic. They concluded that particulate organic matter played a dominant role in

Table 13
TRACE ELEMENT CONCENTRATIONS IN FECAL PELLETS PRODUCED BY A PLANKTONIC EUPHAUSIID[a]

Element	Concentration ($\mu g\ g^{-1}$)
Ag	2.1
Cd	9.6
Co	3.5
Cr	38
Cu	226
Fe	24,000
Mn	243
Ni	20
Pb	34
Zn	950
Ce	200
Cs	6.0
Hg	0.34
Sb	71
Sc	2.8
Se	6.6
Sr	78

[a] Data from Fowler.[92]

controlling the particulate concentrations of Cu, Zn, Pb, Cd, and to a lesser extent Ni and Cr. Certain elements, i.e., Cr, Cu, Zn, Pb, and Cd, were enriched in the particulates relative to crustal material to about the same extent as they were in atmospheric particulates from the same general region (see Table 14). Using a simplistic vertical flux model, the authors were able to show that the atmospheric input of trace metals to the surface waters was of the same approximate magnitude as the rate of removal of particulate trace elements from the mixed layer by sinking in association with POC. They concluded that for Ni, Cr, Cu, Zn, Pb, and Cd atmospherically introduced particulate material may not be the dominant direct source of the particulate forms of these metals in the mixed layer, and that the fate of these atmospherically transported metals may be a function of their solubility in sea water. However, almost all the particulate Al, Fe, and Mn was attributed to an eolian source.

Buat-Menard and Chesselet[83] have made a detailed study of the influence exerted by the atmospheric flux on the trace element composition of North Atlantic TSM. By considering the two principal sources (i.e., biological particles produced within the ocean and crustal weathering products transported to the oceans by rivers, wind, and ice) the authors were able to draw the following conclusion regarding the sources of 17 deep-water particulate trace elements in the oceans: (1) Al, Sc, and Th have solids produced by crustal weathering as their principal source. (2) V, Fe, and Mn are associated with both crustal material and another phase. (3) Co, Cr, Ni, Cu, Zn, Sb, Pb, Au, Hg, and Se are not directly related to solids derived from crustal sources. (4) Most of the trace metals are present in concentrations significantly higher than would be expected from marine plankton. It is apparent from this work, therefore, that sources other than the sinking of crustal weathering products and plankton control the particulate concentrations of Co, Cr, Ni, Cn, Zn, Sb, Pb, Au, Hg, and Se in the deep

Table 14
MEAN PARTICULATE TRACE METAL CONCENTRATIONS AND ENRICHMENT FACTORS IN SARGASSO SEA SURFACE WATER AND BERMUDA AIR[a]

Element	Sargasso Sea (ng l^{-1})	Bermuda Air (ng m^{-3})	Enrichment factor	
			Sargasso Sea	Bermuda Air
Al	83	130	1.0	1.0
Mn	1.6	1.5	1.7	1.5
Fe	62	97	1.1	1.1
Ni	$\leqslant$0.5	0.08	$\leqslant$7	1.1
Cr	2.9	1.0	29	5
Cu	2.4	1.0	43	11
Zn	2.6	2.5	37	22
Pb	1.4	3.4	110	180
Cd	0.04	0.2	200	480

[a] Data from Wallace et al.[73]

Table 15
ENRICHMENT FACTORS FOR ATLANTIC DEEP AND INTERMEDIATE WATER PARTICULATES AND ATMOSPHERIC PARTICULATES[a]

Element	Geometric mean $EF_{(Al)}$; deep and intermediate water particulates	Geometric mean $EF_{(Al)}$; atmospheric particulates
Co	6.9	1.6
Cr	19	2.4
Ni	20	4.0
Cu	41	7.0
Zn	160	28
Pb	510	390
Sb	560	250
Se	2690	4020
Hg	3740	450
Au	3930	570

[a] Data from Buat-Menard and Chesselet.[83]

ocean. For these elements the E.F.s$_{(Al)}$ in the deep-water and atmospheric particulates increased in the same general manner (see Table 15), and the authors attempted to estimate the flux balance for the mixed layer between the atmospheric input and particle sinking. In their model they considered both the small and large particle populations (see above), and by the use of various atmospheric fall-out and particle sinking rates the following fluxes were estimated: Fl, atmospheric deposition; F2A, "small particle" vertical flux; F2B, "large particle" vertical flux; F2, total primary particulate flux (i.e, F2A + F2B). Since they did not find any significant trends in the concentrations of the particulate trace metals from depths of 300 m to 5000 m, it was assumed that the fluxes apply to the whole water column. The results are given in Table 16, from which it can be seen that for all the trace metals flux F2B is the dominant contri-

Table 16
PARTICULATE FLUX BALANCE OF TRACE METALS FOR THE MIXED LAYER BETWEEN ATMOSPHERIC INPUT AND PARTICLE SINKING ($ng\ cm^{-2}\ year^{-1}$)[a]

Element	F1	F2A	F2B	F2
Al	5,000	900	4,300	5,200
Sc	1.1	0.2	0.7	0.9
V	17	3	32	35
Mn	70	38	202	240
Fe	3,200	1,600	12,700	14,300
Co	2.7	1.8	7.2	9
Cr	14	20	180	200
Ni	20	16	101	117
Cu	25	24	210	234
Zn	130	120	920	1,040
Ag	0.9	1.1	5.8	6.9
Au	0.1	0.2	0.6	0.8
Hg	2.1	3.2	23	26.5
Sb	3.5	1.3	10	11.3
Se	14	1.5	12	13.5
Pb	310	70	260	330
Th	0.9	0.15	0.6	0.75

Note: F1 = atmospheric deposition. F2A = "small particle" vertical flux — see text. F2B = "large particle" vertical flux — see text. F2 = total primary particulate flux (F2A + F2B).

[a] Data from Buat-Menard and Chesselet.[83]

bution to the total primary flux F2. The overall conclusions reached may be summarized as follows: (1) Al, Sc, and Th have a primarily atmospheric source in North Atlantic deep water, excluding coastal waters (from which river input is important) and bottom waters (from which sediment resuspension is dominant). (2) An atmospheric flux is important for Fe, V, Mn, Co, and Cr, but it is apparent from Table 16, columns 1 and 4, that another flux component is also involved. This component probably results partly from the association of these elements with particulate organic matter and biogenous debris, and partly from the *in situ* formation of iron oxides. (3) The elements Ni, Cu, Zn, Au, Ag, and Hg are enriched in the particulates relative to crustal material. However, the atmospheric input makes a relatively small contribution (i.e., ≲10%) to the total primary flux of these elements, and their particulate concentrations are probably controlled by their involvement in the marine biological cycle. This conclusion is supported by a number of recent studies (see, e.g., Boyle et al.,[97] Sclater et al.,[98] Bruland et al.[89]), which have shown that dissolved Ni, Cu, and Zn in sea water behave like the nutrients. (4) The elements Sb, Se, and Pb are also enriched in the particulates relative to crustal material, and for these the atmospheric input makes up a significant fraction (i.e., ≳30%) of the total primary flux in sea water. An important fraction of Pb in the marine atmosphere has an anthropogenic source, and the authors concluded that this is reflected in the distribution of Pb in the water column, with ≳90% of the total concentration of this element in Atlantic deep waters having a similar source. The atmospheric and marine geochemistries of Se and Sb are somewhat different to that of Pb. According to Buat-Menard and Chesselet[83] the particulate concentrations of these two elements are identical in deep waters from the North Atlantic, South Atlantic and South Pacific, despite the fact that anthropogenic concentrations of trace metals in the marine atmosphere are probably one order of magnitude lower in the southern than in the northern hemisphere. Because of this, the enrichments of Se and Sb in oceanic particulate matter from the southern hemisphere cannot originate from a direct anthropogenic atmospheric input into southern latitudes, since this would result in lower enrichments than those found in the northern hemisphere. The authors also argued that the time which has elapsed since the industrial revolution is too short to account for a world-wide contamination of oceanic particulate matter by a northern hemisphere atmospheric flux transported north to south within the oceans themselves. It was concluded, therefore, that the oceanic particulate concentrations of Se and Sb are probably the result of a natural input from the atmosphere.

It is apparent from this important investigation that the deposition of atmospheric material over open-ocean areas may play a significant role in controlling the particulate chemistries of some elements (e.g., Al, Sc, Th, Fe, V, Mn, Co, Cr, Pb, Se, Sb). However, for those trace metals which could have an anthropogenic origin (e.g., Pb, Se, Sb), the atmospheric input to the ocean surface would have to be one order of magnitude higher than at present to have a detectable effect on their concentrations in deep water TSM; an exception to this being Pb in the North Atlantic water column.

One of the most important recent advances in our understanding of the role played by TSM in the oceanic trace element budget has resulted from the use of sediment traps. These sampling devices are positioned in the water column and collect material which is undergoing downward transport. Their use permits a *direct* estimate to be made of the downward flux of chemical elements, and they have been employed in a number of investigations (see, e.g., Berger and Soutar,[100] Wiebe et al.,[101]). Spencer et al.[35] positioned a pair of sediment traps at 5367 m and 5581 m within the bottom nepheloid layer of the Sargasso Sea and carried out chemical analyses on the collected material (Table 17). From data obtained over a period of 75 days they estimated that the mean total particulate flux is 1.68 ± 0.79 mg cm^{-2} $year^{-1}$. The material collected consisted of ~50% clays, ~20% calcium carbonate, ~20% silica and ~5% organic matter. The authors attempted to separate the primary flux from the resuspended sediment contribution and concluded that, of the total flux, only ~5% of the clay, most of the calcium carbonate, and ~90% of the organic matter were contributed by rapidly settling large particles derived directly from the surface waters. The principal portion of the total flux appeared to have originated from the resuspension of bottom sediments that may have been aggregated and sedimented from the nepheloid layer. Two distinctly different types of fecal pellet were collected in the traps (see also, Honjo,[36]); "green pellets" which originate at the ocean surface, and "red pellets", the origin of which is uncertain but which are thought to be formed either by organisms which inhabit the nepheloid layer or by coprophagic zooplankton. The elemental composition of the two types of pellets are given in Table 17, together with the concentrations normalized to Al. It is apparent from the normalized concentrations that, relative to Al, the "green" pellets (i.e., those formed in the surface layers and subsequently transported downwards as the primary flux) have an excess of Fe, Mn, and Co; according to the authors this may imply that a distinct nonaluminosilicate phase, e.g., a hydroxide, is incorporated into the primary flux.

Spencer et al.[35] made a first order estimate of the primary elemental flux delivered by the "green" fecal pellets and the results are listed in Table 18, together with the total primary particulate flux given by Buat-Menard and Chesselet[83] for the deep waters of the North Atlantic. These two sets of data represent two different approaches

Table 17
ELEMENTAL COMPOSITIONS OF FECAL PELLETS COLLECTED BY SEDIMENT TRAP IN THE DEEP SARGASSO SEA[a]

		"Green" fecal pellets		"Red" fecal pellets	
Element	Units	Element concentration	Normalized to Al	Element concentration	Normalized to Al
Al	%	2.08	1	7.49	1
Fe	%	2.16	1.04	4.36	0.58
Mg	%	0.81	0.38	2.73	0.37
Ca	%	23.2	11.2	5.69	0.76
Ti	%	0.13	0.06	0.31	0.041
Mn	ppm	2110	0.10	768	0.010
Ba	ppm	192	0.007	526	0.009
Sr	ppm	1430	0.00067	50	0.069
V	ppm	76	0.0036	114	0.0015
I	ppm	496	0.024	83	0.0011
La	ppm	24	0.0012	49	0.00065
Sc	ppm	4	0.002	15	0.0002
Cu	ppm	650	0.032	308	0.004
Sb	ppm	<5	<0.0002	<5	>0.00007
Zn	ppm	<20	<0.001	<20	>0.00027
Co	ppm	15	0.0007	10	0.00013
Org. C	%	15	7.2	5	0.67

[a] Data from Spencer et al.[35]

to estimating the primary flux of material to deep waters from the surface layers in the Atlantic Ocean and despite the uncertainties involved in the calculations and the assumptions made in the models employed, there is remarkably good agreement between them. We now have, therefore, a reasonable first order estimate of this primary particulate flux.

Over some parts of the oceans special conditions prevail which may affect particulates in the water column. Anoxic environments, such as that found at depths $\gtrsim$200 m in the Black Sea, are examples of these special conditions under which precipitation-dissolution phenomena can play an important role in controlling the concentrations of some elements in sea water. In this context, Spencer and Brewer[102] have shown that in the Black Sea deep waters Mn is transported upwards by mixing and precipitates as MnO_2 at the interface between oxygenated and sulfide-containing waters. These particles then sink and are redissolved in the anoxic zone, with the result that the deep water acts as a Mn trap via this precipitation—dissolution cycle.

VII. THE GREAT PARTICLE CONSPIRACY

We can now turn again to Turekian's "great particle conspiracy" which was the starting point of this treatment of the importance of TSM in the oceanic geochemical cycle. In his stimulating paper Turekian[1] pointed out that as analytical and collection techniques have improved it has become apparent that most trace elements are present at very low levels in the oceans and that variations in their concentrations are usually unspectacular. Further, theoretical calculations suggest that most trace metals should be present in ocean waters at concentrations which are many orders of magnitude higher than those found using the best modern analytical techniques. Turekian then posed the question, "Why are the oceans so depleted in these trace metals?" He con-

Table 18
ESTIMATES OF THE PRIMARY PARTICULATE FLUX TO DEEP OCEAN WATERS (ng cm^2 $year^{-1}$)

Element	Primary flux; Atlantic Ocean[a]	Primary flux; Sargasso Sea[b]
Al	5,200	5,000
Fe	14,300	5,200
Mg	—	1,900
Ca	—	55,800
Ti	—	300
Mn	240	500
Ba	—	46
Sr	—	344
V	35	18
I	—	120
La	—	6
Sc	0.9	9.6
Cu	234	160
Sb	11.3	<1
Zn	1,040	<5
Co	9	3.6
Cr	180	—
Ni	117	—
Ag	6.9	—
Au	0.8	—
Hg	26.5	—
Se	13.5	—
Pb	330	—
Th	0.75	—

[a] Calculated total primary flux; Buat-Menard and Chesselet.[83]

[b] Primary flux estimated from the sinking of "green" fecal pellets (see text); Spencer et al.[35]

cluded that the answer lies in the role played by particles as sequestering agents for reactive elements during every step of the transport process from the continents to the ocean floor.

The concept of trace metal sequestration by particles is not new. It has been known for many years that sea water is markedly undersaturated, with respect to any likely solid, for most trace metals (see, e.g., Goldsmidt[103]).

It became apparent, therefore, that the concentration of most trace metals in sea water are not controlled by the solubilities of the least soluble compounds which they form with one of the anionic species in the water, and since the supply of metals during geological time has been more than sufficient to attain saturation, some other removal mechanism must have been operative. Krauskopf[104] used both theoretical and experimental techniques to examine the factors which control the concentrations of a series of trace elements in sea water. Although his work can be criticized on a number of grounds (see Brewer[80]), it still remains one of the most comprehensive studies of this kind ever undertaken. Adsorption was one of the removal mechanisms he considered as a possible controlling factor on trace metal concentrations. He concluded that removal onto adsorbents such as ferric oxide, manganese dioxide, apatite, montmorillonite, dried plankton, and peat moss is a possible control on the concentrations of

Zn, Cu, Pb, Bi, Cd, Hg, Ag, and Mo in sea water. In many subsequent investigations adsorption (or scavenging) onto suspended particulates in the water column has been the most often invoked principal inorganic control on the removal of trace elements from sea water. A review of the processes involved in adsorption in the marine environment has been given by Parks,[105] and there have been numerous reports of laboratory experiments designed to study the characteristics of adsorption (see, e.g., O'Connor and Kester,[106] Murray,[107,108] van der Weijden,[109] van der Weijden and Kruissink[110]). However, the importance of these adsorption processes is difficult to evaluate on an ocean-wide scale, because the processes which occur in sea water are complex and interrelated, whereas those in laboratory experiments are usually much more simple. One investigation in which an attempt was made to simulate natural conditions was carried out by Kharkar et al.[111] These authors examined adsorption-desorption phenomena in a laboratory river-estuarine-marine system and concluded that when a trace metal is adsorbed from solution in river water by a clay mineral, it may be released, at least to some extent, on contact with waters of increasing salinity. For example, they found that the supply of Co desorbed from river detritus exceeded that originally in solution in river water by a factor of two. They also found that there was considerably less desorption of some metals from iron and manganese oxides than from clay minerals on contact with sea water. Humic metal colloids, which may be considered as part of the dissolved river load, undergo flocculation at the fresh water-sea water boundary. According to Coonley et al.[112] and Sholkovitz,[113] Fe can be flocculated in this mixing region. It is probable that during this process trace metals may also be removed from solution. This may overwhelm the effects of the desorption of trace metals from clay minerals, thus highlighting the complexity of the estuarine environment. However, Turekian, in his treatment of trace metal removal processes, followed the transport cycle through all its various phases and concluded that although there is a continuous movement of some trace metals in and out of solution in estuaries, only a small proportion of these "mobile" metals are actually lost from the estuarine system itself. This leads to the interesting concept that although estuaries are an extremely important stage in the geochemical cycle, they are mainly self contained, with release and deposition occurring within the system itself (i.e., they act as traps by removing reactive elements onto particles, many of which are deposited in coastal regions). There is a flux from estuaries out to the open-sea, but Turekian has concluded that it is mainly in the form of fine particles.

By following the transport process through weathering, stream transport, the estuarine cycle, and the oceans, Turekian has shown that particles clean the aqueous system of natural radionuclides and, by analogy, of trace metals. The particles range from active organic suspensions on land and in the sea, to flocculating metal-organic colloids at the river-ocean boundary, to freshly precipitated manganese and iron oxides which are found throughout the aqueous system. Together, these particles act as scrubbing agents which dominate the behavior of dissolved species, and Turekian has concluded that it is therefore no surprise that the oceans are so free of metals.

REFERENCES

1. **Turekian, K. K.,** The fate of metals in the oceans, *Geochim. Cosmochim. Acta,* 41, 1139, 1977.
2. **Sackett, W. M.,** Suspended matter in sea water, in *Chemical Oceanography,* Vol. 7, 2nd ed., Riley, J. P. and Chester, R., Eds., Academic Press, London, 1978, 127.
3. **Jerlov, N. G.,** Particle distribution in the ocean, *Rep. Swed. Deep-Sea Exped.,* 3, 73, 1953.
4. **Lisitzin, A. P.,** Sedimentation in the World Ocean, *Soc. Econ. Paleo. Min. Spec. Pub. 17,* Tulsa, 1972, 218.

5. **Jacobs, M. B. and Ewing, M.**, Suspended particulate matter: concentrations in the major oceans, *Science,* 163, 380, 1969.
6. **Bassin, N. J., Harris, J. E., and Bouma, A. H.**, Suspended matter in the Caribbean Sea: a gravimetric analysis, *Mar. Geol.,* 12, 171, 1972.
7. **Chester, R. and Stoner, J. H.**, Concentration of suspended particulate matter in surface sea water, *Nature,* 240, 552, 1972.
8. **Ewing, M. and Thorndike, E.**, Suspended matter in deep ocean water, *Science,* 147, 1291, 1965.
9. **Hunkins, K., Thorndike, E. M., and Mathieu, G.**, Nepheloid layers and bottom currents in the Arctic Ocean, *J. Geophys. Res.,* 74, 6995, 1969.
10. **Eittriem, S., Ewing, M., and Thorndike, E. M.**, Suspended matter along the continental margin of the North American Basin, *Deep-Sea Res.,* 16, 1969.
11. **Eittriem, S. and Ewing, M.**, Suspended particulate matter in the deep waters of the North American Basin, in *Studies in Physical Oceanography,* Gordon, A. L., Ed., Gordon and Breach, London, 1972, 123.
12. **Eittriem, S., Gordon, M., Ewing, M., Thorndike, E. M. and Bruchhausen, P.**, The nepheloid layer and observed bottom currents in the Indian-Pacific Antarctic Sea, in *Studies in Physical Oceanography,* Gordon, A. L., Ed., Gordon and Breach, London, 1972, 19.
13. **Ewing, M. and Connary, S.**, Nepheloid layer in the North Pacific, *Geol. Soc. Am. Memoir,* 126, 41, 1970.
14. **Ewing, M., Eittriem, S., Ewing, J., and Le Pichon, X.**, Sediment transport and distribution in the Argentine Basin. III. Nepheloid layer and processes of sedimentation, in *Physics and Chemistry of the Earth,* Vol. 8, Ahrens, L. H., Press, F., Runcorn, S. K., and Vrey, H. C., Eds., Pergamon Press, Oxford, 1971, 49.
15. **Connary, S. C. and Ewing, M.**, The nepheloid layer and bottom circulation in the Guinea and Angola Basins, in *Studies in Physical Oceanography,* Gordon, A. L., Ed., Gordon and Breach, London, 1972, 169.
16. **Baker, E. T., Sternberg, R. W., and McManus, D. A.**, Continuous light-scattering profiles and suspended matter over Nitinat deep-sea fan, in *Suspended solids in Sea Water,* Gibbs, R. J., Ed., Plenum Press, New York, 1974, 155.
17. **Bassin, N. J.**, Ph.D. dissertation, Texas A and M University, 1975; quoted in Sackett.[2]
18. **Feely, R. A.**, Major-element composition of the particulate matter in the near-bottom nepheloid layer of the Gulf of Mexico, *Mar. Chem.,* 3, 121, 1975.
19. **Brewer, P. G., Spencer, D. W., Biscaye, P. E., Hanley, P. E., Sachs, A., Smith, P. L., Kadar, S., and Fredericks, J.**, The distribution of particulate matter in the Atlantic Ocean, *Earth Planet Sci. Lett.,* 32, 393, 1976.
20. **Biscaye, P. E. and Eittreim, S. L.**, Suspended particulate loads and transports in the nepheloid layer of the abyssal Atlantic Ocean, *Mar. Geol.,* 23, 155, 1977.
21. **Gibbs, R. J.**, The suspended material of the Amazon Shelf and tropical Atlantic Ocean, in *Suspended Solids in Water,* Gibbs, R. J., Ed., Plenum Press, New York, 1974, 203.
22. **Manheim, F. T., Meade, R. H., and Bond, G. C.**, Suspended matter in surface waters of the Atlantic continental margin from Cape Cod to the Florida Keys, *Science,* 166, 371, 1970.
23. **Meade, R. H., Sachs, P. L., Manheim, F. T., Hathaway, J. C., and Spencer, D. W.**, Sources of suspended matter of the Middle Atlantic Bight, *J. Sediment. Petrol.,* 45, 171, 1975.
24. **Spencer, D. W. and Sachs, P. L.**, Some aspects of the distribution, chemistry and mineralogy of suspended matter in the Gulf of Maine, *Mar. Geol.,* 9, 117, 1970.
25. **Lal, D.**, The oceanic microcosm of particles, *Science,* 198, 997, 1977.
26. **Holeman, J. N.**, The sediment yield of major rivers of the World, *Water Resour. Res.,* 4, 737, 1968.
27. **Cawse, P. A. and Peirson, D. H.**, An analytical study of trace elements in the atmospheric environment, *U.K.A.E.A.,* H. M. Stationery Office, London, 1970.
28. **Delany, A. C., Delany, Audrey, C., Parkin, D. W., Griffin, J. J., Goldberg, E. D., and Reimann, B. E. F.**, Airborne dust collected at Barbados, *Geochim. Cosmochim. Acta,* 31, 885, 1967.
29. **Aston, S. R., Chester, R., Johnson, L. R., and Padgham, R. C.**, Eolian dust from the lower atmosphere of the eastern Atlantic and Indian Oceans, China Sea and Sea of Japan, *Mar. Geol.,* 14, 15, 1973; Plank, W. S., Pak, H., and Zaneveld, J. R. V., Light scattering and suspended matter in nepheloid layers, *J. Geophys. Res.,* 77, 1689, 1972.
30. **Chester, R., Baxter, G. G., Behairy, A. K. A., Connor, K., Cross, D., Elderfield, H., and Padgham, R. C.**, Soil-sized dusts from the lower troposphere of the Eastern Mediterranean Sea, *Mar. Geol.,* 24, 201, 1977.
31. **Prospero, J. M.**, Mineral and sea salt aerosol concentrates in various ocean regions, *J. Geophysical Res.,* 84, 725, 1979.
32. **Goldberg, E. D.**, Marine Pollution, in *Chemical Oceanography,* Vol. 3, 2nd ed., Riley, J. P. and Skirrow, G., Eds., Academic Press, London, 1975, 39.
33. **Goldberg, E. D.**, Marine geochemistry: chemical scavengers of the sea, *J. Geol.,* 62, 249, 1954.

34. **Jacobs, M. B. and Ewing, M.,** Mineralogy of particulate matter suspended in sea water, *Science,* 149, 179, 1965.
35. **Spencer, D. W., Brewer, P. G., Fleer, A., Honjo, S., Krishnaswami, S., and Nozaki, Y.,** Chemical fluxes from a sediment experiment in the deep Sargasso Sea, *J. Mar. Res.,* 36, 493, 1978.
36. **Honjo, S.,** Sedimentation of materials in the Sargasso Sea at a 5,367 m deep station, *J. Mar. Res.,* 36, 469, 1978.
37. **Behairy, A. K., Chester, R., Griffiths, A. J., Johnson, L. R., and Stoner, H. J.,** The clay mineralogy of particulate material from some surface seawaters of the Eastern Atlantic Ocean, *Mar. Geol.,* 18, M 45, 1975.
38. **Chester, R., Stoner, J. H., and Johnson, L. R.,** Montmorillonite in surface detritus, *Nature,* 249, 335, 1974.
39. **Krishnaswami, S., Lal, D., Somayajulu, B. L. K., and Craig, H.,** Investigations of grain quantities of Atlantic and Pacific surface particulates, *Earth Planet. Sci. Lett.,* 32, 403, 1976.
40. **Chester, R. and Hughes, M. J.,** A chemical technique for the separation of ferro-manganese minerals, carbonate minerals and adsorbed trace elements from pelagic sediments, *Chem. Geol.,* 2, 249, 1967.
41. **Elderfield, H.,** Hydrogenous material in marine sediments; excluding manganese nodules, in *Chemical Oceanography,* Vol. 5, 2nd ed., Riley, J. P. and Chester, R., Eds., Academic Press, London, 1976, 137.
42. **Chesselet, R., Jedwab, J., Darcourt, C., and Dehans, F.,** Barite as discrete suspended particles in the Atlantic Ocean, *Am. Geophys. Union Trans.,* 57, 255, 1976.
43. **Krishnaswami, S. and Sarin, M. M.,** Atlantic Surface particulates; composition, setting rates and dissolution in the deep sea, *Earth, Planet. Sci. Lett.,* 32, 430, 1976.
44. **Bergher, W. H.,** Biogenous deep-sea sediments: production, preservation and interpretation, in *Chemical Oceanography,* Vol. 5, 2nd ed., Riley, J. P. and Chester, R., Eds., Academic Press, London, 1976, 265.
45. **Freiling, E. C. and Ballou, N. E.,** Nature of nuclear debris in sea water, *Nature,* 195, 1283, 1962.
46. **Joseph, A. B. Gustafson, P. F., Russell, I. R., Schuert, E. A., Volchok, H. L., and Tamplin, A.,** in *Radioactivity in the Marine Environment,* National Academy of Science, Washington, D. C., 1971, 6.
47. **Bowen, V. T. and Sugihara, T. T.,** Oceanographic implications of radioactive fall-out distributions in the Atlantic Ocean: from 20°N to 25°S, from 1957 to 1961, *J. Mar. Res.,* 23, 123, 1965.
48. **Osterberg, C., Cutshell, N., and Cronin, J.,** Chromium-51 as a radioactive tracer of Columbia River water at sea, *Science,* 150, 1585, 1965.
49. **Osterberg, C., Carey, A. G., and Herbert, C.,** Acceleration of sinking rates of radionuclides in the ocean, *Nature,* 200, 1270, 1963.
50. **Fowler, S.,** Biological transfer and transport processes, in *Pollutant Transfer and Transport in the Sea,* Vol. II, Kullenberg, G., Ed., CRC Press, Boca Raton, Fla., in press.
51. **Burton, J. D.,** Radioactive nuclides in the marine environment, in *Chemical Oceanography,* Vol. 3, 2nd ed., Riley J. P. and Skirrow, G., Eds., Academic Press, London, 1975, 91.
52. **Morris, B. F., Butler, J. N., Sleeter, T. D., and Cadwallader, J.,** Transfer of particulate hydrocarbon material from the ocean surface to the water column, in *Marine Pollutant Transfer,* Windom, H. L. and Duce, R. A., Eds., Lexington Books, Lexington, 1976, 213.
53. **Horn, M. H., Teal, J. M., and Backus, R. H.,** Petroleum lumps on the surface of the sea, *Science,* 168, 245, 1970.
54. **Butler, J. N., Morris, B. F., and Sass, J.,** Pelagic Tar from Bermuda and the Sargasso Sea, *Bermuda Biological Station For Research, Spec. Pub. No. 10,* 1975, 346.
55. **Morris, B. F.,** Petroleum: tar quantities floating in the northwestern Atlantic taken with a new quantitative neuston net, *Science,* 173, 1971.
56. **Burns, K. A. and Teal, J. M.,** Hydrocarbons in the pelagic Sargassum community, *Deep-Sea Res.,* 20, 207, 1973.
57. **Lee, R. F., Sauerheber, R., and Benson, A. A.,** Petroleum hydrocarbons: uptake and discharge by the marine mussel, *Mytilus edulis, Science,* 177, 344, 1972.
58. **Conover, R. J.,** Some relations between zooplankton and Bunker C oil in Chedabucto Bay following the wreck of the tanker Arrow, *J. Fish. Res. Board Can.,* 28, 1327, 1971.
59. **Cowell, E. B.,** Oil pollution of the sea, in *Marine Pollution,* Johnston, R., Ed., Academic Press, London, 1976, 353.
60. **Risebrough, R. W., de Lappe, B. W., and Walker, W.,** Transfer of higher molecular weight chlorinated hydrocarbons to the marine environment, in *Marine Pollutant Transfer,* Windom, H. L. and Duce, R. A., Eds., Lexington Books, Lexington, 1976, 261.
61. **Bidleman, T. F., Rice, C. P., and Olney, C. E.,** High molecular weight chlorinated hydrocarbons in the air and sea: rates and mechanisms of air/sea transfer, in *Marine Pollutant Transfer,* Windom, H. L. and Duce, R. A., Eds., Lexington Books, Lexington, 1976, 323.

62. **Hom, W., Risebrough, R., Soutar, A., and Young, D. R.,** Deposition of DDE and PCB in dated sediments of the Santa Barbara Basin, *Science,* 184, 1197, 1974.
63. **Harvey, G. R. and Steinhauer, W. G.,** Biogeochemistry of PCB and DDT in the North Atlantic, in *Environmental Biogeochemistry,* Nriagu, J. O., Eds., Ann Arbor Science, Ann Arbor, 1976.
64. **Painter, H. A.,** Chemical, physical, and biological characteristics of wastes and waste effluents, in *Water and Water Pollution Handbook,* Vol. 1, Ciaccia, L. L., Ed., Marcel Dekker, New York, 1971, 329.
65. **Topping, G.,** Sewage and the sea, in *Marine Pollution,* Johnston, R., Ed., Academic Press, London, 1970, 303.
66. **Chow, T. J., Brutland, K. W., Bertine, K. K., Soutar, A., Koide, M., and Goldberg, E. D.,** Lead pollution: records in southern California coastal sediments, *Science,* 181, 1973.
67. **Brutland, K. W., Koide, M., Bowser, C., Maher, L. J., and Goldberg, E. D.,** Lead-210 and pollen chronologies on Lake Superior sediments, *Quarternary Research,* 5, 89, 1975.
68. **Perkins, E. J., Gilchrist, J. R. S., Abbott, O. J., and Halcrow, W.,** Trace metals in Solway Firth sediments, *Mar. Pollut. Bull.,* 4, 59, 1973.
69. **Skei, J. M., Price, N. B., and Calvert, S. E.,** Particulate metals in waters of Sörfjord, West Norway, *Amkio,* 2, 122, 1972.
70. **Chester, R. and Stoner, J. H.,** Trace elements in sediments from the lower Severn Estuary and Bristol Channel, *Mar. Pollut. Bull.,* 6, 92, 1975.
71. **Chester, R., Cross, D., Griffiths, A., and Stoner, J. H.,** The concentrations of "aluminosilicates" in particulates from some surface waters of the World Ocean, *Mar. Geol.,* 22, M59, 1976.
72. **Sackett, W. M. and Arrhenius, G. O. S.,** Distribution of aluminium species in the hydrosphere. I. aluminium in the oceans, *Geochim. Cosmochim. Acta,* 26, 955, 1962.
73. **Wallace, G. T., Hoffman, G. L., and Duce, R. A.,** The influence of organic matter and atmospheric deposition on the particulate trace metal concentration of northwest Atlantic surface seawater, *Mar. Chem.,* 5, 143, 1977.
74. **Feely, R. A., Sackett, W. M., and Harris, J. E.,** Distribution of particulate aluminium in the Gulf of Mexico, *J. Geophys. Res.,* 76, 5893, 1971.
75. **Joyner, T.,** The determination and distribution of particulate aluminium and iron in the coastal waters of the Pacific Northwest, *J. Mar. Res.,* 22, 259, 1964.
76. **Brewer, P. G., Spencer, D. W., and Bender, M. L.,** Elemental composition of suspended matter from the northern Argentine Basin, *Trans. Am. Geophys. Union,* (abstr.), 56, 308, 1974.
77. **Parsons, T. R.,** Particulate Organic Carbon in the Sea, in *Chemical Oceanography,* Vol. 2, 2nd ed., Riley, J. P. and Skirrow, G., Eds., Academic Press, London, 1975, 365.
78. **Chester, R. and Stoner, J. H,.** The distribution of particulate organic carbon and nitrogen in some surface waters of the World Ocean, *Mar. Chem.,* 2, 263, 1974.
79. **Riley, G. A.,** Particulate organic matter in sea water, in *Advances in Marine Biology,* Vol. 8, Russell, F. S. and Yonge, C. M., Eds., Academic Press, London, 1970, 1.
80. **Brewer, P. G.,** Minor elements in sea water, in *Chemical Oceanography,* Vol. 1, 2nd ed., Riley, J. P. and Skirrow, G., Eds., Academic Press, London, 1975, 415.
81. **Chester, R., Griffiths, A., and Stoner, J. H.,** The concentration of Al, Mn, V, Cu, Pb and Zn in surface water particulates from the China Sea, the North Eastern Atlantic, and open-ocean areas of the South Atlantic and Indian Oceans, in preparation, 1979.
82. **Chester, R. and Stoner, J. H.,** Trace elements in total particulate material from surface sea water, *Nature,* 255, 1975.
83. **Buat-Menard, P. and Chesselet, R.,** Variable influences of the atmospheric flux on the trace metal chemistry of oceanic suspended matter, *Earth Planet. Sci. Lett.,* 1979, in press.,
84. **Griffin, J. J., Windom, H., and Goldberg, E. D.,** The distribution of clay minerals in the World Ocean, *Deep-sea Res.,* 15, 433, 1968.
85. **Chester, R., Elderfield, H., Griffin, J. J., Johnson, L. R., and Padgham, R. C.,** Eolian dust along the eastern margins of the Atlantic Ocean, *Mar. Geol.,* 13, 91, 1972.
86. **Lerman, A., Lal, D., and Dacey, M. F.,** Stokes settling and chemical reactivity of suspended particles in natural waters, in *Suspended Solids in Water,* Gibbs, R. J., Ed., Plenum Press, New York, 1974, 17.
87. **McCave, I. N.,** Verticle flux of particles in the ocean, *Deep-Sea Res.,* 22, 491, 1975.
88. **Grun-Cotton, J. C.,** Stokes settling and dissolution rate model for marine particles as a function of size distribution, *J. Geophys. Res.,* 81, 1601, 1976.
89. **Bishop, J. K. B., Edmond, J. M., Kellen, D. R., Bacon, M. P., and Silker, W. B.,** The chemistry, biology, and vertical flux of particulate matter from the upper 400 m of the equatorial Atlantic Ocean, *Deep-Sea Res.,* 24, 511, 1977.
90. **Elder, D. L. and Fowler, S. W.,** Polychlorinated biphenyls: penetration into deep ocean by zooplankton fecal pellet transport *Science,* 197, 459, 1977.

91. **Higgo, J. J. W., Cherry, R. D., Heyrand, M., and Fowler, S. W.,** Rapid removal of plutonium from the ocean surface layer by zooplankton fecal pellets, *Nature,* 266, 623, 1977.
92. **Fowler, S. W.,** Trace elements in zooplankton particulate products, *Nature,* 269, 51, 1977.
93. **Cherry, R. D., Fowler, S. W., and Higgo, J. J. W.,** A possible relationship between element concentrations in zooplanktonic faecal pellets and oceanic residence times, in preparation, 1979.
94. **Small, L. F., Fowler, S. W., and Keckes, S.,** Flux of zinc through a macroplanktonic crustacean, in *Radioactive Contamination of the Marine Environment,* I.A.E.A., Vienna, 1973, 437.
95. **Soutar, A., Kling, S. A., Crill, P. A., Duffrin, E., and Bruland, K. W.,** Monitoring the marine environment through sedimentation, *Nature,* 266, 136, 1977.
96. **Bishop, J. K. B., Kellen, D. R., and Edmond, J. M.,** The chemistry, biology and vertical flux of particulate matter from the upper 400 m of the Cape Basin in the Southeast Atlantic Ocean, *Deep-Sea Res.,* 25, 1121, 1978.
97. **Boyle, E. A., Sclater, F. R., and Edmond, J. M.,** The distribution of dissolved copper in the Pacific, *Earth Planet. Sci. Lett.,* 37, 38, 1977.
98. **Sclater, F. R., Boyle, E. A., and Edmond, J. M.,** On the marine geochemistry of nickel, *Earth Planet. Sci. Lett.,* 31, 119, 1976.
99. **Bruland, K. W., Knauer, G. A., and Martin, J. H.,** Zinc in north-east Pacific water, *Nature,* 271, 741, 1978.
100. **Berger, W. H. and Soutar, A.,** Planktonic foraminifera: field experiment on production rate, *Science,* 156, 1495, 1967.
101. **Wieke, P. H., Boyd, S. H., and Winget, C.,** Particulate matter sinking to the deep-sea floor at 2000 m in the tongue of the Ocean, Bahamas, with a description of a new sedimentation trap, *J. Mar. Res.,* 34, 341, 1976.
102. **Spencer, D. W. and Brewer, P. G.,** Vertical advection diffusion and redox potentials as controls on the distribution of manganese and other trace metals dissolved in waters of the Black Sea, *J. Geophys. Res.,* 76, 5877, 1971.
103. **Goldschmidt, V. M.,** The principles of distribution of chemical elements in minerals and rocks, *J. Chem. Soc.,* 1937, 655, 1937.
104. **Krauskopf, K. B.,** Factors controlling the concentrations of thirteen rare metals in sea-water, *Geochim. Cosmochim. Acta,* 9, 1, 1956.
105. **Parks, G. A.,** Adsorption in the Marine Environment, in *Chemical Oceanography,* Vol. 1, 2nd ed., Riley, J. P. and Skirrow, G., Eds., Academic Press, London, 1975, 241.
106. **O'Connor, T. P. and Kester, D. R.,** Adsorption of copper and cobalt from fresh and marine systems, *Geochim. Cosmochim. Acta,* 39, 1531, 1975.
107. **Murray, J. W.,** The interaction of metal ions at the manganese dioxide-solution interface, *Geochim. Cosmochim. Acta,* 39, 505, 1975.
108. **Murray, J. W.,** The interaction of cobalt with hydrous manganese dioxide, *Geochim. Cosmochim. Acta,* 39, 635, 1975.
109. **Van der Weijden, C. H.,** Experiments on the uptake of zinc and cadmium by manganese oxides, *Mar. Chem,.* 4, 377, 1976.
110. **Van der Weijden, C. H., and Kruissink, E. C.,** Some geochemical controls on lead and barium concentrations in ferromanganese deposits, *Mar. Chem.,* 5, 93, 1977.
111. **Kharkar, D. P., Turekian, K. K., and Bertine, K. K.,** Stream supply of dissolved silver, molybdenum, antimony, selenium, chromium, cobalt, rudidium and cesium to the oceans, *Geochim. Cosmochim. Acta,* 32, 285, 1968.
112. **Coonley, L. S., Baker, E. C., and Holland, H. D.,** Iron in the Mullica River and in Great Bay, New Jersey, *Chem. Geol.,* 7, 51, 1971.
113. **Sholkovitz, E. R.,** Flocculation of dissolved organic and inorganic matter during the mixing of river water and seawater, *Geochim. Cosmochim. Acta,* 40, 831, 1976.
114. **Martin, J. M. and Maybeck, M.,** Elemental mass-balance of material carried by world major rivers, in preparation, 1979.
115. **Martin, J. H. and Knauer, G. A.,** The elemental composition of plankton, *Geochim. Cosmochim. Acta,* 37, 1639, 1973.
116. **Slowey, J. F. and Hood, D. W.,** Copper, manganese and zinc concentrations in Gulf of Mexico waters, *Geochim. Cosmochim. Acta,* 35, 121, 1971.
117. **Chester, R. and Aston, S. R.,** The geochemistry of deep-sea sediments in *Chemical Oceanography,* Vol. 8, 2nd ed., Riley, J. P. and Chester, R., Eds., Academic Press, London, 1976, 281.
118. **Chester, R., Griffiths, A. G., and Hirst, J. M.,** The influence of soil-sized atmospheric particulates on the elemental chemistry of the deep-sea sediments of the northeastern Atlantic, *Mar. Geol.,* in press, 1979.

Chapter 3

SEDIMENTS AND TRANSFER AT AND IN THE BOTTOM INTERFACIAL LAYER*

Egbert K. Duursma and Maarten Smies**

TABLE OF CONTENTS

* Communication No. 188 of the Delta Institute for Hydrobiological Research, Yerseke, The Netherlands.

** On secondment from Shell Internationale Research Maatschappij B.V., Group Toxicology Division, The Hague, The Netherlands.

I. INTRODUCTION

The transfer processes of pollutants in the watercolumn of the sea depend on a great number of properties of these substances. When dissolved, the substances are diluted in the sea water, but when in particulate form and having a specific gravity higher than sea water, they will be precipitated to the seafloor. Equally, substances adsorbed to sedimenting materials will ultimately arrive at the seawater/seafloor interface. Sorption of dissolved compounds directly from the supernatant water to the bottom interfacial layer is another process which exists. Its relevance compared to the sedimentation of pollutants depends on factors such as time, transport over the seafloor, composition of the bottom interfacial layer, and the physical and chemical properties of the pollutant.

Both in the deep sea and in shallow coastal waters, the bottom sediments can accumulate many of the known pollutants of domestic, agricultural, and industrial origin. Practically all heavy metals, a great deal of the radionuclides of nuclear effluents and many organic substances, ultimately find their way into the sediments: sorption, desorption, chemical conversion, and other processes determine how these substances are bound to or mixed with the bottom sedimentary particles. Concentration factors as large as 10^4 or 10^5 may occur, representing the concentration per gram or milliliter of sediment relative to that in 1 mℓ of sea water. Thus potentially, the bottom interfacial layer can accumulate a great deal of pollutant from the watercolumn. One millimeter of bottom layer can for example accumulate the same amount that is contained by approximately 1.2×10^5 mm = 120 m of water column for a concentration factor of 10^5. For the same factor 2.4 km watercolumn equals 20 cm of bottom layer, a layer in which not unusually pollutants are found.

These figures illustrate clearly the potential role of the seafloor in the distribution of pollutants in the oceans, coastal waters, and estuaries. The actual role, however, is connected with the transfer processes, since these determine the thickness of the sediment layer in which the pollutants can penetrate. Some of these processes are very time dependent.

II. TRANSFER MEDIA AND DRIVING FORCES

In the seafloor three media can be distinguished by which transfer processes can occur: pore water, sedimentary particles, and living organisms. In the first 20 cm potentially all these media play a role, but deeper in the seafloor transfer processes are mainly limited to occur in the pore water.

The "driving forces" are numerous: currents, wave action, biological activities, gradients of concentration, dissolution, precipitation, decomposition, pressure, and density gradients. Each of these "forces" results in a different kind of transfer process in and at the seabed.

A. Media

1. Pore Water

Sea-bottom sediments usually have a pore-water content in the range of 10 to 80% volume per volume depending on the packing of the sedimentary particles. The loosely packed sediment are those freshly deposited or redeposited in areas where currents or waves have a direct influence on the bottom structure. In more quiescent regions, the packing and thus also the pore-water content may be rather constant over several meters in the sediment. Closer packing occurs deeper due to compaction over a great range of years — centuries or longer.

The problem is whether both the content and the composition of the pore-water

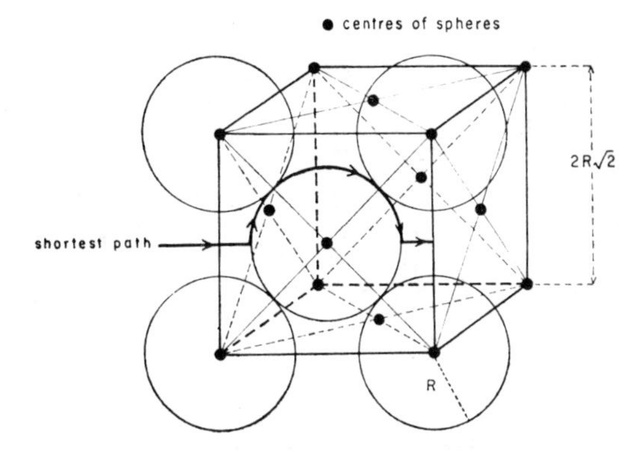

FIGURE 1. Cubic close-packing of identical spheres. Centers of spheres are on the angular points and centers of the lateral faces of the cube. R = radius of spheres. *Interspace volume.* Volume of cubic unit: $16R^3\sqrt{2}$. Volume of spheres in one cubic unit (4 spheres): $16/3\ \pi R^3$. Relative interspace volume:

$$\frac{16R^3\sqrt{2} - \frac{16}{3}\pi R^3}{16R^3\sqrt{2}} \times 100\% = 26\%$$

Free path. For the free path is taken the shortest way around the spheres, as is presented by the thick line in the figure. Free path per cubic unit: $2R\sqrt{2} - 2R + \pi R = 3.97$ R. Free path: $3.97\ R/2R\sqrt{2} = 1.40 \times$ straight length. *Cross section.* Surface area of one lateral face of the cubic unit: $8R^3$. Area of circles intersected by a lateral face: $2\pi R^3$. Relative area between the circles: $8R^2 - 2\pi R^2/8R^2 \times 100\% = 21.5\%$. Number of circles per lateral face of the cubic unit: 2. Number of circles per cm^2: $1/4R^2$. *Surface area of the spheres.* Per cubic unit ($16R^3\sqrt{2}$) the surface area of the spheres is: $16\pi R^3$ which is per cm^3: $\pi/R\sqrt{2}$ (proportional to R^{-1}). (From Duursma, E. K. and Bosch, C. J., *Neth. J. Sea Res.*, 4, 395, 1970. With permission.)

influence transfer processes of substances in the pore water. These factors cannot be separated completely since the pore-water content may be of influence on some physicochemical characteristics of the pore water which may change its composition.

a. Content

Although seabeds are generally not composed of single-sized particles, it is worthwhile to start this discussion with the pore-water content of a sediment with an idealized particle distribution. As demonstrated schematically in Figure 1 identical spheres can be packed in close-packing conditions. The cubic or hexagonal close-packing systems[1] contain 26.0 and 32.8% void volumes, respectively. A calculation has been made (Figure 1)[2] for the shortest free path, the cross-section, and for the surface area of the spheres in the case of cubic close packing.

The result shows that the void space is independent of the radius of the spheres, as is the shortest free path. The cross section (number of circles per unit surface) is pro-

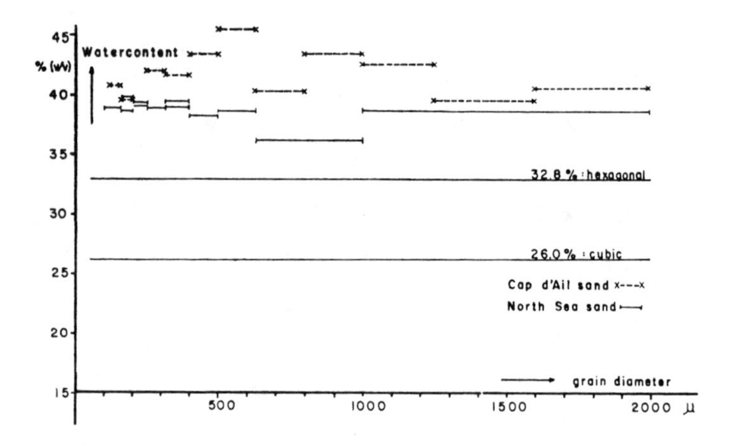

FIGURE 2. Interspace water content (porosity) in % volume to volume of sieved sand fractions in μm compared with models of close-packed spheres. (From Duursma, E. K. and Bosch, C. J., *Neth. J. Sea Res.*, 4, 395, 1970. With permission.)

portional to R^{-2} (R being the radius of the spheres). The surface area of the spheres per volume of "sediment" is proportional to R^{-1}.

Marine sediments are neither composed of spherical nor identically sized particles. Nevertheless, the porosities of sieved sand fractions may be fairly identical when experimentally brought to close packing[2] (Figure 2). Since sorting processes exist also in the sea, it is not surprising that sometimes deep-sea clays have similar porosities as beach sand.

The effect of the porosity on the transfer processes may be illustrated from some diffusion experiments with sieved quartz sand fractions. Both the diffusion coefficients (D) of NaCl and $K^{36}Cl$ (Figure 3)[2] show a relation with the grain diameter. For small grain sizes the diffusion coefficients are lower, but a relation could be demonstrated with the specific surface of the spheres. The explanation given by Duursma and Bosch[2] supposes a surface action between solids and liquid. This was based on the effect found for capillary rise (Figure 4). The experimental results were in accordance with the theoretical formula of Glasstone[1]:

$$h = \frac{2\gamma \cos \Theta}{g\rho} \times \frac{1}{r} \qquad (1)$$

where h is the capillary rise in cm, g the acceleration due to gravity (0.98×10^3 cm/sec^2), ϱ the specific weight of the liquid, r the radius of the tube, Θ the solid-liquid angle of contact, and γ the surface tension solid-liquid (interface tension). Assuming now that $\cos \Theta/g\varrho$ is the same for the different grain-size fractions, and r is proportional to the diameter of the sand particles, the formula becomes:

$$h = \frac{A\gamma}{d} \qquad (2)$$

in which A is a constant and d the diameter of the particles.

Diffusion in liquid is dependent on the viscosity, and the next question is whether an equal porosity but with smaller pores would affect the viscosity of the interstitial water. This can be demonstrated with a simple flow experiment with sieved sand fractions below 1000 μm (Figure 5).[2] The flow rate (v) is proportional to the square of the diameter of the grains, according to Darcy's law[3] for soils:

$$\frac{v}{t} = \frac{A\,d^2}{\eta} \qquad (3)$$

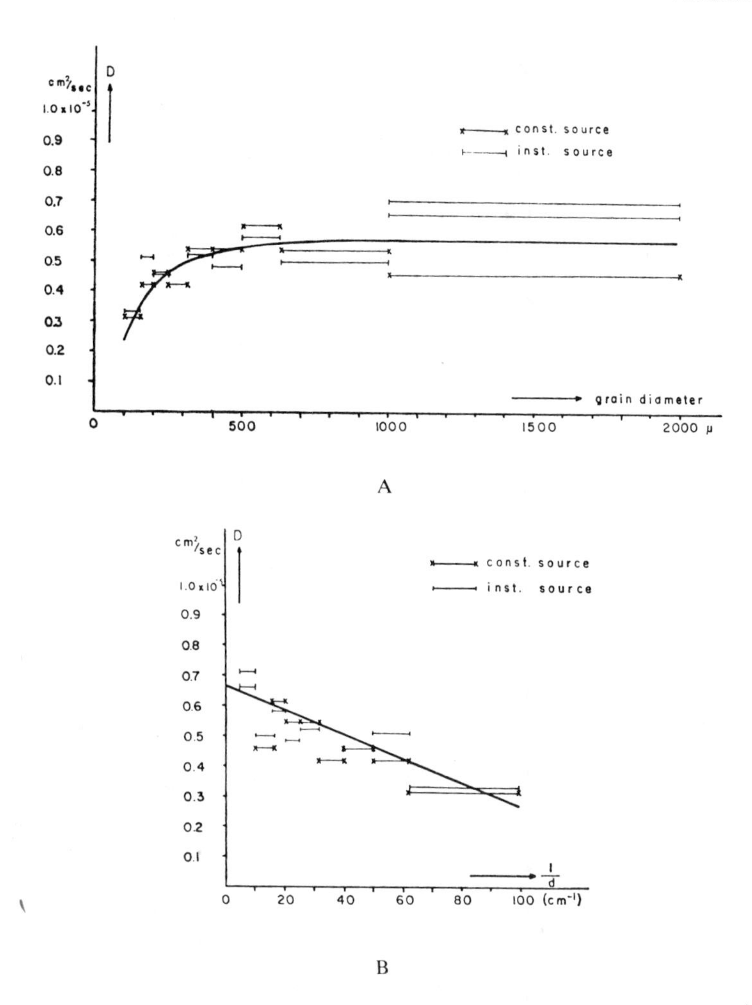

FIGURE 3. (A) Diffusion coefficients for NaCl and $K^{36}Cl$, diffusing from a constant source and an instantaneous source, respectively, in sieved North Seas and fractions. (B) The same diffusion coefficients as in (A) related to the specific surface area of the sand particles, assumed to be cubic close-packed spheres. The specific surface area is $\pi\sqrt{2}/d$ cm^{-1} in which d is the diameter. (From Duursma, E. K. and Bosch, C. J., *Neth. J. Sea Res.*, 4, 395, 1970. With permission.)

where A is a constant if the driving pressure and the length of the "tubes" are constant; d is the diameter of the grains, t is the time of the experiment, and η the viscosity coefficient. The results show a constant viscosity coefficient for different sizes.

b. Composition

Migration in pore water cannot be regarded independently from the composition of the sedimentary particles. However, some physicochemical properties of the water influence the diffusion of the dissolved components.

First of all, cations or anions can only diffuse together since, otherwise electrical potentials are built up which counteract the migration. For example, the diffusion coefficient of chloride in the presence of its counter cations can differ by a factor of 3 (Table 1).[2] The one which is only related to its ionic radius is found if K^+ as cation is used. This ion has the same ionic radius which results also in an identical equivalent conductance at infinite dilution. Also partial dissociation of an electrolyte is effective. This can be demonstrated for strontium chloride (Table 2)[2] in sea water. Diffusion is slower at higher dissociation (higher dilution).

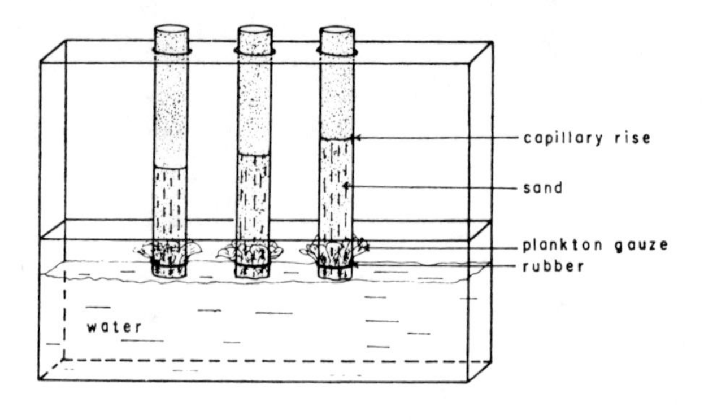

A

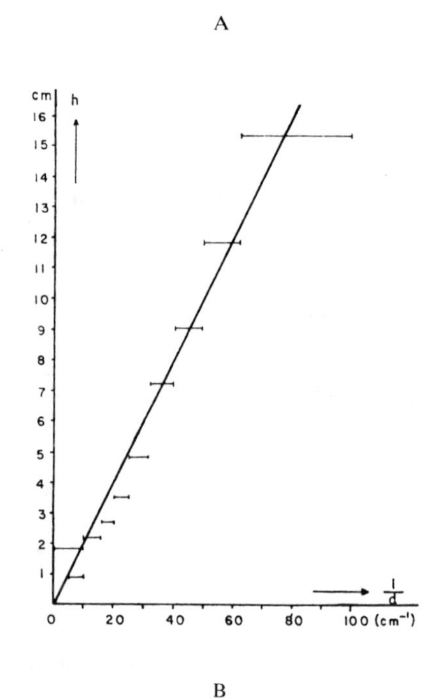

B

FIGURE 4. (A) Technique used in the experiment to determine the capillary rise of water in sieved North Sea sand fractions. (B) Capillary rise (h) after 22 hr as function of the reciprocal of the grain-size diameter (d) of sieved North Sea sand fractions. (From Duursma, E. K. and Bosch, C. J., *Neth. J. Sea Res.*, 4, 395, 1970. With permission.)

In marine interstitial waters this situation is complicated because of the presence of a great many ions. On the other hand, however, these pore waters have a rather constant ionic strength and composition in all deep ocean sediments. Only the pH varies slightly, while the dominant variations concern the physicochemical alterations caused by redox-potential changes due to the O_2/H_2S-oxic/anoxic regimes. As long as these alterations do not cause changes in solubility, transfer by diffusion is not greatly influenced. When, however, reactions are involved which cause adsorption from the pore water to the sedimentary particles the changes will be very large. They will be discussed later in Section II.B.5.

2. Sedimentary Particles

Although sedimentary particles as solid material contribute (as a transfer medium) little to transfer processes in and at the bottom interfacial layer, they participate largely

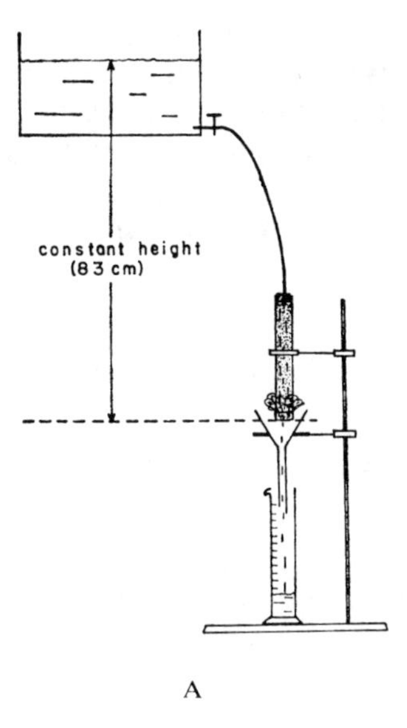

A

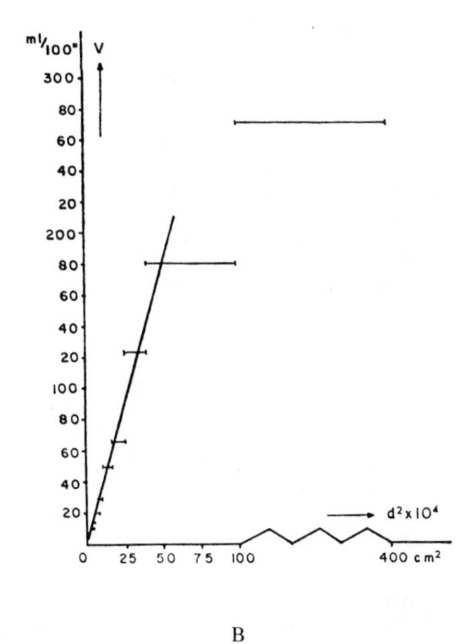

B

FIGURE 5. (A) Technique used to determine the viscosity of the interstitial water in sieved North Sea sand fractions. (B) Flow rate per 100 sec in relation to the square of the diameter measured for sieved North Sea sand fractions. (From Duursma, E. K. and Bosch, C. J., *Neth. Sea Res.*, 4, 395, 1970. With permission.)

when the particles are moved vertically or horizontally. Migration into particles is regulated by diffusion having[2,4] diffusion coefficients of 10^{-15} to 10^{-19} cm^2 s^{-1}. Relative to any other process in the pore water this process can be neglected, which is not the case for the binding of certain elements by sedimentary particles (Figure 6),[5] where binding means chemical reaction with, adsorption to, absorption by, and precipitation on sedimentary particles. The result is that sediments may be a greater sink to such elements, and transfer with the sedimentary particles becomes more effective.

Table 1
DIFFUSION COEFFICIENTS OF THREE CATIONS AND CHLORIDE IN SAND, AS DETERMINED FROM THE DIFFUSION OF KCl, $CaCl_2$ and HCl (1 mℓ OF CONCENTRATED SOLUTIONS) WITH EITHER SEA WATER OR DISTILLED WATER AS INTERSTITIAL WATER. NORTH SEA SAND OF 250—315 μm OR CLEANED MEDITERRANEAN SAND OF 125 μm

Source	Interstitial water	Sediment	Equivalent conductance at infinite dilution $ohm^{-1}cm^2$ (25°C)	Diffusion coefficients, $10^{-5}cm^2s^{-1}$	
				Cation	Chloride
KCl	Distilled water	North Sea	K^+:74.5 Cl^-:75.5	0.79	0.75
KCl	Sea water	North Sea		0.77	0.78
$CaCl_2$	Distilled water	North Sea	½Ca^{++}:60	0.28	0.45
$CaCl_2$	Sea water	North Sea		0.32—0.50	0.38—0.72
HCl	Distilled water	Mediterranean	H^+:350	1.20	1.17
HCl	Sea water	Mediterranean		1.30	1.21

From Duursma, E. K. and Bosch, C. J., *Neth. J. Sea Res.*, 4, 395, 1970. With permission.

Table 2
DIFFUSION COEFFICIENTS OF $SrCl_2$ IN NORTH SEA SAND (250 to 375 μm) USING A HIGH AND AN EXTREMELY LOW CONCENTRATED INSTANTANEOUS SOURCE

Source	Interstitial water	Diffusion coefficients $10^{-5}cm^2s^{-1}$	
		Cation	Anion
1 mℓ conc $SrCl_2$	Sea water	0.38	0.47—0.79
1 mℓ $^{90}SrCl_2$	Sea water	0.15	0.64

From Duursma, E. K. and Bosch, C. J., *Neth. J. Sea Res.*, 4, 395, 1970. With permission.

In order to evaluate these processes it is worthwhile to understand which properties of marine sediments play a role in this context. In some publications[6] it has been postulated that binding of pollutants should be inversely proportional to the grain size of the sedimentary particles (i.e., proportional to the surface-per-unit weight). However, natural marine sediments are not composed of grains of identical composition and therefore the reactions with the sediment particles might have a different relation. This causes a wide spread in the results which is difficult to explain.

The specific surface of major ocean sediments, collected for radionuclide sorption experiments (Figure 7)[7] and determined with ethyleneglycol according to the method of Dyal and Hendricks,[8] shows almost no relationship with the reciprocal of the median grain size (Figure 8).[7] There is a better relationship between this specific surface and the base-exchange capacity (Figure 9).[7] This property is empirically determined after percolation of the sediment with alcohol to remove sea salts and subsequently

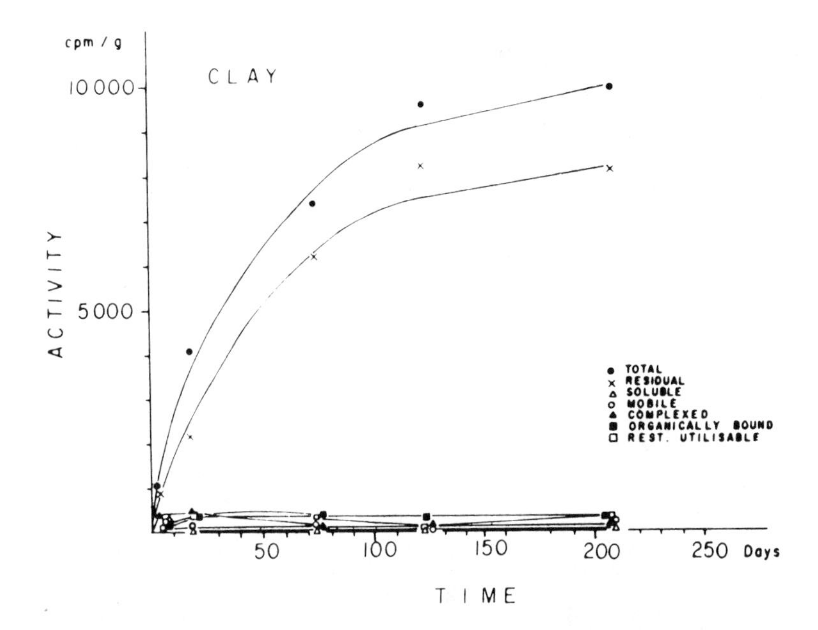

FIGURE 6. ^{65}Zn concentrations in different leachable fractions of a marine sediment as function of time of sorption, as compared to the total and residual (non-leachable) amount. (From Duursma, E. K., Dawson, R., and Ros Vicent, J., *Thalassia Jugosl.*, 11, 47, 1975. With permission.)

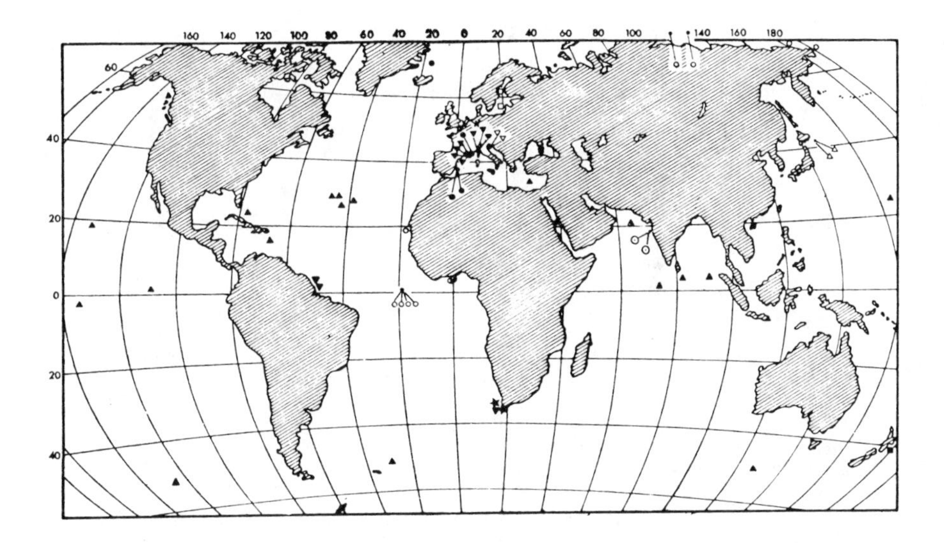

FIGURE 7. Bottom samples taken by oceanographic vessels of various countries for subsequent investigation of the sorption capacity of radionuclides. (From Duursma, E. K. and Eisma, D., *Neth. J. Sea Res.*, 6, 265, 1973. With permission.)

with 1 *N* NaCl, so that all exchange sites are filled by Na^+. A percolation with 1 *N* NH_4Cl removes again this Na^+, which can be determined and gives the base-exchange capacity in milliequivalents per 100 g.

Radionuclide sorption reactions involving ion exchange could be better understood in terms of specific surface and base-exchange capacity than other properties of the sediment, including the clay minerals. For other reactions than ion exchange there was no relationship with any of the sediment properties. Binding or precipitation occurred independent of these properties.

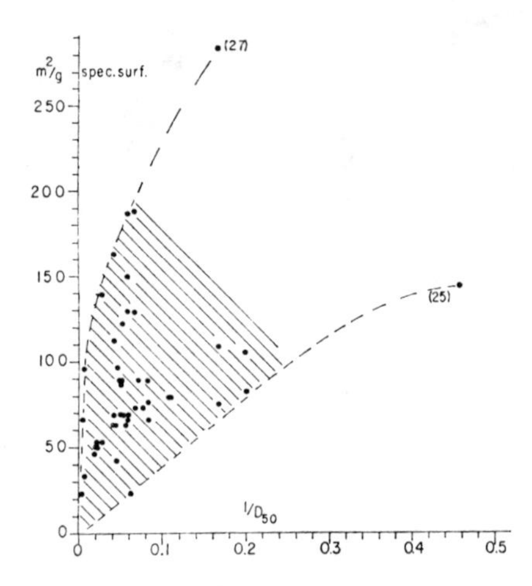

FIGURE 8. Relationship between the specific surface of marine sediments and the reciprocal of the median grain size (D_{50}) in μm^{-1}. (From Duursma, E. K. and Eisma, D., *Neth. J. Sea Res.*, 6, 265, 1973. With permission.)

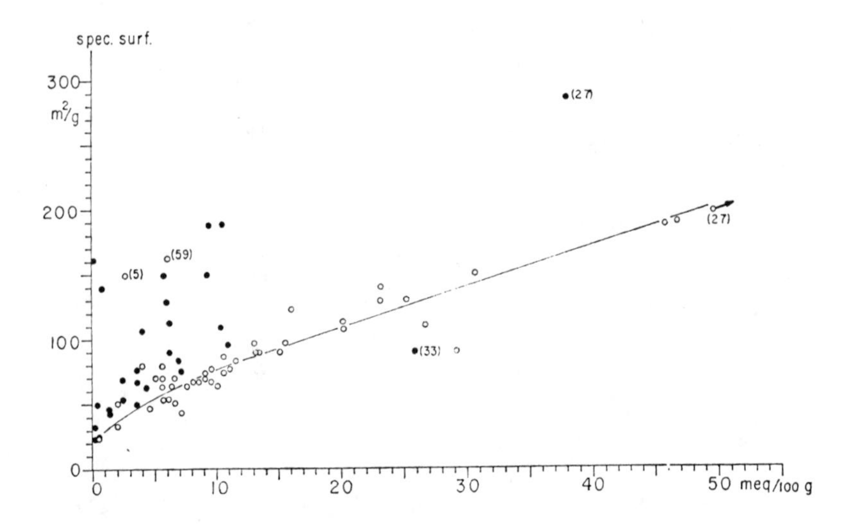

FIGURE 9. Relationship between the specific surface of marine sediments and their base-exchange capacity (o), and their exchange capacity, (•), as calculated from the clay-mineral composition. (From Duursma, E. K. and Eisma, D., *Neth. J. Sea Res.*, 6, 265, 1973. With permission.)

3. *Living Organisms*

Organisms living in marine sediments can be divided into three groups, the macrofauna, the meiofauna, and the microfauna. Roughly, these groups are determined by the size of the organisms being larger than 1 mm, between 30 μm and 1 mm and smaller than 30 μm, respectively. All of them can act as a medium for pollutants by accumulation and having them transported in and at the bottom interfacial layer.

The most well-known benthic species are reviewed by Rhoads[9] with respect to their bioturbation activities in the sea floor. They generate transfer processes and are media at which transfer occurs.

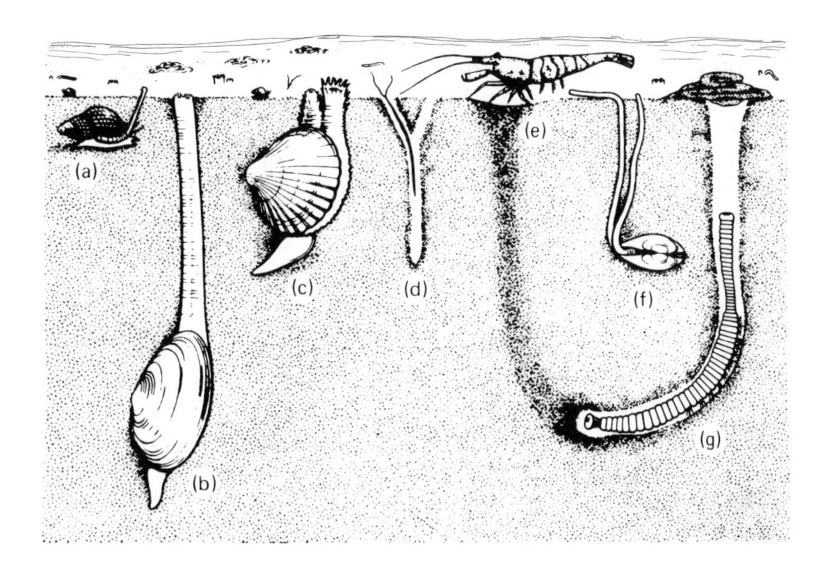

FIGURE 10. Vertical section through a *Macoma* community. (a) The tiny mud snail *(Hydrobia)*; (b) a small whelk *(Nassarius)*; (c) a cockle *(Cardium)*; (d) a bristle worm *(Pygospio)* in its tube; (e) the common shrimp *Crangon*; (f) the bivalve *Macoma baltica*; and (g) the lug worm *(Arenicola)*. (After Thorson, G., *Life in the Sea*, World University Library, London, 1971. With permission.)

a. Macrofauna

Although an abundant literature exists on the accumulation of pollutants by marine organisms, the relevant information for bottom organisms is limited. On benthic fish, in particular flatfish, a comparison can be made with other species, but the question is to what extent these organisms can be classified as transfer media at and in the bottom layer. This is different for the real bottom dwellers like polychaetes, some bivalves, snails, and crustaceans. Since these organisms move, they transfer material including pollutants. A typical distribution of such animals is given in Figure 10 (representative for an estuary in a temperate zone).[10]

Most of the benthic macrofauna species live from living or dead organic matter which arrives at the bottom interface. For these organisms, there is no reason for horizontal or large vertical displacements inside the bottom, which makes them less effective as medium than as a driving force. Some of them, e.g., *Arenicola* sp. return their feces to the sediment surface. The polychaete group of nereids, however, migrates through the sediment, being able to accumulate pollutants from the sediment and releasing them at other locations.

b. Meiofauna

This is a group composed of many species, the most well-known of which are the harpacticides (copepods), the nematodes, and the tartigrades. They live in the pores of the sediment and have a certain liberty of displacement which is from some millimeters to some centimeters per minute. The knowledge on their behavior in coastal zones and estuaries is rapidly growing, but with regard to their function as a transfer medium very little is known.

Nevertheless they cannot be neglected since they may exist in numbers up to a few millions per square meter, mostly distributed over a depth of a few to about 20 cm. For such numbers their biomass can be estimated at 1 to 3 g m^{-2} (dry weight).[11]

c. Microfauna

This group contains the bacteria which occur in sediments in numbers of 10^5 to 10^8 per gram wet sediment with a biomass of approximately 10^{-5} to 10^{-2} mg wet weight per gram wet sediment.[12]

Although little is known about their migration patterns in sediments, their distribution is mostly connected with the redox conditions in the sediment and the presence of organic matter. They have the tendency to zonate in certain horizons and thus have little function as transfer media.

B. Driving Forces

Transfer at and in the bottom interfacial layer cannot occur without a cause or a number of causes. These causes have been classified here as "driving forces" which for some of those listed is correct, but for others is actually not the exact description. Nevertheless they will be treated under this term, while at the same time the media will be mentioned through which the particular "driving force" is effective as far as the transfer processes are concerned.

1. Currents

Currents displace sediments and cause small or large displacements at or in the bottom interfacial layer. As long as there is no pressure difference, currents are probably less effective in causing displacements of pore water, at least if the sediment layer remains untouched. When sedimentary particles are picked up by currents, the pore water of the affected layer is also mixed with the supernatant water.

Generally these currents cause large horizontal transfer of pollutants, picked up together with the sedimentary particles at one place and deposited somewhere else. For tidal currents a back-and-forward transport occurs which results, depending on the grain size of the sedimentary particles, in a residual migration inward or outward from the estuaries. The salt-wedge effect from outflowing fresh water is superposed on this system, which causes a residual inward flow over the bottom.

The classical information on these effects[13] is presented in Figures 11 and 12. The consequences for pollutants attached to these sediments are identical as for the sedimentary particles as long as no sorption-desorption is occurring. Measurements of de Groot et al.,[14] Duinker and Nolting,[15] and Müller and Förstner[16] indicate that a number of heavy metals, for example, show conservative properties in association with estuarine sediments, so that mixing with marine sediments can be distinguished.

The mostly fine-grained "salt-wedge" sediments have the greatest chance to be enriched with metals precipitating from river water which mixes with sea water, but also here the river itself supplies additional sedimentary material which can adsorb these quantities. For iron[17] this is illustrated in Figure 13.

Nevertheless, particles loaded with high amounts of metals can be observed during measurements over a whole tidal cycle at one station which was explained by Duinker et al.[18] in the following way: "A very fine fraction of particulate matter exists, probably of colloidal iron and manganese oxides including trace metals, occurring at low concentrations in the sea. At low suspension loads these colloids are the cause of relatively high particulate metal concentrations (as μg metal per g particulate matter), while at high suspension loads these colloids are not distinguishable."[19]

The impact of currents over the deep-sea floor is certainly of another scope, although according to recent investigations, the so-called nepheloid layer contains higher amounts of, for example, Zn than the particulate matter in the overlying water column. A resuspension from the bottom interface is suspected, thus causing horizontal transport[20] along the seafloor. This last effect is considered very essential in the evaluation of the impact of dumped pollutants with respect to recycling in the oceans.

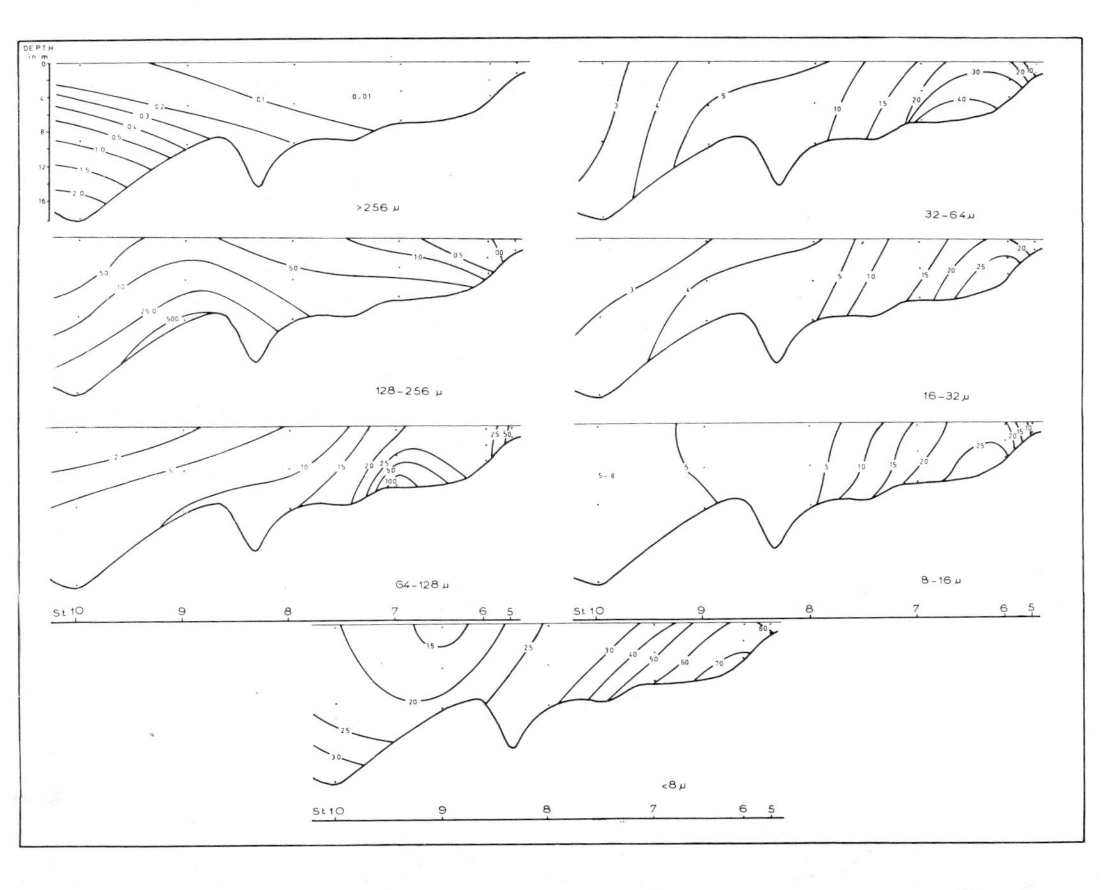

FIGURE 11. Distribution of suspended matter in a vertical profile along the Ameland estuary. The values relate to the average concentrations, expressed in mg/ℓ, of the various grainsize fractions throughout a tidal period.[13] (From Postma, H., *Neth. J. Sea Res.*, 1,448, 1961. With permission.)

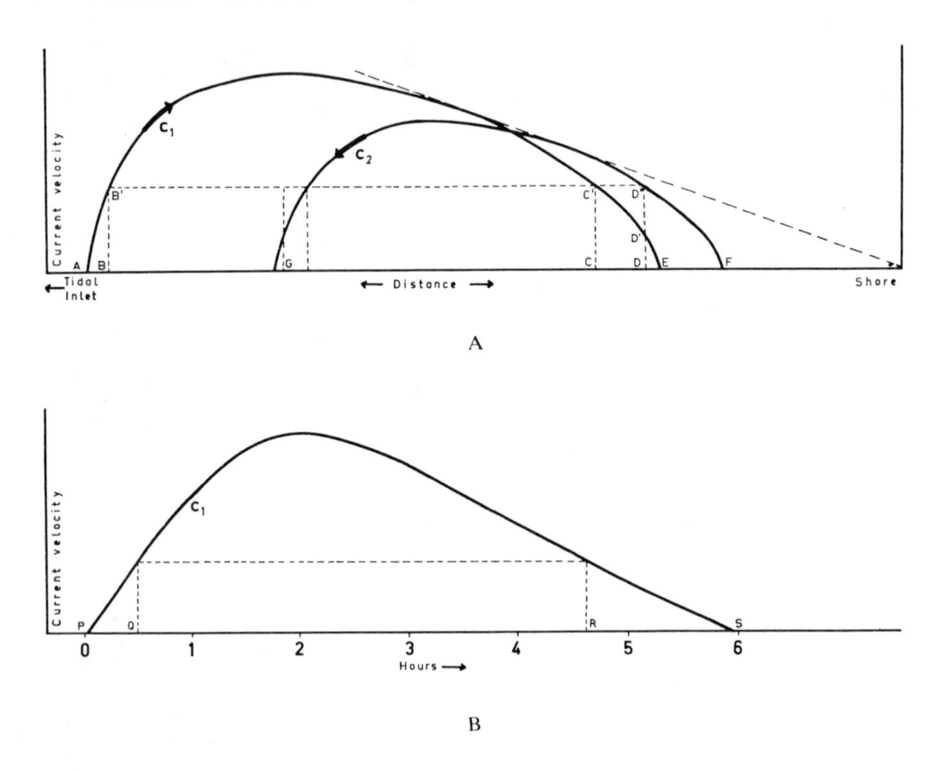

FIGURE 12. (A) Distance-current velocity diagrams of water masses moving from a tidal inlet towards the shore and back, demonstrating the inward shift of sediment particles; C_1 denotes flood, C_2 ebb. (B) Time-current velocity diagram for the water mass represented by curve C_1 in (A). (From Postma, H., *Neth. J. Sea Res.*, 1, 448, 1961. With permission.)

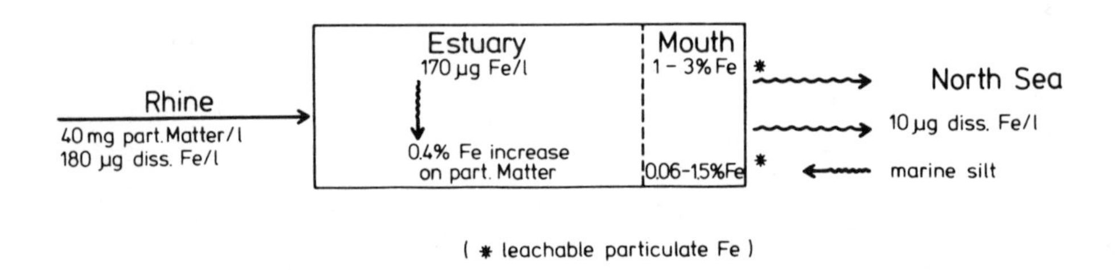

FIGURE 13. Schematic representation of iron precipitation in the Rhine estuary. (Duursma, E. K., Vegter, F., and Kelderman, P., *Hydrobiol. Bull.*, 12, 215, 1978. With permission.)

A last point reflects the indirect effect of tidal currents on pore-water transport in tidal flats. In particular, on the borders of tidal flats with the gullies, there will be a loss of pore water due to the water-level difference at low tide. For sandy flats this process will occur to a larger extent than for silty or muddy flats. Pollutants present in the flats will leak out with the pore water as far as they are dissolved in the pore water. The process is clearly illustrated by iron which leaks out as Fe^{++} and precipitates afterwards as Fe^{+++}, which gives brown stainings at the border of the tidal flats.

With rising water the process is not reversed, since the loss of pore water at the border is mainly replenished from above. Thus, as a whole, this process causes a kind of percolation from above to the borders of the tidal flats, being effective to some tens of meters from the gully and particularly for tidal flats with steep borders.

2. Wave Actions

The action of waves on beaches is well documented and large amounts of sand can be reworked by breakers, in particular during storms. This causes vertical as well as horizontal transport in and along the beaches where the tidal currents will displace these loads of sedimentary particles. Waves also cause a selection of grain sizes on the beaches, since fine materials are kept in suspension and large-grained particles are transported to deeper parts. Thus sands with little sorting can occur on beaches which are exposed to rather constant types of waves.

Percolation of pore water is a second effect that occurs, since a part of the uprunning wave water is returning through the sand which causes a regular flow of water through the top 20 to 50 cm of the water line of the beach. Pollutants in the water may thus contaminate beaches over a thickness of this size in a very rapid way.

Wave action also has effects on the seafloor up to approximately 25 times the wave heights.[21] This process is held responsible for the well-known ripples on the sea floor and thus causes sediment transport over short distances and reworking over a few tens of centimeters or the height of the ripples. Probably, pore-water currents of circular types will occur, thus giving opportunities for exposing a certain layer of the sea bottom to pollutants.

3. Biological Activities

The benthic organisms, divided into three size groups as stated before, can cause a variety of activities which result in transfer processes. These are (1) reworking of the sediment, (2) holes along which transport can occur, (either passively or stressed by the organisms), and (3) enlarging the random motion of the pore water.

Data have been published on the total effect that these processes have on the distribution of radionuclides in the seafloor, with impact on the determination of sedimentation rates. These showed anomalies in the top tens of centimeters. Already in 1958 Goldberg and Arrhenius[22] made an evaluation of these findings and suggested that biological activities were the cause.

For deep-sea sediments with a relatively scarce benthic population, the activities are effective over very long periods of time. Since, however, the sedimentation rates are also only a few centimeters per thousand year, the total result is detectable in cores (Figure 14).

On the continental shelf and in coastal areas the benthic fauna is more abundant, and data from Livingston and Bowen[24] for Pu-239, from Templeton and Preston[25] for Ru-106, and from Billen[26] for NO_3^- showed vertical distributions which suggested diffusion coefficients higher than could be estimated on the basis of the properties of these substances towards sedimentary particles and in the pore water. As calculated for the first two examples, according to diffusion models of Duursma and Hoede,[27] Duursma[28] estimated 10^{-8} cm^2 s^{-1} for Pu-239; Duursma and Gross[29] estimated 10^{-7} cm^{2}s^{-1} for Ru-106, and Billen[26] deducted 10^{-4} cm^2 s^{-1} for NO_3^-. This last figure points to a dispersion that is more rapid than the molecular diffusion in pore water (0.3×10^{-5} cm^2 s^{-1}), while the coefficients for Pu and Ru should be in the range of 10^{-9} to 10^{-10} cm^2 s^{-1} on pure physical arguments (see Sections II.B.5 and III.B).

Experiments in shallow brackish stagnant waters showed that colored sand could be reworked for two months with a diffusion coefficient of 10^{-7}cm^{2}s^{-1} where the abundant macropopulation were worms (*Arenicola*) and cockles (Cardiidae). This process occurred only in a distinct top layer of about 20 cm, and a diffusion model could only be applied for the period mentioned.[30]

As reviewed by Rhoads[9] and Cadée,[31] the bioturbation of recognized species can be expressed also in other terms, such as the annual thickness of layer reworked for a number of organisms, where this thickness may represent a thin layer that is reworked frequently. An alternative measure gives the amount of sediment displaced per animal.

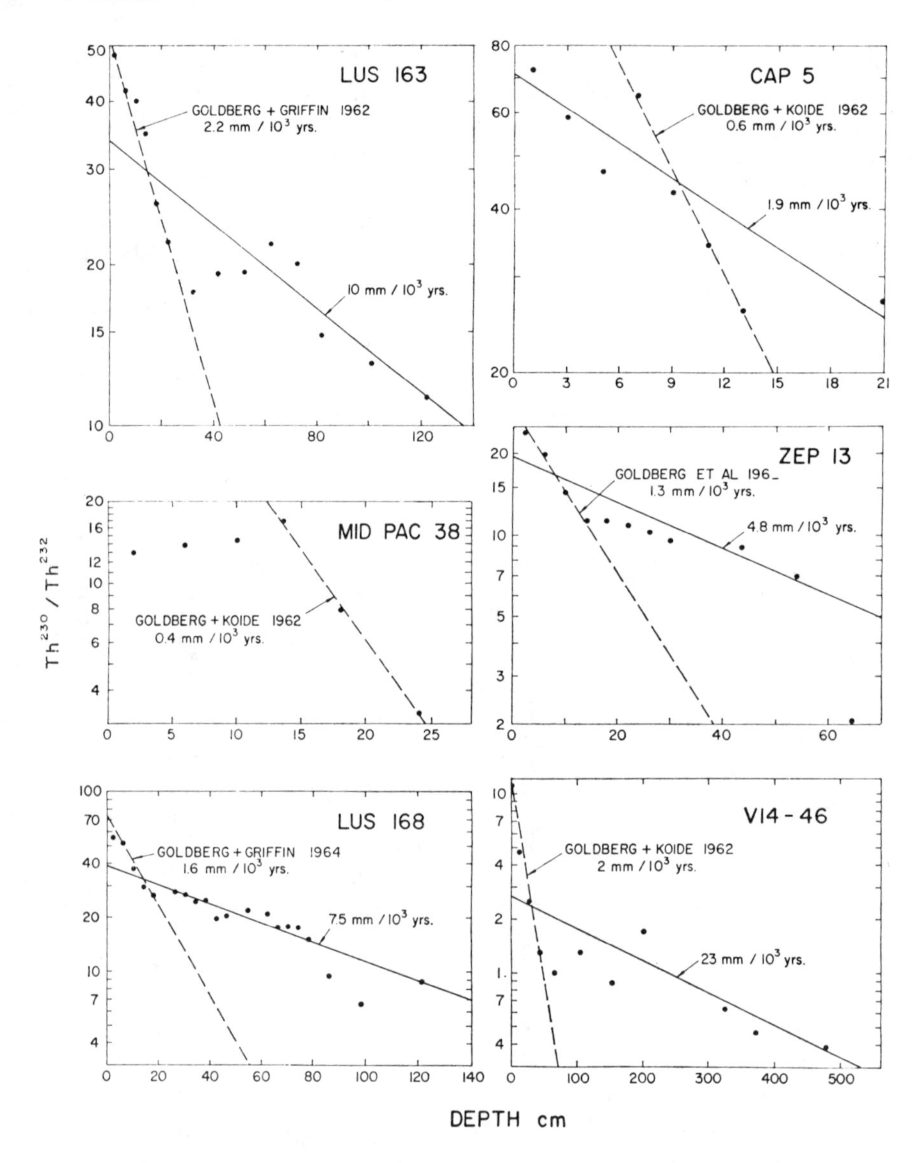

FIGURE 14. Plots of the logarithm of the ^{232}Th activity against depth for six of the cores studied by Goldberg and co-workers. The dotted line is that used by Goldberg to establish the rate of sedimentation for the last 100,000 years. The solid line represents the best fit to the data. The cores represent examples of cases where the difference in interpretation leads to quite different sedimentation rates. (After Ku, T. L., Broecker, W. S., and Opdyke, N., *Earth Planet. Sci. Lett.*, 4, 1, 1968. With permission.)

It is very hard to convert one result into the other. For pollutants, the overall transfer is of interest for the geochemist, while for the biologist, the actual transport by certain species or the contact of the organisms with polluted sediment is of interest.

However, since sediments in shallow areas (but also from the deep sea) are potentially great adsorbers of pollutants, the reworking of sediments by biota enlarges the speed of total accumulation in the bottom, while the recycling to the aquatic system is similarly intensified.

In the case of the NO_3^- distribution, meiofauna may be effective since it concerns a dissolved electrolyte. Molecular diffusion is based on the random molecular movements in the pore water. The meiofauna can probably enlarge such movements by one or two orders of magnitude. However, little is known on this effect, but it is quite possible that owing to the meiofauna the transport of the oxygen into the sediment is also enhanced, so that these organisms create their own optimal environment. This has its repercussions for the breakdown or transformation processes of pollutants either chemically, enzymatically, or by bacteria.

4. Gradients of Concentration

When pollutants arrive at the bottom interface, this creates a gradient of concentration vertically into the sediment. Exceptions might occur when organisms filter the overlying sea water and deposit a layer of feces inside the sediments, as is the case for a number of polychaetes such as nereids.

Another system occurs where the top 10- or 20-cm layer is rapidly reworked by currents, waves, or organisms, and where a gradient thus exists from there deeper into the sediment. Locally, also man can cause special situations when harbor or sewage sludges are dumped so that contaminated packages are lying on the seafloor. This also causes gradients to the interface and loss to the supernatant water. In all occasions transformation processes in the sediment may change the physicochemical behavior of the pollutants, thereby changing their solubilities.

The redox potential and bacterial activities play a great role in this system, since these predominantly cause horizons of equal properties, where the oxygen or $SO_4^{=}$ flowing into the sediment are regulatory factors. Thus the transfer of the pollutants involved depends also on the transfer of other substances involved in the reactions.

5. Dissolution/Precipitation/Sorption/Desorption

Most pollutants such as oil, heavy metals, radionuclides, chlorinated hydrocarbons and other organic compounds can be attached to settling sedimentary particles and arrive or are scavenged to the bottom.

For metals and radionuclides certain physicochemical reactions can be assumed, but the problem of attachment is less known for oil, chlorinated hydrocarbons, and other organic compounds. As already mentioned in Section II.B.2, the first supposition is that the reactions with the sediments should be surface dependent. There is not always a simple relationship as already suggested, the question being how the active surface is to be determined.

Another point, which is problematic for all groups of pollutants, is the distinction between dissolution and desorption and between precipitation and sorption. In and at the bottom interfacial layer, these processes are mostly mixed and are difficult to describe in thermodynamically defined terms.

Nevertheless, Duursma and Eisma[7] tried to characterize some of the radionuclide sorption reactions on marine sediments in two groups as ion exchange and precipitation/chemical compound formation reactions. In Table 3 the possible sorption reactions are evaluated for 14 radionuclides, taking into account (for 12 of them) the half-time of sorption found with the so-called thin-layer sorption technique of Duursma and Bosch.[2] This technique applies 10 mg of sediment on a 1-in. filter on which the radionuclide is adsorbed from a solution for a period of 1 month. Short half-times are supposed to be related to ion-exchange reactions, while longer half-times suggest precipitation and/or binding reactions of another kind than ion exchange.

From these results it is obvious that for the ion-exchange reacting radionuclides, a better relation can be found between the sorption-distribution coefficients and the base-exchange capacities of the sediments, which are supposed to represent the number of exchange sites per unit of sediment. The distribution coefficients (K) are representing the apparent equilibrium distribution after one month as

$$K = \frac{\text{amount of radionuclide/m}\ell\text{ sediment (dry)}}{\text{amount of radionuclide/m}\ell\text{ sea water}}$$

Table 3

AVERAGE DISTRIBUTION COEFFFICIENTS (K) DETERMINED BY THE THIN-LAYER METHOD RELATIVE TO THE BASE-EXCHANGE CAPACITY OF MAJOR MARINE SEDIMENTS ACCORDING TO DUURSMA AND EISMA;[7] THE HALF-TIME PERIOD OF SORPTION, $t_{1/2}$, AND THE POSSIBLE SORPTION REACTIONS ARE EVALUATED FROM THE RADIONUCLIDE BEHAVIOR IN THE EXPERIMENTS

	Stable element in sea water			K/base-exchange capacity (100 g/meq)			
Isotope	Chemical form	Conc (μg/ℓ)	log K^a	Average	Range	$t_{1/2}$ (days)	Possible sorption reaction
^{90}Sr	Sr^{++}	8000	1.9 ± 0.3	1	0.5—1.5	2.3 ± 1.7	Isotopic exchange
^{137}Cs	Cs^+	0.3	2.8 ± 0.4	50	25—90	0.7 ± 0.4	Ion exchange
^{65}Zn	Zn^{++}, $ZnOH^+$, $ZnCO_3$, (organic ?)	5—10	3.4 ± 0.4	150	100—400	1.9 ± 1.3	Ion exchange, chemical binding
^{60}Co	Co^{++}	0.4	3.9 ± 1.0	?	100—4000	8.2 ± 5.1	Co-OH-CO_3 compounds?
$^{95}Zr/Nb$	$ZrOH_n^{(4-n)+}$ n = 1–4	0.03	3.9 ± 0.4	1000	200—1600	5.7 ± 2.3	Precipitation
^{54}Mn	Mn^{++}, $MnCl^+$	2	3.9 ± 1.0	1000	200—2000	6.1 ± 4.6	Mn insoluble oxides
^{59}Fe	Fe^{+++}, $Fe(OH)^{++}$, $Fe(OH)_2^+$, $Fe(OH)_3^0$	3	4.0 ± 0.5	1250	400—2500	18.0 ± 6.5	Precipitation
^{106}Ru	cations and anions	10^{-4}	4.4 ± 0.5	2850	1400—5000	8.9 ± 3.3	Ion exchange Ru-Fe compounds
^{147}Pm	$Pm(OH)_3^0$	—	4.7 ± 0.4	3300	2800—6700	11.4 ± 3.5	Precipitation and "aging"
^{144}Ce	$Ce(OH)_3^0$	10^{-3}	4.8 ± 0.4	4000	2000—8000	10.5 ± 4.4	Precipitation and "aging"
$^{239,240}Pu$	$PuO_2(CO_3)_3^{4-}$	—	4 ± ?	52	?	n.d.	Binding to carbonates?
^{45}Ca	Ca^{++}	4 × 10^5	2 ± ?	45	?	<1	Isotopic exchange
^{86}Rb	Rb^+	120	3 ± ?	620	?	<0.5	Ion exchange
^{210}Pb	Pb^{++}, $PbCl^+$, $PbCl_3^-$, $PbCO_3^0$, $Pb(CO_3)_2^{2-}$	0.03	4.4 ± ?	2700	?	n.d.	Pb-OH-CO_3 compounds?

[a] Thirty marine sediments; ± = standard deviation; ? = only one sediment; n.d. = not determined

From Duursma, E. K. and Eisma, D., *Neth. J. Sea Res.*, 6, 265, 1973. With permission.

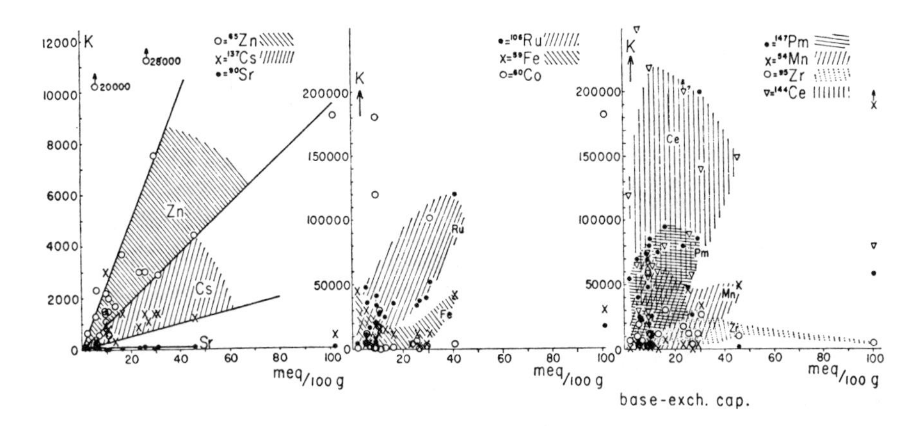

FIGURE 15. Relationship of the sorption distribution coefficients, (k), as determined by the thin-layer technique, with the base-exchange capacity of the sediments. (From Duursma, E. K. and Eisma, D., *Neth. J. Sea Res.*, 6, 265, 1973. With permission.)

In Figure 15 it is shown that for Zn, Cs, and Sr the relations with the base-exchange capacity are indeed much better than for the other isotopes.

The distribution of these radionuclides over the different grain sizes, as determined by a suspension experiment for two different sediments, is given in Table 4. It is clear from this table that there exists a great variety of distribution over the different fractions which between two sediments is not necessarily consistent for one isotope. It is also difficult to find from this inconsistency a relationship with the clay composition.

For oil, chlorinated hydrocarbons, and other organics very little fundamental knowledge is available, except for the fact that accumulation in sediments can occur. At oil-spill sites, such as the Gulf of Maine, it is after a number of years that the residual fractions attached to the sediment are migrating into or out of the bottom, thus causing carcinogenic effects upon bottom organisms.[32] PCBs are found in deep Mediterranean sediments.[33]

6. Decomposition

The constant flow of organic matter from the watercolumn to the seafloor supplies a source of food for benthic organisms and heterotropic bacteria. In the majority of shallow marine sediments, this causes anoxic conditions below the bottom interface. For deep-sea sediments this is different, since decomposition occurs already during the downward transport in the watercolumn. The residual material is decomposed at the surface layer of the sediment. Cores from deep-sea sediment are mainly oxic, although for some they may become anoxic when stored at higher temperatures. The temperature at the deep-sea floor is very low.

The decomposition in shallow-water sediments has an impact on the bottom to supernatant water exchange of nutrients, which can be demonstrated by data on the PO_4^{-} distribution in an enclosed sea arm,[34] the Grevelingen in the Netherlands. Since 1971 the PO_4^{-} in the created stagnant saline lake has increased by leakage of PO_4^{-} from the bottom, a process which could not be distinguished when the Grevelingen was still an estuary in contact with the North Sea (Figure 16).

Also redox conditions may cause alterations of some metals which become either more or less mobile in the sediment. Lower and higher distribution coefficients have been found for some radionuclides.[29] Indirectly, decomposition probably results in the creation of a number of "driving forces" for transfer processes.

Table 4
DISTRIBUTION OF RADIONUCLIDES SORBED BY DIFFERENT GRAIN SIZE FRACTIONS OF DUTCH WADDEN SEA SEDIMENT AND MEDITERRANEAN SEDIMENT

Size fractions		Distribution coefficients × 10^2									
(μm)	(% wght)	^{90}Sr	^{137}Cs	^{106}Ru	^{59}Fe	^{65}Zn	^{60}Co	^{147}Pm	^{54}Mn	^{95}Zr/Nb	^{144}Ce
					Dutch Wadden Sea sediment						
<4	3.7 ± 0.8	26	6.2	4.7	540	112	220	280	97	670	124
4—8	8.4 ± 1.9	0	16	5.2	510	380	430	320	480	1220	540
8—16	5.6 ± 2.0	0	12	7.3	370	490	59	0	8.0	1040	950
16—32	9.5 ± 3.2	0	5.4	0	53	260	65	0	35	66	120
32—64	21.7 ± 4.5	0	1.6	3.1	76	0	3.7	0	1.4	42	3.3
>64	51.1 ± 4.9	0	0	0	0	0	5.8	0	0	0.3	0.4
					Mediterranean sediment						
<4	6.2 ± 1.9	5.9	0.5	7.8	63	97	140	130	23	160	41
4—8	39.1 ± 7.5	9.3	2.3	34	183	140	820	101	76	310	147
8—16	36.5 ± 8.6	7.6	1.4	22	118	150	730	4.5	61	150	82
16—32	15.2 ± 4.2	2.5	1.3	11	101	140	540	0.8	41	290	73
32—64	3.0 ± 1.3	15.0	2.8	41	15	68	380	24	19	117	32
>64	0	—	—	—	—	5.2	—	—	—	—	—

Note: The sediments were suspended in radionuclide-enriched sea water for 1 month (30 g sediment in 20ℓ sea water). Size fractions were separated by sedimentation technique using a 100-cm column of 20ℓ content. For ^{65}Zn another Mediterranean sediment was used with 2.3% <4 μm, 12.7% 4—8 μm, 24.9% 8—16 μm, 49.2% 16—32 μm, 7.6% 32—64 μm, and 3.3% >64 μm.

From Duursma, E. K. and Eisma, D., *Neth. J. Sea Res.*, 6, 265, 1973. With permission.

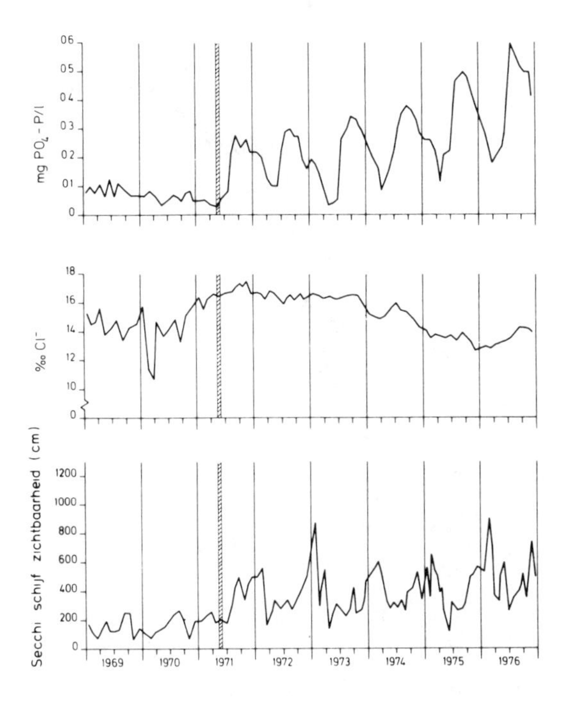

FIGURE 16. Orthophosphate, chlorinity, and Secchi-disc visibility in the Grevelingen estuary, closed in 1971 from the North Sea by a dam. (From Nienhuis, P. H., *Vakbl. Biol.*, 57, 271, 1977. With permission.)

7. Compaction

Compaction of marine sediments is a result of sedimentation and causes an upward movement of pore water, compensating the downward packing of the sedimentary particles. For regulary reworked sediments, this process occurs in periods when these reworking processes slow down, for example, with seasonally controlled bioturbation or quiet weather conditions after storms.

For deep-sea sediments the compaction is a very slow process ranging over thousands of years. For areas with only a sedimentation rate of 1 cm/1000 years, the porosity (pore water content in percentage of volume to volume) can be stabilized only at a 3000-m depth in the bottom (Figure 17).[35, 36]

The main point of interest is in how far and to what extent the slow pore-water upwelling, due to compaction in the deep-sea sediments, is effective relative to diffusion of substances in the pore water system. Relative to the upward moving sediment/water interface, water *and* sediment are precipitating. Relative to a fixed point deep inside the bottom, water is moving upward (due to compaction) at a smaller velocity than the upward movement of the sediment/water interface due to sedimentation. Hoede and van Beckum[37] recalculated Berner's[35] models of these processes, using a fixed origin as well as the traditionally employed moving origin. Thus, the waterfronts in the bottom are receding from the upward moving water/sediment interface with the result that dissolved substances are not pushed out of a compacting ocean sediment. This model calculation is based on the assumption of steady-state diagenesis, so that the porosity profile with depth is constant when viewed from the sediment/water interface.

Thus, upward diffusion of any substance in the pore water, out of marine sediments, has to counteract the sedimentation rate of water (and sediment). For deep-ocean sediments with very low sedimentation rates, this will be, however, only competitive for very slow diffusion.

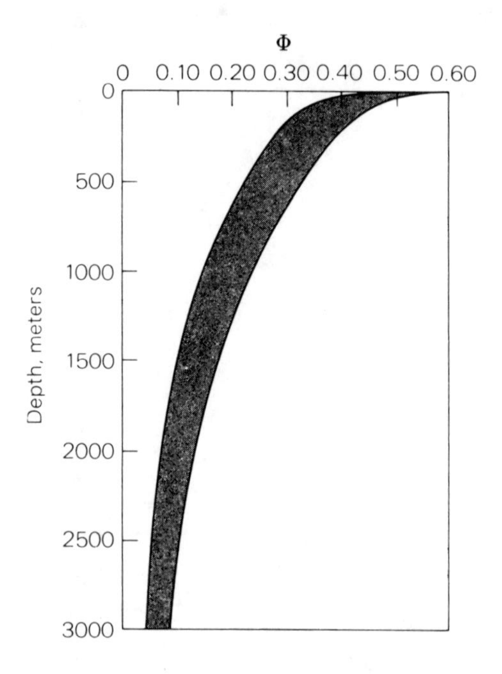

FIGURE 17. Plot of porosity ϕ vs. depth for some fine-grained marine terrigenous sediments. (After Hedberg, H. D., *Am. J. Sci.*, 31, 241, 1936. With permission.)

8. Density Gradients

Density gradients due to, for example, differences in salinity of pore waters cause transfer processes which can be distinguished from molecular diffusion. The effect depends, however, on the freedom of these density currents in the system, which will be larger in sand than in sand mixed with silt (Figure 18).[2]

Such processes certainly occur in estuaries where due to tidal currents the bottom is exposed to regularly changing waters of different salinity: fresh or brackish water at low tide and salt water at high tide. Transfer processes will be induced which might overlap molecular diffusion, having a diffusion coefficient of about $0.3 \times 10^{-5} cm^2 s^{-1}$.

Different from these processes is the process of upwelling of fresh water from continental origin through deeper layers when a fresh-water wedge exists underneath the coastal sediments. This upwelling is, however, much less distinct or homogeneous over the bottom concerned, since as frequently can be observed by divers, single wells of fresh water exist, leaving the rest of the seafloor undisturbed.[38]

III. DIFFUSION MODELS OF CHEMICAL BEHAVIOR IN THE BENTHIC BOUNDARY LAYER

A. Preliminaries

The application of Fick's second law of diffusion[39] to the rate of change in the concentration of a dissolved chemical in the interstitial water at a given depth z below the sediment surface will give a description of the dispersion of the chemical in the sediment column:

$$\frac{\partial \phi C}{\partial t} = -\frac{\partial F}{\partial z} = D \frac{\partial}{\partial z}\left(\phi \frac{\partial C}{\partial z}\right) \tag{4}$$

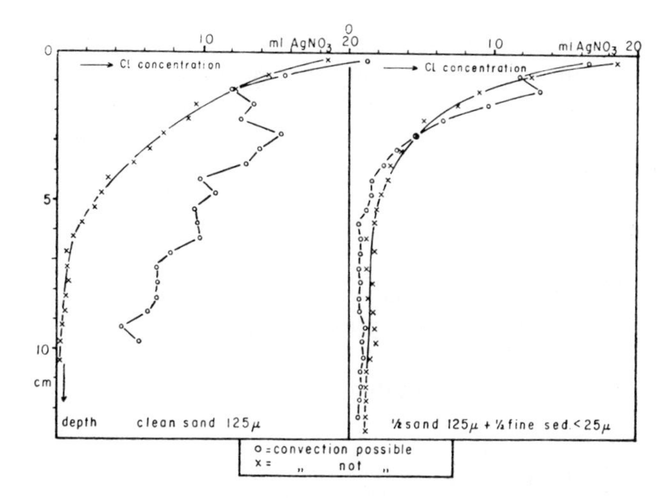

FIGURE 18. Graph showing the influence of convection due to density differences on the diffusion of chloride in two sediments. Constant source experiment with sea water as source and distilled water as interstitial water of the sediment. Tubes hanging upright O——O or upside down x——x. (From Duursma, E. K. and Bosch, C. J., *Neth. J. Sea Res.*, 4, 395, 1970. With permission.)

with C the concentration of the chemical in the pore water, ϕ the (fractional) sediment porosity, F the flux in the vertical, D the molecular diffusion constant (corrected for sediment tortuosity, i.e., the actual pore-water pathlength relative to the thickness of the sediment layer) and z the depth (positive) below the sediment surface.

If the flux due to the vertical advection of pore water is also taken into account, then

$$\frac{\partial \phi C}{\partial t} = D \frac{\partial}{\partial z}\left(\phi \frac{\partial C}{\partial z}\right) - W \frac{\partial \phi C}{\partial z} \qquad (5)$$

with W the velocity of the mass transfer of the interstitial water. Thus, if porosity is constant with time and depth, the equations reduce to the standard Fick's second law and diffusion-advection equation, respectively. Many authors have made use of the above equations, and especially Equation 5 to describe the behavior of chemicals in sediments. In that context it is usually called the *diagenetic equation.*[35] Often terms are added to the diagenetic equation to allow for processes that may add a chemical to or remove it from solution in the pore water. Similar equations can be formulated to describe the dispersion of chemicals in other phases of the sediments (e.g., in the adsorbed layer on the particles) and the equations may be combined (e.g., in the form of compartment models) to take the exchange processes between the different phases into account.

Steady-state diffusion models have been applied to problems of chemical diagenesis, and a number of analytical solutions of model equations under certain boundary conditions have been reported in the literature.[40,41] By contrast, transient-state models[41-43] are used seldom, not only because of the more intricate mathematics, but also because steady-state models have in many instances been shown to give a satisfactory fit to experimental data, so that the need for transient-state models is relatively small.

Besides true molecular diffusion, other mechanisms of dispersion of chemicals in the sediment layer can sometimes be described adequately in terms of diffusion models. Examples are in the mixing of pore water and sediment grains by physical and biological processes that disperse dissolved or adsorbed chemicals by turbulent mass transfer.

Many chemicals that are of interest as pollutants are closely associated with the sediment particles and with the suspended material in the overlying water. Models that not only describe the behavior of dissolved chemicals but also take into account the behavior in particulate-bound form, are therefore of special interest for a description of the transfer of the chemicals in the benthic boundary layer.

B. Models for Dissolved Species

1. Basic Equations

The simplest possible model for the vertical diffusion of dissolved compounds in pore waters is Fick's second law of diffusion (Equation 4) with the assumption of constant porosity with time and depth. For $C = 0$ at $t < 0$ and $C = C_0$ at $z = 0$ at $t \geq 0$ (constant source), the solution of Equation 4 is:

$$C = C_0 \operatorname{erfc}\left(\frac{z}{2\sqrt{Dt}}\right) \tag{6}$$

with erfc denoting the complementary error function.[39] (As ϕ is taken constant with time and depth, it no longer appears in the equations.)

For an instantaneous source of amount s at ($z = 0$; $t = 0$), the analytical solution of Equation 4 is[27]

$$C = \frac{s}{2\sqrt{\pi Dt}} \exp\left(-\frac{z^2}{4Dt}\right) \tag{7}$$

while at steady state (i.e., $\partial c/\partial t = 0$) with $C = C_0$ at $z = 0$ and $C = C_h$ at $z = h$,

$$C = C_0 - z\,\frac{(C_0 - C_h)}{h} \tag{8}$$

and the flux of material

$$F = -D\,\frac{dC}{dz} = D\,\frac{(C_0 - C_h)}{h} = \text{constant} \tag{9}$$

The incorporation of addition/removal terms in this model may be desirable. Although such terms may be of any order, we select for sake of simplicity a first-order removal term, making the change in concentration by that process dependent upon the actual concentration:

$$\frac{\partial C}{\partial t} = D\frac{\partial^2 C}{\partial z^2} - kC \tag{10}$$

with k the removal rate constant (positive, with dimension $time^{-1}$). The steady-state solution of Equation 7 with $C = C_0$ at $z = 0$ and $C = 0$ at $z = \infty$ is

$$C = C_0 \exp\left(-z\sqrt{\frac{k}{D}}\right) \tag{11}$$

If the concentration of the chemical at $z = 0$ now changes instantaneously from C_0 to C_1, then the solution of Equation 10 for the transient state is[40]

$$C = C_1 \exp\left(-z\sqrt{\frac{k}{D}}\right) + \frac{1}{2}(C_0 - C_1)\left\{\exp\left(-z\sqrt{\frac{k}{D}}\right) \times \right.$$

$$\operatorname{erfc}\left(\sqrt{kt} - \frac{z}{\sqrt{4Dt}}\right) - \exp\left(z\sqrt{\frac{k}{D}}\right) \times$$

$$\left.\operatorname{erfc}\left(\sqrt{kt} + \frac{z}{\sqrt{4Dt}}\right)\right\} \qquad (12)$$

with the initial conditions

$C = C_0 \exp(-z\sqrt{k/D})$ at $z = 0$, $t \leq 0$; $C = C_0$ at $z = 0$, $t \leq 0$; and $C = C_1$ at $z = 0$, $t > 0$.

The time required to reach the new steady state with $C = C_1 \exp(-z\sqrt{k/D})$ can now be calculated by successive approximation.

Usually an advection term is included in the basic equation, so that

$$\frac{\partial C}{\partial t} = D\frac{\partial^2 C}{\partial z^2} - W\frac{\partial C}{\partial z} - kC \qquad (13)$$

At steady state and with boundary conditions $C = C_o$ at $z = 0$ and $C = 0$ at $z = \infty$; its analytical solution is

$$C = C_0 \exp\left\{z\left(\frac{W}{2D} - \sqrt{\frac{W^2}{4D^2} + \frac{k}{D}}\right)\right\} \qquad (14)$$

2. Applications

The application of models of the type described is confined to chemicals that do not undergo appreciable partitioning between the dissolved and particulate phases of the sediment by adsorption, precipitation, etc. Chemicals that fall into this category are primarily a number of macronutrient ions and dissolved gases. Some of these macronutrients may reach polluting concentrations where estuaries and coastal seas carry high domestic or agricultural effluent loads. In highly industrialized areas fluorides may be of local importance as aquatic pollutants that occur primarily in solution.

The principal model is Equation 4 with possible extensions to allow for advection, removal, etc. If we assume a pollutant to be introduced into the overlying water at a concentration of C_0, then according to Equation 6 and taking $D = 10^{-5}\ cm^2s^{-1}$, it will take 7 hr for the concentration of the pollutant at 1 cm in the sediment to reach 15% of C_0 and 29 days to reach 90%. At 10 cm below the sediment surface, these values are 29 and 2900 days, respectively. However, as Vanderborght and Wollast[44] have pointed out, the turbulent structure of the sediment-water interface in coastal waters causes a faster turbulent mass transfer with apparent diffusion coefficients in the order of $10^{-4}\ cm^2\ s^{-1}$ in the upper sediment layer. In that case, it will take only ¾ hr to reach 15% of C_0 at a depth of 1 cm and 2.9 days to reach 90%. At a depth of 3.5 cm, which in muddy near-shore sediments is the lower boundary of a poorly consolidated top layer, these values are 8.5 hr and 35 days, respectively.

Table 5
DIAGENETIC EQUATION FOR DISSOLVED SILICA, OXYGEN, SULFATE, NITRATE, AND AMMONIUM IN MUDDY SEDIMENTS IN THE SOUTHERN NORTH SEA[45,46]

General equation: $$D\frac{d^2C}{dz^2} - U\frac{dC}{dz} + R = 0$$

	Upper layer ($z<z_n$)	Lower layer ($z>z_n$)
Diffusion constant	D_1	D_2
Reaction term R		
Silica	$+k(C_{z=\infty} - C)$	$+k(C_{z=\infty} - C)$
Oxygen	$-k$	$-k'C$
Sulfate	0	$-k_{SO_4}C$
Nitrate	$+k_{NO_3}$	$-k'C$
Ammonium	$-k_{NO_3} + k_{NH_4}$	$+\alpha k_{SO_4}C_{SO_4}$

Table 6
APPROXIMATE PARAMETER VALUES FOR SILICA, OXYGEN, SULFATE, NITRATE, AND AMMONIUM DIAGENETIC EQUATIONS[45,46]

		Upper layer		Lower layer
D		10^{-4} cm^{2}s^{-1}		10^{-6} cm^{2}s^{-1}
U		10^{-9} cm s^{-1}		10^{-9} cm s^{-1}
Silica	$C_{z=\infty}$	400 μmol ℓ$^{-1}$		400 μmol ℓ$^{-1}$
	k	5×10^{-7} s^{-1}		5×10^{-7} s^{-1}
Oxygen	k	5×10^{-6} μmol mℓ$^{-1}$ s^{-1}	k′	1.5×10^{-3} s^{-1}
Sulfate			k	2.5×10^{-8} s^{-1}
Nitrate	k_{NO_3}	1.5×10^{-6} μmol mℓ$^{-1}$ s^{-1}	k′	5×10^{-6} s^{-1}
Ammonium	k_{NH_4}	2.0×10^{-6} μmol mℓ$^{-1}$ s^{-1}	α	0.18

The above calculations show that the downward transfer of chemical species by molecular diffusion in the pore water is a very slow process, albeit it can be speeded up considerably by turbulent diffusion near the sediment-water interface. Considerable deep downward diffusional transport of solutes in pore water only takes place on very long time scales. However, in the sedimentary environment, sedimentation itself can cause a (virtual) downward transport of sediment and pore water by raising the sediment-water interface. Nevertheless, molecular and turbulent diffusion still play a significant role at rates of sedimentation of up to, say, 1 cm yr^{-1} ($\simeq 3 \times 10^{-8}$ cm s^{-1}). This results in a scale length (D/W) of about 3 cm for a molecular diffusion coefficient of 10^{-5} cm^2 s^{-1} in sediment, while in the sediment layer the scale length is 30 cm if an apparent diffusion coefficient of 10^{-4} cm^2 s^{-1} is assumed.

Vanderborght et al.[45,46] studied the diagenesis of silica, oxygen, sulfur, and nitrogen in muddy sediments in the southern North Sea. They proposed a two-layer model and included in their diagenetic equations a number of chemical reaction terms. The diagenetic equations derived for the chemical species in question are shown in Table 5. From simulations with the models for the different species they obtained estimates for the parameters in these equations that are shown in Table 6. As can be seen from Table 5, the models used are steady-state equations, and they give a good fit to concentration-depth profiles from the field. Vanderborght and Billen[40] showed that with regard to nitrate diagenesis, new steady states of the concentration-depth profile were established in a matter of days in response to changes in the nitrate concentration in the overlying water. Therefore, steady-state models are in many cases acceptable even if they appear strictly not to be wholly applicable.

C. Models That Include Sorption

1. Basic Equations

Many of the chemicals in the sediment will become associated with the inorganic and organic sediment particles to a considerable degree the inclusion of sorption-distribution terms in diagenetic models necessary. In which way such terms are to be formulated depends partly on the kinetics of the distribution processes between pore water and sediment particles. If distribution (e.g., by adsorption) proceeds at a slower or at an equal rate as the physical transfer processes such as diffusion, dispersion, or pore-water advection), there will be no distribution equilibrium for a dissolved chemical that is introduced into the pore water. In this transient state the removal of the chemical to the sediment particles obeys normal chemical reaction kinetics, so that it can be incorporated in the removal term (Equation 10). In the steady state, however, there will be a dynamic equilibrium between the dissolved and particulate-bound forms of the chemicals. In that case and if distribution by adsorption, ion exchange, precipitation/dissolution is fast relative to physical transfer, then

$$C_s = f(C) \tag{15}$$

the concentration of the chemical on the solid phase (C_s), expressed in terms of mass per unit volume of pore water, being a function (f) of its pore water concentration. Duursma and Hoede[27] have indicated that under certain conditions such as ion exchange or another reversible reaction with the solid phase, the linear distribution law applies

$$C_s = KC \tag{16}$$

with K a constant of proportionality. In that case the total diffusive transport in the vertical direction may be described by

$$\frac{\partial C}{\partial t} = D\frac{\partial^2 C}{\partial z^2} \tag{17}$$

which denotes the diffusion in the pore water, and by

$$\frac{\partial C_s}{\partial t} = D_s\frac{\partial^2 C_s}{\partial z^2} \tag{18}$$

denoting the diffusion in the solid phase, with D_s the applicable diffusion coefficient. Substituting $K \cdot C$ for C_s in Equation 16 and summing Equations 17 and 18 yields the total diffusion

$$\frac{\partial C}{\partial t} = \frac{D + KD_s}{K + 1}\frac{\partial^2 C}{\partial z^2} \tag{19}$$

which can be reduced to Fick's second law by defining

$$D' \equiv \frac{D + KD_s}{K + 1} \tag{20}$$

If $D_s = 0$ (i.e., if there is no diffusion in the solid phase) then Equation 20 reduces to

$$D' \equiv \frac{D}{K + L} \tag{21}$$

Rather than using the simple linear distribution $C_s = K \cdot C$, the use of the adsorption equations of Freundlich or Langmuir,[1] or other equations, may be preferred to describe the partitioning of a chemical between the dissolved and solid phases. The Freundlich equation is the simpler one,

$$C_s = KC^n \tag{22}$$

with n usually taking on values in the range 0.1 to 1. The Langmuir equation

$$C_s = \frac{K_1 C}{1 + K_2 C} \tag{23}$$

(with K_1 and K_2 constants) and others can be incorporated into the diagenetic equation, but since this results in its becoming nonlinear, no analytical solutions are available.

As pointed out by Lerman,[47] diffusion occurring in the solid phase enhances the diffusional flux. On the other hand, if there is no diffusion in the solid phase, then at steady state diffusion takes place with a diffusion coefficient D (cf. Equation 17) and not with D′ (Equation 21), as has been pointed out by Berner.[48]

2. Applications

Diffusion models that include sorption are usually of the same form as those used to describe dissolved transport only. The diffusion-advection equation (Equation 13) applies with minor modifications, provided linear adsorption distribution is assumed. Then, by analogy with Equations 17, 18, and 19:

$$\frac{\partial C}{\partial t} = \frac{D + KD_s}{K + 1}\frac{\partial^2 C}{\partial z^2} - \frac{W + KW_s}{K + 1}\frac{\partial C}{\partial z} - \frac{k + Kk_s}{K + 1} C \tag{24}$$

Where the s-subscripted variables denote diffusion, advection, and removal for the absorbed phase. Similarly, as for Equation 20, Equation 24 can be reduced to

$$\frac{\partial C}{\partial t} = D' \frac{\partial^2 C}{\partial z^2} - W' \frac{\partial C}{\partial z} - k'C \tag{25}$$

which is the same as equation 13, so that the same solutions are applicable.

As the difference between Equations 24 and 13 lies in the effective parameter values of D′, W′, and k′; these may be considered more closely. The values of D′, W′, and k′ relative to D, W, and k, depend on the degree of adsorption and on the value of D_s, W_s, and k_s relative to D, W, and k. To illustrate this point Figure 19 shows D′/D for different values of K and D_s/D. The graph has been extended to cover ratios of D_s/D greater than one that are meaningless in the case of diffusion, but that may be of interest for other parameters (e.g., k) since for W′/W and k′/k the same graph applies. As can be seen from Figure 19, strong adsorption and little diffusive transport in the adsorbed layer cause an appreciable decrease of the effective diffusion constant.

Typical values of D, D_s, and K for compounds that we know as pollutants would be 10^{-6} cm^2s^{-1}, 10^{-8} cm^2s^{-1}, and 10^3, in which case D′ becomes 1.1×10^{-8} cm^2s^{-1}. By contrast, if $D_s = 0$, then D′ would be 1×10^{-9} cm^2s^{-1}.

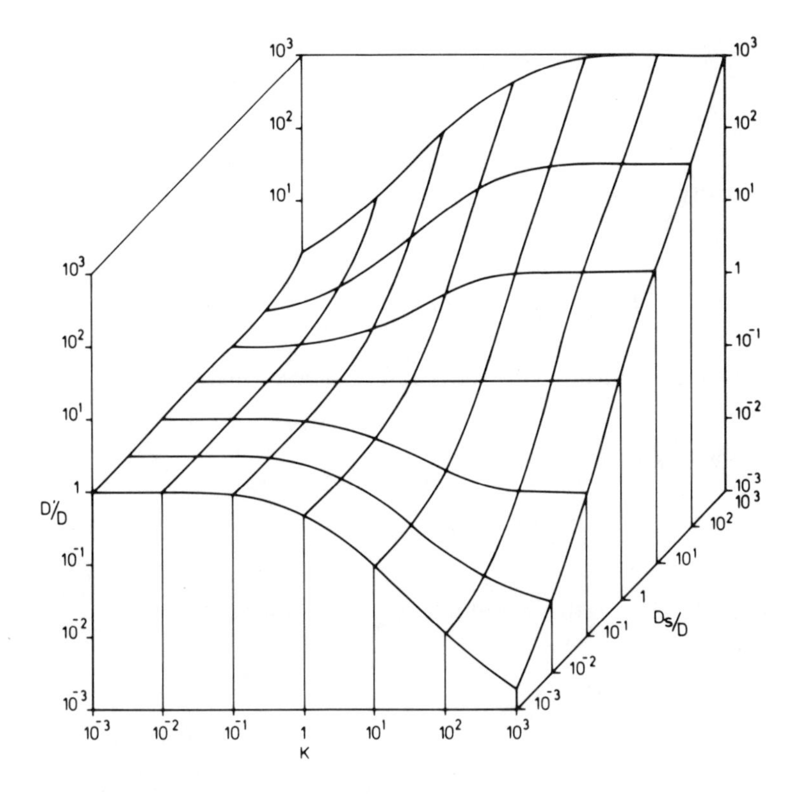

FIGURE 19. The effective diffusion constant (D′) as affected by adsorption (K) and diffusion in the adsorbed layer (D_s).

The situation becomes rather more complicated if porosity is considered not to be constant with depth, which is nearly always the case in the upper sediment layer. The diagenetic equation for the dissolved phase must then include ϕ and becomes

$$\frac{\partial \phi C}{\partial t} = \frac{\partial}{\partial z}\left(D\phi \frac{\partial C}{\partial z}\right) - W_w \frac{\partial \phi C}{\partial z} - k\phi C \tag{26}$$

and the equation for the solid phase

$$\frac{\partial \phi C_s}{\partial t} = \frac{\partial}{\partial z}\left(D_s\phi \frac{\partial C_s}{\partial z}\right) - W_s \frac{\partial \phi C_s}{\partial z} - k_s\phi\, C_s \tag{27}$$

In these equations W_w is the advection rate of the pore water, while the other variables have their customary meaning. Assuming $C_s = K \cdot C$ and adding equations 26 and 27

$$\frac{\partial \phi C}{\partial t} = \frac{D + KD_s}{K+1} \frac{\partial}{\partial z}\left(\phi \frac{\partial C}{\partial z}\right) - \frac{W_w + KW_s}{K+1} \frac{\partial \phi C}{\partial z} - \frac{k + Kk_s}{K+1} \phi\, C \tag{28}$$

Now at $z = 0$, $W_w = W_s = W$, with W being the local deposition rate. Therefore at any depth z

$$W_s + W_w = 2W \tag{29}$$

and

$$W_w = W \frac{\phi}{\phi_0} \tag{30}$$

with ϕ_0 the porosity at $z = 0$.

Consequently,

$$W_s = W \left(\frac{2\phi_0 - \phi}{\phi_0}\right) \tag{31}$$

Therefore, the advection parameter $W_w + KW_s$ can be evaluated as follows:

$$\begin{aligned} W_w + KW_s &= \frac{W}{\phi_0} (\phi + 2 K \phi_0 - K \phi) \\ &= \frac{W}{\phi_0} (2 K \phi_0 + (1 - K) \phi) \end{aligned} \tag{32}$$

As K is large for those cases that are of interest, Equation 3 reduces to:

$$\frac{KW}{\phi_0} (2\phi_0 - \phi) \tag{33}$$

Thus, if we take $D' = D + KD_s$, $W' = KW/\phi_0$, and $k' = k + Kk_s$, then at steady state Equation 28 becomes:

$$D' \left(\phi \frac{d^2 C}{dz^2} + \frac{d\phi dC}{dz^2}\right) - W' (2\phi_0 - \phi) \frac{d\phi C}{dz} - k'\phi C = 0 \tag{34}$$

It is clear that this type of equation calls for a numerical rather than an analytical solution.

Models that seek to incorporate all possible diagenetic parameters tend to become intangible. It should therefore be stressed that very simple diagenetic models often adequately fit the data, provided one is ready to accept that the diagenetic parameters are being included in the specific model parameters, e.g., in the form of an apparent diffusion coefficient. Those parameters only have a meaning for the environment in which they were determined.

Aston and Stanners[49] have used the simple diffusion-advection equation together with estimates of the apparent diffusion coefficients for radionuclides from the literature to calculate the sedimentation rate in the River Esk estuary (U.K.), and this approach was very successful indeed.

D. Bioturbation as an Apparent Diffusion Process

1. Background

Bioturbation is the reworking of the sediment-water interface by benthic organisms and more specifically, the reworking of sediment particles. Considering the relative ineffectiveness of diffusion in the solid phase of the sediment in transporting chemicals, bioturbation may be an additional factor in distributing chemicals that are closely associated with sediment particles through the sediment column. Provided the rate of sediment reworking is faster than the local net sedimentation rate, and this is necessarily the case if benthic organisms are not to be buried alive, bioturbation must play a role in redistributing strongly bound chemicals through the upper sediment layer.

Viewed on a longer time scale, bioturbation is a mixing process and therefore it has been described in terms of diffusion, commonly using a diffusion-advection equation with the advection term representing sedimentation. In the short term, however, bioturbative transport itself has a definite advective element, since burial does not only take place by sedimentation, but also by surface deposition of (pseudo) feces, transported from the subsurface by the infauna. As buried particles can, in time, be brought to the surface again, the associated chemicals pass through a layer where fast changes in physicochemical parameters, notably the redox-potential, occur.

In the coastal and intertidal environment bioturbation rates are much greater than those of local net sedimentation. Risk and Moffat[50] estimated that a high-density population ($3500\ m^{-2}$) of *Macoma balthica* could rework some 28 cm of sediment per year, and Cadée[31] found that dense populations of *Arenicola marina* ($85\ m^{-2}$) reworked up to 33 cm $year^{-1}$. This compares with typical net sedimentation rates in coastal areas in the order of 1 cm $year^{-1}$.

2. Models Used

The type of equation commonly used to describe the vertical transport under the influence of bioturbation is the now familiar diffusion-advection equation:

$$\frac{\partial \phi C}{\partial t} = D \frac{\partial}{\partial z}\left(\phi \frac{\partial C}{\partial z}\right) - W \frac{\partial \phi C}{\partial z} \qquad (35)$$

with D being redefined to include bioturbative mixing. This equation then applies only to the depth to which the sediment is being perturbed. Since the models were first used to describe the distribution of radionuclides in sediments, a decay term was normally included as well. Schink and Guinasso[51] extended the simple diagenetic equation to allow for adsorption and bioturbation. In their diagenetic equation in corrected form:

$$\phi (K+1) \frac{\partial C}{\partial t} = \frac{\partial}{\partial z}\left(D_T \frac{\partial C}{\partial z} - W_T C\right) + R \qquad (36)$$

D_T and W_T are the redefined diffusion constant and advection rate, respectively, while R represents all reactions that take place. The redefined diffusion constant contains the following terms:

$$D_T = \phi (D + D_A) + KD_B \qquad (37)$$

with D the molecular diffusion constant of the dissolved compound, D_A a coefficient of biological mixing of the pore water, and D_B a coefficient of biological mixing of the sediment particles. The advection rate W_T is defined

$$W_T = \phi W_w + KW_s - D_B \frac{\partial K}{\partial z} \qquad (38)$$

with W_W the pore water and W_S the sediment particle advection rate, respectively.

3. Applications

As pointed out by Aller,[52] bioturbation is not necessarily described adequately by apparent diffusion only. Amiard-Triquet[53] shows that the short-term bioturbation by *Arenicola marina* has an important advective component, although her presentation of the data does not render them amenable to mathematical analysis. Our own work[54] with sediment cores from intertidal flats in the Dutch Delta region shows the same feature. Application of Equation 35 to the vertical transport of painted sediment grains in such cores yielded a mean apparent diffusion coefficient of 2×10^{-8} cm^2s^{-1} and a mean downward advection of particles of 1.5×10^{-7} cm s^{-1} caused by burial of the colored grains by (pseudo)feces of the infauna. The average depth of biogenic sediment reworking in these cores of 3.5 to 4 cm agrees well with the thickness estimate of the upper sediment layer in the southern North Sea by Vanderborght et al.[45]

Calculation of the scale length for our cores yields a distance in the order of 10^{-1} cm, which shows that initially the advective component of bioturbation is much more important than mixing, i.e., outright burial is the essential process. However, by the time sediment particles are being brought up to the surface again in the casts of worms and molluscs a more even spread of particles in the upper sediment layer is achieved.

Apparent diffusion coefficients for bioturbation of sediments are generally found to be in the order of 10^{-6} to 10^{-8} cm^2 s^{-1} in estuaries and coastal seas, while being much smaller in abyssal sediments. Guinasso and Schink[55] have defined a mixing parameter for bioturbation

$$G = D_B/LW \tag{39}$$

with D_B the biological mixing constant, L the depth of bioturbative activity, and W the rate of sedimentation.

With D_B having values of 10^{-6} to 10^{-8} cm^2 s^{-1}, L being 3 to 10 cm and W being in the order of 10^{-8} cm s^{-1}, G takes on values varying from 0.1 to 10. At the former value of G, bioturbation hardly plays a role in downward transport of sediment while at the latter value, the deposited sediment grains become uniformly mixed through the zone of bioturbation before being transported below the layer of bioturbative activity by sedimentation.

Recalling Schink and Guinasso's[51] diagenetic equation (Equation 36), which can be written as follows:

$$\frac{\partial C}{\partial t} = \frac{D_T}{\phi(K+1)} \frac{\partial^2 C}{\partial z^2} - \frac{W_T}{\phi(K+1)} \frac{\partial C}{\partial z} + \frac{R}{\phi(K+1)} \tag{40}$$

its effective diffusive term becomes (cf. Equation 37)

$$D'_T = \frac{\phi D}{\phi(K+1)} + \frac{K\, D_B}{\phi(K+1)} + \frac{\phi\, D_A}{\phi(K+1)} \tag{41}$$

Similarly, the effective advective term (cf. Equation 38) becomes

$$W'_T = \frac{\phi W_w}{\phi(K+1)} + \frac{K\, W_s}{\phi(K+1)} + \frac{D_B}{\phi(K+1)} \frac{\partial K}{\partial z} \tag{42}$$

As in soft sediments, ϕ is always in the order of 0.4 to 0.9 and as Equation 36 is only of interest for chemicals that exhibit considerable adsorption, with K taking on values of say 100 or more, the effective terms may be somewhat simplified to:

$$D'_T = \frac{D}{K} + \frac{D_B}{\phi} + \frac{D_A}{K} \tag{43}$$

and

$$W'_T = \frac{W_w}{K} + \frac{W_s}{\phi} + \frac{D_B}{\phi(k+1)} \frac{\partial K}{\partial z} \tag{44}$$

As can be seen from these expressions, bioturbative mixing and advection of sediment particles become progressively more important in distributing chemicals when K increases.

IV. RESIDENCE TIMES OF CHEMICALS IN THE BENTHIC BOUNDARY LAYER

A. Background

In a reservoir with equal input and output the amount of material present does not change with time. Material passing through will therefore spend a certain amount of time in the reservoir, the mean residence time of the material, τ, which can be defined as[41]

$$\tau \equiv \frac{M}{\sum_i F_i} \tag{45}$$

with M the amount of material in the reservoir and $\sum_i F_i$ the sum of all input *or* removal fluxes, expressed in units of material per time. Alternatively, fractional mean residence times may be defined for individual input or removal processes:

$$\tau_i \equiv \frac{M}{F_i} = \frac{1}{k_i} \tag{46}$$

with k_i the rate constant (in $time^{-1}$) for the process.
Then

$$\tau = \frac{1}{\sum_i k_i} \tag{47}$$

and

$$\frac{1}{\tau} = \sum_i \frac{1}{\tau_i} \tag{48}$$

For a reservoir without equal input and output, Equation 45 denotes the "instantaneous" residence time of material, which can be compared with the mean residence time for the steady state.

To illustrate the preceding points, let us take the following example.[28] We have a reservoir with volume V, material enters the reservoir at a rate G ($g\ s^{-1}$) and is removed from the reservoir by a first-order process with rate constant k (s^{-1}). Within the reservoir complete mixing takes place. The rate of change of the concentration of material C is described by

$$\frac{dC}{dt} = \frac{G}{V} - kC \tag{49}$$

the analytical solution of which is

$$C = \frac{G}{Vk}(1 - e^{-kt}) \tag{50}$$

At steady state

$$\frac{G}{V} - kC = 0 \tag{51}$$

and

$$\tau = \frac{CV}{G} = \frac{1}{k} \tag{52}$$

The time it takes to reach the steady state can be calculated with Equation 50. However, it is convenient to express this time relative to the steady-state residence time of the material, as this gives us an indication of the usefulness of steady-state models. In order to attain steady state, the term e^{-kt} in Equation 50 must become zero. As $\tau = 1/k$ it can easily be calculated that 95% of the steady state concentration will be reached at $t = 3\,\tau$ and 99% at $t = 4.6\,\tau$. For cases where more elaborate models are appropriate, calculation of material fluxes will normally be the first step in determining the mean residence time of the material.

B. Residence Time in the 0- to 10-cm Sediment Layer

1. Dissolved Species

Fluxes for dissolved species can be calculated using the equations given in Section III.B. Considering Fick's first law of diffusion $F = -D(dC/dz)$ this relationship can be used if $C = f(z)$ is known. If the flux is known, the residence time can be calculated.

Let us take as an example Equation 10, of which the steady-state solution was

$$C = C_0 \exp\left(-z\sqrt{\frac{k}{D}}\right)$$

the flux at $z = 0$ is:

$$F = D\left.\frac{dC}{dz}\right|_{z=0} = C_0\sqrt{kD} \tag{53}$$

Vanderborght and Billen[40] have used this model to describe the vertical distribution of nitrate in sediment pore water. Taking $D = 10^{-5}$ cm^2 s^{-1} and $k = 2 \times 10^{-6}$ s^{-1}, they calculated using Equation 12 that if C_0 changed instantaneously from 150 to 30 μmol ℓ^{-1} nitrate, the initially resulting concentration peak in the sediment had disappeared after 55 hr. Calculating the flux across the sediment-water interface with Equation 4 and assuming a mean concentration of nitrate in the upper 1 cm of pore water of 125 μmol ℓ^{-1}, the mean residence time of nitrate in the upper 1 cm of the sediment is found to be

$$\tau = \frac{M}{F} = \frac{125}{150\sqrt{2 \times 10^{-6} \times 10^{-5}}}\ \text{s} = 52\ \text{hr}$$

At a depth of 10 cm, the mean residence time is longer. Here the flux equation is

$$F = -D\frac{dC}{dz} = C_0\sqrt{kD}\ \exp\left(-z\sqrt{\frac{k}{D}}\right) \tag{54}$$

and the mean residence time

$$\tau = \frac{M}{F} = \frac{2.2}{150\sqrt{2\times10^{-6}\times10^{-5}}\ \exp\left(-10\sqrt{2\times10^{-6}/10^{-5}}\right)}\ s \simeq 80\ hr$$

Leaving out removal/addition terms, the flux of dissolved material becomes constant and independent of depth within the sediment (Equation 9). The mean residence time then becomes

$$\tau = \frac{M}{F} = \frac{Mh}{D\,(C_0 - C_h)} \tag{55}$$

with C_0 the concentration at $z = 0$ and C_h the concentration at depth $z = h$. The concentration at depth $z = h$ may be determined by horizontal ground water flow, removing the diffusing substance.

2. Adsorbed Species and Bioturbation

For chemical species that become to a considerable extent bound to sediment particles by adsorption, the flux of the compound through the sediment-water interface and downward in the sediment is likely to depend more strongly upon local deposition than on diffusion, if compared with the diagenesis of dissolved chemicals. In consequence, diffusion-advection equations are more appropriate descriptions, since the scale length D/W decreases if D is replaced by $D' = (D + KD_s)/(K + 1)$ (Equation 20) and W by $W' = (W + KW_s)/(K + 1)$, because $D_s \ll D$ and $W_s \simeq W$ in general, so that $D' < D$ and $W' \simeq W$. If we also assume that $k_s = k$ in $k' = (k + Kk_s)/(K + 1)$ (Equation 24) them $k' = k$ and Equation 25 becomes

$$\frac{\partial C}{\partial t} = D'\frac{\partial^2 C}{\partial z^2} - W\frac{\partial C}{\partial z} - kC \tag{56}$$

and its steady-state solution with $C = C_0$ at $z = 0$ and $C = 0$ at $z = \infty$ is

$$C = C_0 \exp\left\{z\left(\frac{W}{2D'} - \sqrt{\frac{W^2}{4D'^2} + \frac{k}{D'}}\right)\right\} \tag{57}$$

and the flux

$$F = -D'\frac{dC}{dz} + WC$$

$$= \left\{-D'\left(\frac{W}{2D'} - \sqrt{\frac{W^2}{4D'^2} + \frac{k}{D'}}\right) + W\right\} C_0 \times$$

$$\exp\left\{z\left(\frac{W}{2D'} - \sqrt{\frac{W^2}{4D'^2} + \frac{k}{D'}}\right)\right\} \tag{58}$$

The mean residence time can be calculated. Let

$$a = \frac{W}{2D'} - \sqrt{\frac{W^2}{4D'^2} + \frac{k}{D}} \tag{59}$$

then

$$C = C_0 \exp(za)$$

and

$$F = (-D'a + W) C_0 \exp(za)$$

If M denotes the mass of the compound in the sediment per unit of surface then in the sediment layer z_1 to z_2:

$$M = \int_{z_1}^{z_2} C_0 \exp(za)\, dz = \frac{C_0}{a} \exp(za) \Big|_{z_1}^{z_2} \tag{60}$$

The mean residence time in (z_1 to z_2) is:

$$\tau = \frac{\frac{C_0}{a} \{\exp(z_2 a) - \exp(z_1 a)\}}{(-D'a + W) C_0 \exp(z_1 a)}$$

$$= \frac{1}{a(-D'a + W)} \{\exp(z_2 - z_1) a - 1\} \tag{61}$$

if the flux is measured at z_1 and

$$\tau = \frac{1}{a(-D'a + W)} \{1 - \exp(z_1 - z_2) a\} \tag{62}$$

if the material flux is measured at z_2.

Thus for a typical value of W of 1 cm year^{-1} we can calculate the mean residence time in the 0- to 10-cm layer of sediment for, e.g., ^{134}Cs with a decay rate of 0.33 year^{-1} if we assume $D' = 10^{-8}$ cm^{2}s^{-1}. With these data a = 0.301 cm, and

$$\tau = 9.09 \times 10^7 \text{ s} = 2.9 \text{ year}$$

if the flux is measured at z = 0 cm and

$$\tau = 1.85 \times 10^9 \text{ s} = 59 \text{ year}$$

if the flux is measured at z = 10 cm.

Such estimates of residence time become more accurate if the calculation is carried out for a thinner sediment layer, and for any layer of a 1-cm thickness in the above example

$$\tau = 2.49 \times 10^7 \text{ s} = 0.79 \text{ year}$$

if the flux is measured at the top of the layer and

$$\tau = 3.36 \times 10^7 \text{ s} = 1.06 \text{ year}$$

if the flux is measured at the bottom of the layer, so that the residence time in a 10-cm layer is

$$7.9 < \tau < 10.6 \text{ year}$$

If there is no removal (k = 0) and if the scale length (D′/W) is smaller than the thickness of the sediment layer under consideration, the mean residence time of the chemical is

$$\tau = z/W \quad (63)$$

since advection in this case outpaces the apparent diffusion. For the previous example of ^{134}Cs with k = 0, the mean residence time then becomes

$$\tau = 10/1 = 10 \text{ year}$$

In the case that bioturbation plays a role as well, similar arguments apply and calculation procedures will be the same. The parameter values, however, will differ.

REFERENCES

1. **Glasstone, S.,** *Textbook of Physical Chemistry,* 2nd ed., McMillan and Co. Ltd., London, 1960.
2. **Duursma, E. K. and Bosch, C. J.,** Theoretical, experimental and field studies of radioisotopes concerning diffusion in sediments and suspended particles in the sea. B. Methods and experiments, *Neth. J. Sea Res.,* 4, 395, 1970.
3. **Bird, R. B., Stewart, W. E., and Lightfood, E. N.,** *Transport Phenomena,* John Wiley & Sons, London, 1960.
4. **Lavav, H. and Bolt, G. H.,** Self-diffusion of Ca-45 into certain carbonates, *Soil Sci.,* 97, 293, 1964.
5. **Duursma, E. K., Dawson, R., and Ros Vicent, J.,** Competition and time of sorption of various radionuclides and trace metals by marine sediments and diatoms, *Thalassia Jugosl.,* 11, 47, 1975.
6. **Jones, R. F.,** The accumulation of nitrosyl ruthenium by fine particles and marine organisms, *Limnol. Oceanogr.,* 5, 312, 1960.
7. **Duursma, E. K. and Eisma, D.,** Theoretical, experimental, and field studies concerning reactions of radioisotopes with sediments and suspended particles of the sea. C. Applications to field studies, *Neth. J. Sea Res.,* 6, 265, 1973.
8. **Dyal, R. S. and Hendricks, S. B.,** Total surface of clays in polar liquids as a characteristic index, *Soil Sci.,* 62, 421, 1950.
9. **Rhoads, D. C.,** Organism-sediment relations on the muddy seafloor, *Oceanogr. Mar. Biol. Annu. Rev.,* 12, 263, 1974.
10. **Thorson, G.,** *Life in the Sea,* World University Library, London, 1971.
11. **Willems, K. and Sandee, A. J. J.,** Meiozoobenthos: density and biomass, *Verh. K. Ned. Akad. Wet. Afd. Natuurk, Reeks 2,* 73, 168, 1979.
12. **Sepers, A. B. J.,** personal communication, 1979.
13. **Postma, H.,** Transport and accumulation of suspended matter in the Dutch Wadden Sea, *Neth. J. Sea Res.,* 1, 448, 1961.
14. **de Groot, A. J., Salomons, W., and Allersma, E.,** Processes affecting heavy metals in estuarine sediments, in *Estuarine Chemistry,* Burton, J. D. and Liss, P. S., Eds., Academic Press, London, 1976, chap. 5.
15. **Duinker, J. C. and Nolting, R. F.,** Mixing, removal and mobilization of trace metals in the Rhine estuary, *Neth. J. Sea Res.,* 12, 205, 1978.
16. **Müller, G. and Förstner, U.,** Heavy metals in sediments of the Rhine and Elbe estuaries: mobilization or mixing effect? *Environ. Geol.,* 1, 33, 1975.
17. **Duursma, E. K., Vegter, F., and Kelderman, P.,** Aspects of hydrochemical water quality of the Delta waters, *Hydrobiol. Bull.,* 12, 215, 1978.

18. **Duinker, J. C., van Eck, G. T. M., and Nolting, R. F.** On the behaviour of copper, zinc, iron, and manganese, and evidence for mobilization processes in the Dutch Wadden Sea, *Neth. J. Sea Res.*, 8, 214, 1974.
19. **Aston, S. R. and Duursma, E. K.**, Concentration effects on ^{137}Cs, ^{65}Zn, ^{60}Co and ^{106}Ru sorption by marine sediments with geochemical implications, *Neth. J. Sea. Res.*, 6, 225, 1973.
20. **Fukai, R. and Huynh-Ngoc, L.**, Trace metals in offshore Mediterranean waters, *Thalassia Jugosl.*, 13, 1, 1977.
21. **Webb, J. E. and Theodor, J. L.**, Wave-induced circulation in submerged sands, *J. Mar. Biol. Assoc. U.K.*, 52, 903, 1972.
22. **Goldberg, E. D. and Arrhenius, G. O. S.**, Chemistry of Pacific pelagic sediments, *Geochim. Cosmochim. Acta*, 13, 153, 1958.
23. **Ku, T. L., Broecker, W. S., and Opdyke, N.**, Comparison of sedimentation rates measured by paleomagnetic and the ionium methods of age determination, *Earth Planet. Sci. Lett.*, 4, 1, 1968.
24. **Livingston, H. D. and Bowen, V. T.**, Pu and ^{137}Cs in coastal sediments, *Earth Planet. Sci. Lett.*, 43, 29, 1979.
25. **Templeton, W. L. and Preston, A.**, Transport and distribution of radioactive effluents in coastal and estuarine waters of the United Kingdom, in *Disposal of Radioactive Wastes into Seas, Oceans, and Surface Waters*, International Atomic Energy Agency, Vienna, 1966, 267.
26. **Billen, G.**, *Etude écologique des Transformations de l'Azote dans les Sédiments Marins*, Ph.D. thesis Universite Libre, Bruxelles, 1976.
27. **Duursma, E. K. and Hoede, C.**, Theoretical, experimental and field studies concerning molecular diffusion of radioisotopes in sediments and suspended solid particles of the sea, A. Theories and mathematical calculations, *Neth. J. Sea Res.*, 3, 423, 1967.
28. **Duursma, E. K.**, Migration in the seabed, some concepts, in *Biogeochemistry of Estuarine Sediments*, Proc. UNESCO/SCOR Workshop, Melreux, UNESCO, Paris, 1978, 179.
29. **Duursma, E. K. and Gross, M. G.**, Marine sediments and radioactivity, in *Radioactivity in the Marine Environment*, National Academy of Sciences, Washington, D.C., 1971, 147.
30. **Sips, H.** personal communication, 1979.
31. **Cadée, G. C.**, Sediment reworking by *Arenicola marina* on tidal flats in the Dutch Wadden Sea, *Neth. J. Sea Res.*, 10, 440, 1976.
32. **Yevitch, P. D.**, personal communication, 1979.
33. **Elder, D. L. and Villeneuve, J. P.**, Polychlorinated biphenyls in marine air, deep sediments and water of the Mediterranean Sea, *Rapp. P. V. CIESM Monaco*, 24(8), 59, 1977.
34. **Nienhuis, P. H.**, Veranderingen in de Grevelingen na de afdamming, *Vakbl. Biol.*, 57, 271, 1977.
35. **Berner, R. A.**, *Principles of Chemical Sedimentology*, McGraw-Hill, New York, 1971.
36. **Hedberg, H. D.**, Gravitational compaction of clays and shales, *Am. J. Sci.*, 31, 241, 1936.
37. **Hoede, C. and Van Beckum, F. P. M.**, Calculations on the compaction of sediments, unpublished, 1981; cf. Berner, R. A., Early diagnosis, Princeton University Press, 1980.
38. **Cousteau, J. Y.**, personal communication, 1973.
39. **Carslaw, H. S. and Jaeger, J. C.**, *Conduction of Heat in Solids*, Oxford University Press, Oxford, 1959.
40. **Vanderborght, J. P. and Billen, G.**, Vertical distribution of nitrate concentration in interstitial water of marine sediments with nitrification and dentrification, *Limnol. Oceanogr.*, 20, 953, 1975.
41. **Lerman, A.**, *Geochemical Processes: Water and Sediment Environments*, John Wiley & Sons, New York, 1979.
42. **Imboden, D. M.**, Interstitial transport of solutes in non-steady state accumulating and compacting sediments, *Earth Planet. Sci. Lett.*, 27, 221, 1975.
43. **Lasaga, A. C. and Holland, H. D.**, Mathematical aspects of non-steady-state diagenesis, *Geochim. Cosmochim. Acta*, 40, 257, 1976.
44. **Vanderborght, J. P. and Wollast, R.**, Mass transfer properties in sediments near the benthic boundary layer, in *Bottom Turbulence*, Nihoul, J. C. J., Ed., Elsevier, Amsterdam, 1977, 209.
45. **Vanderborght, J. P., Wollast, R., and Billen, G.**, Kinetic models of diagenesis in disturbed sediments. I. Mass transfer properties and silica diagenesis, *Limnol. Oceanogr.*, 22, 787, 1977.
46. **Vanderborght, J. P., Wollast, R., and Billen, G.**, Kinetic models of diagenesis in disturbed sediments. II. Nitrogen diagenesis, *Limnol. Oceanogr.*, 22, 794, 1977.
47. **Lerman, A.**, Migrational processes and chemical reactions in interstitial waters, in *The Sea*, Vol. 6, Goldberg, E. D., McCave, I. N., O'Brien, J. J., and Steele, J. H., Eds., John Wiley & Sons, New York, 1978, 695.
48. **Berner, R. A.**, Inclusion of adsorption in the modelling of early diagenesis, *Earth Planet. Sci. Lett.*, 29, 333, 1976.
49. **Aston, S. R. and Stanners, D. A.**, The determination of estuarine sedimentation rates by $^{134}Cs/^{137}Cs$ and other artificial radionuclide profiles, *Estuarine Coastal Mar. Sci.*, 9, 529, 1979.

50. **Risk, M. J. and Moffat, J. S.,** Sedimentological significance of fecal pellets of *Macoma balthica* in the Minas Bay of Fundy, *J. Sediment. Petrol.*, 47, 1425, 1977.
51. **Schink, D. R. and Guinasso, N. L.,** Redistribution of dissolved and adsorbed materials in abyssal marine sediments undergoing biological stirring, *Am. J. Sci.*, 278, 687, 1978.
52. **Aller, R. C.,** The effects of animal-sediment interactions on geochemical processes near the sediment-water interface, in *Estuarine Interactions,* Wiley, M., Ed., Academic Press, New York, 1978, 157.
53. **Amiard-Triquet, C.,** Etude expérimentale de la contamination par le cérium 144 et le fer 59 d'un sédiment à *Arenicola marina* L. (Annélide Polychète), *Cah. Biol. Mar.*, 15, 483, 1974.
54. **Smies, M. and Francke, J. W.,** unpublished data, 1979.
55. **Guinasso, N. L. and Schink, D. R.,** Quantitative estimates of biological mixing rates in abyssal sediments, *J. Geophys. Res.*, 80, 3032, 1975.

Chapter 4

ESTUARIES AND FJORDS

Herman G. Gade

TABLE OF CONTENTS

I. INTRODUCTION

Whenever sea water is partially enclosed in coastal basins such as lagoons, estuaries, or fjords, the free motion of the water is restricted and both the lateral and vertical exchanges are affected. Usually the water is considerably more stratified than in the open ocean, another factor which contributes to inhibiting exchange between the surface layers and the deeper water.

The vertical limitation of the wind-induced mixing in such stratified coastal water bodies has biological consequences which may affect the entire basin. Pollutants can greatly enhance these effects, often with drastic consequences.

By and large we shall deal with pollutants in dissolved or suspended form, and unless otherwise stated, the pollutant is assumed to be conservative. For the many cases in which this assumption is not permissible, the diffusion equations have to include decay terms such as are dealt with in Volume II, Chapter 1, Section IV. Furthermore, it will be assumed that the pollutant is passive, i.e., with no effect on the density of the ambient water other than that related to the inherent discharge of fresh water.

In an estuary or fjord, discharge of a pollutant will be subject to dispersion which partly is of a turbulent diffusive nature and partly rests on the natural or induced circulation of the basin. In the case of the former, small-scale mixing usually combines with eddies on a larger scale, including that of tidal motion, to cause lateral (specifically longitudinal) dispersion.

Of the circulatory processes, on the other hand, the estuarine circulation is the most prevalent and often a nearly stationary feature, whereas wind-induced circulation and renewals related to density currents associated with a varying offshore density field are more variable or intermittent phenomena. Insofar as a conservative pollutant is associated with the discharge or run-off of fresh water, its transport to the sea can be described by the same methods as those pertaining to the fresh water. Otherwise one is obliged to resort to individual approaches often based upon field experiments of suitable kinds.

The transports related to the exchange processes of importance in estuarine environments are basically additive in the sense that advection and eddy diffusion contribute to the total as if each process were operating independently. The distinction between advection and diffusion is, however, often vague, depending also on the relevant time scales involved. One should also bear in mind that advective processes are often included in some overall diffusion terms, in which case the addition is already performed. The topic is dealt with further in Volume I, Chapter 4, Section II. In the present section a discussion on the principles of the basic transfer mechanisms active in estuaries will be presented.

In accordance with customary usage we shall define an estuary as any partly enclosed sea-water body which has a noticeable discharge of fresh water. The discussion will, however, be limited to systems with residence times less than 100 years. Thus the analysis extends to systems as large as the Baltic Sea whereas the Black Sea is outside the scope of the discussion.

The material covered in the present chapter is mainly based upon individual papers selected from the literature. In some cases the reader is referred to other relevant articles[3-5] and more comprehensive treatises on the subject.[1,2]

II. CLASSIFICATION OF ESTUARIES

Estuaries have traditionally been divided into three geomorphologically different categories: the fjord type, the bar built type (lagoon), and the coastal plain estuary. Estuaries not belonging to one of these groups may be considered separately. Among the latter we have primarily the tectonically formed estuaries.

Classification has otherwise been according to the degree of salinity stratification, varying from the vertically well mixed to the highly stratified estuary. A stratification parameter $\delta S/S_0$ is used, where δS stands for the vertical salinity difference in a two-layered system or refers to the salinity defect ($= S_0 - S$) of the surface water in continuously stratified systems and S_0 stands for the bottom or source water salinity or in some cases, the cross-sectional average salinity.[6]

In the case of two-layered systems the stratification parameter is a direct measure of the proportion of fresh water in the brackish flow and is thus an indicator of the effect of vertical entrainment and mixing. This follows from application of conservation equations for salt and mass. With a mass m_1 of source water (salinity S_0) mixing with a mass m_2 of fresh water to form a mass m_3 of brackish water (salinity S), the appropriate budget equations are $m_1 + m_2 = m_3$ and $m_1 S_0 = m_3 S$, from which it follows that

$$\frac{\delta S}{S_0} = \frac{S_0 - S}{S_0} = \frac{m_2}{m_3}$$

In estuarine systems the salinity is usually decisive for the density. Therefore in two-layered systems the stratification can often be expressed advantageously by the corresponding density ratio $\delta\varrho/\varrho$ or by the related parameter $\delta\varrho/\delta\varrho_0$ where $\delta\varrho$ is the vertical density jump and $\delta\varrho_0$ the corresponding value at the head of the estuary where the upper layer is assumed fresh.

Estuary classification may be further refined by relating the stratification to principal features of the flow regime. Thus, according to Hansen and Rattray,[6] estuaries are generally described by the rates of river flow and tidal flow relative to the speed of long internal waves. Fundamental in this connection is the interfacial Froude number $F_i = U_b/U_d$, where U_b is the average velocity of the upper layer and U_d the speed of the long internal wave. For two-layered flow in a channel of constant width, U_b is related to the fresh-water discharge R per unit width of the estuary by

$$U_b \cdot h \cdot \delta\rho = U_{b_0} \cdot h_0 \cdot \delta\rho_0 = R\,\delta\rho_0 \qquad (1)$$

where h is the thickness of the upper (brackish) layer and U_{b0}, h_0, and ϱ_0 are the corresponding values of the variables at the head of the estuary. For shallow estuaries R is often replaced by DU_f, where D is the total depth and U_f the equivalent of the vertical average of the fresh-water velocity in the estuary.

The tidal flow is generally described by either the peak or the rms tidal velocity U_t. Traditional representation has been by the flow ratio $B = U_f/U_t$ relevant to estuaries of intermediate and shallow depth.

Whenever wind action on the surface affects the flow regime significantly, a third parameter involving some measure of the wind speed will be needed. In problems where the surface stress is essential, the friction velocity may be a suitable parameter being a direct measure of the wind stress and related to the wind velocity profile through the roughness parameter of the sea surface. In other problems the wind is important in connection with the turbulence set up in the water, in which case the root mean square (rms) turbulent velocity might be a suitable parameter.

Any suitable diagrammatic representation of the stratification vs. the flow parameters can be used for estuary classification. In the method proposed by Hansen and Rattray[6] the stratification is studied by means of a circulation parameter U_s/U_f, where U_s is the surface velocity (see Figure 1).

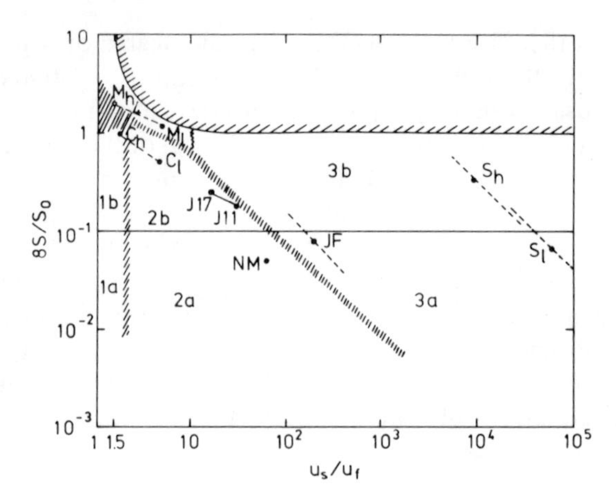

FIGURE 1. Classification diagram with examples. Station code: M, Mississippi River mouth; C, Columbia River estuary; J, James River estuary; NM, Narrows of the Mersey estuary; JF, Strait of Juan de Fuca; S, Silver Bay. Subscripts h and l refer to high and low river discharge; numbers indicate distance (in miles) from mouth of the James River estuary. Shaded zones separate main flow regimes: (1) with predominant zero-order circulation; (2) with predominant first-order circulation, but with appreciable longitudinal eddy flux of salt; (3) with predominant first-order circulation with negligible longitudinal eddy flux of salt; and (4) with pronounced salt wedge. Thin horizontal line indicates arbitrary division between (a) weakly and (b) strongly stratified estuaries (see also definition in Chapter IV.B). (From Hansen, D. V. and Rattray, M., Jr., *Limnol. Oceanogr.*, 11, 319, 1966. With permission.)

The significance of the various lines in Figure 1 connecting observations along the same estuary can be envisaged by the following example. For two-layered flow in a uniform channel it follows from Equation 1 that

$$\frac{U_s}{U_f} \approx \frac{U_b}{U_f} = \left\{\frac{gh \cdot \delta\rho/\rho_0}{gD \cdot \delta\rho_0/\rho_0}\right\}^{-1} \tag{2}$$

in which case the proposed circulation parameter is a nondimensional, inverse measure of the square of the speed of long internal waves. For the case of a uniform channel referred to above it follows (from Equation 2) that

$$\frac{\delta\rho}{\rho_0} = C\left(\frac{U_b}{U_f}\right)^{-1} h^{-1}$$

where C is a constant. Thus, in double logarithmic representation of the expression above, departure from the straight line (slope −1) reflects the influence of a varying thickness of the upper layer.

Recognizing that for deep estuaries the square of the interfacial Froude number F_i can be written:

$$F_i^2 = \frac{U_b^2}{gh\,\delta\rho/\rho_0} = \frac{R^2\left(\dfrac{\delta\rho_0}{\rho_0}\right)^2}{gh^3\left(\dfrac{\delta\rho}{\rho_0}\right)^3} \tag{3}$$

we see that

$$F_i^{-2/3} = \left(\frac{R^2}{g}\left(\frac{\delta\rho_0}{\delta_0}\right)^2\right)^{-1/3} \cdot h\,\delta\rho/\rho_0 \tag{4}$$

varies as $h\delta\varrho$. It also follows from Equations 1 and 4 that

$$U_b = C_1\,F_i^{2/3} \tag{5}$$

where C_1 is a constant.

Recent development in fjord dynamics[7] based upon the model developed by Long[8] has shown that for low-discharge fjords of constant width, $h_2\varrho$ is nearly constant along the fjord in its inner reaches. Application of this result together with Equation 1 gives by elimination of $\delta\varrho$

$$h = C_2\,U_b$$

and by elimination of h

$$\delta\rho = C_3\,U_b^{-2}$$

where C_2 and C_3 are constants. The latter result implies that in double logarithmic representation the Long model leads to a −2 slope of $\delta\varrho$ vs. U_b or employing Equation 5, that $\delta\varrho = C_4F_i^{-4/3}$, leading to a −4/3 slope vs. V_i. These results are at variance with those of the continous model of Hansen and Rattray,[6] where the corresponding line of $\delta\varrho$ vs. U_b (δS vs. U_s) has a slope of −1.

It should be noted, however, that the relations above are highly sensitive to changes of width of the estuary. For these cases it is necessary to compute the complete solution to the Long model (e.g., Gade and Svendsen[7]).

Attempts have been made to relate the stratification energetically to the dominant process of mixing. Ippen and Harleman[9] introduced a stratification number defined as proportion of consumed tidal energy to the increase of potential energy of the stratification in the estuary. The stratification number thus defined is hence some overall inverse flux Richardson number.

For an estuary in the form of a uniform channel of length L and depth D the stratification number suggested by Ippen and Harleman[9] can be shown to be related to the tidal velocity U_t, the fresh-water discharge ratio R and the brackish water to fresh-water admixture ratio P by the following expression

$$M = \frac{U_t^2/D\ L}{g\delta\rho/\rho h/2\ (P-1)\ R/\nu} \tag{6}$$

where h is the characteristic thickness of the brackish layer and ν the kinematic viscosity.

In terms of the flow ratio $B = U_f/U_t$ and a densimetric Froude number

$$F_m = \frac{U_f}{\sqrt{gh\,\delta\rho/\rho}}$$

introduced by Hansen and Rattray,[6] the stratification number can formally be represented by

$$M = \frac{2F_m^2 \ L/D}{B^2 \ (P-1) \ Re} \tag{7}$$

where the net flow per unit width is measured by the corresponding Reynolds number Re.

Another, equivalent form of the stratification number M is obtained by introducing the speed U_d of the long internal wave and the relative thickness $\eta = h/D$ of the upper layer:

$$M = \frac{2 \ U_t^2 \ L/D \ (1-\eta)}{U_d^2 \ Re \ (P-1)} \tag{8}$$

This expression becomes particularly simple where the thickness of the brackish layer is but a small fraction of the total depth. The usefulness of the stratification number M in such cases is, however, doubtful because of the minor role of the tides in providing energy for the estuarine circulation. It is evident from the expressions above that the stratification number M involves parameters characterizing the forcing of the system by the tides and the fresh water discharge as well as dependent parameters measuring the stability and strength of the estuarine circulation.

Attempts to combine the various types of estuaries into a single all-comprising system of classification beyond what is mentioned above have not been successful. Work toward this aim is presently going on and there is hope for some progress in this field in the near future.

III. PRINCIPAL FEATURES OF ESTUARINE MIXING AND CIRCULATION

A. Zero and First-Order Circulation Modes

Precipitation and run-off from land into an estuary cause the level of the surface to rise so as to slope downward toward the sea. Associated with the slope there will be a pressure gradient driving a flow seaward. The cross-sectional average or net flow may appropriately be termed the zero order circulation and accounts for the transport of fresh water through the estuary.

In the general case the pressure gradient resulting from the sloping sea surface in the estuary is balanced by frictional, intertial, and baroclinic pressure forces associated with the internal field of density. For two-dimensional steady flow the mathematical formulation of the force balance in the horizontal direction may be written

$$u \frac{\partial u}{\partial x} + w \frac{\partial u}{\partial z} = \frac{1}{\rho} \frac{\partial \tau}{\partial z} - \frac{g}{\rho} \left[\rho s \frac{\partial \xi}{\partial x} + \frac{\partial}{\partial x} \int_{-z}^{0} \rho dz \right] \tag{9}$$

where $\delta\xi/\delta x$ denotes the longitudinal slope of the surface, τ the frictional shear stress, and ϱ_s the surface water density. For the other symbols see the list of notations following this chapter.

In shallow estuaries, particularly those vertically well mixed, the zero order circulation may dominate other semipermanent circulatory features. This means that the last term on the right hand side of Equation 9, the baroclinic component of the pressure gradient, is likely to be of less magnitude than the second term at all levels in the estuary. Even if this should not be the case in the lower levels, the frictional forces together with the barotropic pressure gradient may counteract the effect of the baroclinic pressure field. This will normally require reversal of the curvature of the velocity profile.

Usually the internal pressure gradient arising from the longitudinally varying salinity is sufficiently strong to establish density currents transporting sea water up the estuary. Such semipermanent counter currents are invariably maintained by redistribution of density by vertical mixing. The principal or normal mode of density currents in an estuary is the compensation current, which flows up the estuary as a consequence of the mixing of fresh or brackish water with more saline water below. The strength of this first-order circulation mode may become quite remarkable, accounting for transports up to 20 times the net fresh-water flow in the estuary.

As the lower part of the first-order estuarine circulation, the compensation current will normally comprise all levels which are affected by mixing with water from the brackish layer. This generally means that the compensation current extends to the bottom of the estuary, even in the deeper ones and fjords. However, in the case of deep estuaries the compensation current has a maximum shortly below the brackish current of the upper layer, the velocity dimishing rapidly to practically zero at greater depths. This means that the pressure gradient associated with the internal density field here more or less completely balances the effect of the sloping surface.

B. Higher-Order Circulation Modes

Multilayered currents are frequently observed in estuarine environments. In most cases we are faced with transient processes responding to changing density conditions offshore and causing intrusions of sea water at various levels. As a stationary feature, multilayered currents appear as second or higher-order circulation modes superimposed on the basic estuarine circulation.

Such higher-order circulation modes develop as results of the formation of new water masses which tend to flow out of the estuary at their appropriate density levels. The mixing itself constitutes a drain on the source waters above and below giving rise to two additional compensation currents. Most often the upper one may not be distinguishable from the normal flow of sea water below the brackish surface current, or from that of the neighboring circulation cell in the case of multilayered currents. In this manner the higher-order circulation pattern is bound to consist of an even number of currents.

Estuarine circulation modes higher than third are rarely identified as stationary features. The velocities tend to be weak and to be masked by the principal estuarine circulation and transient fluctuations. Moreover, a rather dense array of current meters would be required for their detection.

The causes of higher-order circulation are likely to be found in the coupling of tidal currents with an irregular topography. Either through the generation and decay of internal waves or simply by direct interaction with submarine ridges or other irregularities, tidal energy is being transferred to turbulence localized to specific points or levels in the estuary. Similarly, internal seiches arising from sudden winds or atmospheric pressure disturbances may have such effects but without necessarily causing a sustained long-term mixing and circulation.

So far we have been concerned with steady-state features of estuarine flows. In general, estuaries are subject to time-dependent barotropic and baroclinic disturbances entering from the sea. The most prevalent of these are of oscillatory nature as, for example, the tides. The estuarine circulation must then be dealt with in terms of time averages and departures from these. An introduction to the topic is given in Volume I, Chapter 4, Section II.A.

Oscillatory disturbances of the type mentioned above are important sources of energy for mixing in the estuary. This is particularly the case with the tides which give rise to appreciable currents. The tidal velocities are commonly comparable with that of the estuarine circulation, and in many shallow estuaries very much stronger, ranging up several orders of magnitude higher than the mean flow.

The conversion of tidal energy to turbulence derives principally from three different mechanisms. The most universal is the generation of boundary (bottom) turbulence which in shallow water may easily affect the entire water column. Less common, but nevertheless generally well known, are examples of breaking or spilling of the tidal wave as it advances up the estuary. The breaking follows from convergence of tidal energy to the extent that the wave form becomes dynamically unstable. The process is recognized by a steepening of the wave crest which eventually transforms into a highly turbulent front or bore. Such convergence of tidal energy is caused by shoaling and narrowing of the estuary. Both conditions are greatly enhanced by the river flow which tends to slow the propagation of the tide. In the extreme case of supercritical flow the tidal energy becomes more or less completely trapped.

The third mechanism of importance for conversion of tidal energy lies in the generation of internal waves. Internal waves of tidal period develop as a baroclinic response of stratified waters to interaction of the surface tide with the topography. The process is favored by the existence of submarine edges and ridges and particularly by sills in fjords and estuaries. Also, high-frequency internal oscillations are frequently encountered in estuarine environments and are generally attributed to shear flow instability. Short-period oscillations are also sometimes seen coupled to the propagation of the internal tide. This phenomenon appears as stable waves trailing the crest of the internal tide and has been explained as an effect of nonlinear coupling between the internal tide and the mean flow.

As the bottom conditions inside estuaries rarely favor reflection of internal waves, the energy so trapped is eventually transformed into turbulence. The actual processes by which this is accomplished are essentially the same as those mentioned above for the surface wave. Both bottom turbulence and breaking against the shoaling bottom are commonly recognized in connection with internal waves. Furthermore, internal waves tend to interact with the shear flow of the estuarine circulation so as to cause billow turbulence and, occasionally, internal bores.

In most estuarine environments mixing processes also contribute to the horizontal dispersion of matter, particularly longitudinally, either by direct lateral exchange or by the so-called shear effect, a combined effect of steady or oscillatory flow with vertical or transverse eddy diffusion. Although of considerable importance for the dispersion of matter, longitudinal eddy transports tend to reduce the horizontal gradient of salinity and thereby weaken the (first-order) circulation in the estuary. Wherever the degree of stratification is high, longitudinal gradients of salinity are mainly small and the corresponding eddy fluxes of salt of minor importance.

IV. THE COASTAL PLAIN ESTUARY

The coastal plain estuary appears generally as a submerged extension of a river valley opening toward the sea. Quite often the coastal plain estuary may comprise not only the river valley proper, but also a wider depression of the area (Gulf of St. Lawrence), thus forming a wide bay. In both cases coastal processes may have caused the formation of more or less narrow entrances, following sedimentary processes in connection with longshore transports. Tidal currents, if not the estuarine circulation itself, act to maintain the opening to the sea in a steady balance.

Except for specific geomorphological reasons the coastal plain estuary does not have a pronounced sill near the mouth. We shall therefore not be concerned with basins under this heading, but return to the specific problems of basins in connection with fjords and estuaries with dynamic control.

Most coastal plain estuaries are rather shallow with maximum depths generally less than 50 m. The actual horizontal extent of the estuary noticeably influenced by sea water may be up to the order of 100 km, with the James River (Virginia) as a classical example (about 45 km).

Estuaries react to oceanic tides with variations of water level so as to correspond roughly with that of the adjacent open ocean. During the flood, this normally involves the entry of a volume of sea water constituting some fraction of the tidal prism, itself defined as the total volume of water represented by the elevation of the surface from low water to the following high water. The fresh water discharged during the same half tidal cycle makes up the rest. The ratio between the two fractions is decisive for the nature of the estuary. Their relative importance is described by the flow ratio, which expresses the ratio of fresh water discharged during the tidal cycle to the volume of the tidal prism.

Besides being a source of sea water the tides contribute energy to the mixing in the estuary. Since the energy associated with the tidal wave depends upon the square of the elevation, the flow ratio (or rather the inverse of it) is no measure of the energy potentially available per unit volume of fresh water discharged into the estuary. Moreover, the actual amount of tidal energy converted to turbulence (and thereby contributing to the mixing) is of the order (see Equation 6) of

$$\frac{U_t^2}{D} \propto \frac{H^2}{D^3}$$

where H is the height of the tidal prism (tidal height) and D the characteristic depth of the estuary. The topography, including the bottom and shoreline irregularities, is therefore of prime importance in determining the hydrographic characteristics of the estuary. Nevertheless, it has been shown that coastal plain estuaries with flow ratios less than 0.1 are mostly well mixed, having typical vertical salinity differences less than 1 ppt.[10]

The stratification in the estuary and the associated zero and first-order circulation modes are of utmost importance for the dispersion and transporting capacities of coastal plain estuaries. The topic will be dealt with by traditional subdivision in the following.

A. The Vertically Well-Mixed Estuary

It is an open question whether vertically well-mixed estuaries really exist in the sense that there is no detectable first-order circulation. Nevertheless the designation may be appropriate in some applications where the vertical range of salinity is but a small fraction of the total range of salinity in the estuary. Otherwise, the dynamics of the well-mixed estuary does not differ significantly from that of the weakly stratified estuary. The baroclinic flow regime of the residual currents is only weakly determined by the stratification. In both cases the estuary is influenced decisively by the tidal currents of the region and in the study of transports of pollutants we can safely combine the two groups.

B. The Weakly Stratified Estuary

It may be convenient to refer to estuaries with a vertical range of salinity everywhere less than 10 ppt and a vertical salinity gradient generally less than 1 ppt/m as weakly stratified. This arbitrary definition may cause some conflict, but should otherwise be helpful for a quick assessment of a specific estuary. The condition of being weakly stratified is normally brought about by fairly intense vertical mixing inside the estuary. With low discharge rates of river water it is possible to obtain this condition by wind-induced mixing. Otherwise, in areas of tides, tidal currents acting in shallow water are normally held responsible for the necessary generation of turbulence.

In the weakly stratified estuary the strength of the first-order circulation is possibly less than or of the same order of magnitude as that of the zero-order mode. Longitudinal dispersion of a pollutant will therefore depend on both processes as well as on eddy exchange. The latter is greatly enhanced by the so-called shear effect, in which longitudinal dispersion is brought about by vertical or lateral diffusion combined with shear in oscillatory or steady flow. The shear effect has been found to account for up to 90% of longitudinal dispersion in estuaries.

1. Longitudinal Dispersion by the Shear Effect

The importance of shear for the dispersion of a constituent in turbulent flows was first recognized by Taylor[11] who examined the shear effect for flow in pipes. In oceanography the shear effect was first considered by Bowles et al.[12] and further investigated by Bowden,[13] Okubo and Carter[14] and Okubo.[15] The concept of vertical shear has been generalized by Kullenberg[16] with application to coastal water.

The principal mechanisms of the shear effect can be visualized by considering the dispersion of a contaminant in flow in a two-dimensional channel with vertical shear. If the contaminant is in the form of a cloud, as if released from a point source, the vertical shear will cause the cloud to stretch longitudinally while eddy diffusion will tend to homogenize the concentrations along the vertical. With respect to steady flow the shear effect is fairly obvious. The net effect of dispersion as measured by averaging the concentrations in the vertical is, however, dependent on the vertical distributions of the contaminant, of the longitudinal velocity, and of the associated eddy diffusivity. With respect to oscillatory flow the shear effect is far from trivial. For very low values of the vertical diffusivity, as for instance in laminar flow, the longitudinal dispersion is negligible, growing initially linearly, however, with increasing vertical diffusivity. At a certain point, a maximum is reached beyond which the effect of net longitudinal dispersion decreases with increased diffusivity.[13]

Analysis of the shear effect follows from application of a characteristic subdivision of the field variables into time and space averages and departures from these. Thus, if the bar denotes the time average over a tidal period, the tilde the average taken laterally across the estuary, and the bracket the average along the vertical, the instantaneous value of a field variable q can be written

$$q = \langle \tilde{\bar{q}} \rangle + q_1\,(z) + q_2\,(y,z) + q_t\,(y,z,t) \qquad (10)$$

where q_1 and q_2 are the first- and second-order spatial departures and q_t the departure from the time average, as defined by the following equations;

$$q_1\,(z) = \tilde{\bar{q}}\,(z) - \langle \tilde{\bar{q}} \rangle \qquad (11)$$

$$q_2\,(y,z) = \bar{q}\,(y,z) - \tilde{\bar{q}}\,(z) \qquad (12)$$

$$q_t\,(y,z,t) = q\,(y,z,t) - \bar{q}\,(y,z) \qquad (13)$$

Applied to the longitudinal velocity component u and the concentration s of a constituent, the mass flux density Q_s, averaged over a vertical section in the estuary and the tidal period, can be composed of at least four terms as follows:

$$Q_s = <\tilde{\bar{u}}><\tilde{\bar{s}}> + <u_1 \; s_1> + <\widetilde{u_2 \; s_2}> + <\widetilde{\overline{u_t \; s_t}}> \qquad (14)$$

Here, on the right-hand side, the first term represents the flux contribution of the net flow in the estuary. The second and third terms represent the contributions from the vertical (estuarine) and the horizontal circulations, respectively, and the last term is the contribution by oscillatory and turbulent eddy fluxes.

The importance of the shear effect in estuaries has been analysed theoretically by Bowden[13] with respect to vertical shear in oscillatory flow and steady circulation. Fischer[17] has expanded the theory to include horizontal shear associated with the horizontal circulation. Further penetration of the topic of shear in estuaries has been made by Dyer,[18] Talbot and Talbot,[19] Fischer,[20] and Smith.[21,22] For an elementary presentation of the subject the reader is referred to Officer.[2]

2. Practical Approaches to the Study of Transporting Capacities of Weakly Stratified Estuaries

a. Flushing Time Evaluation

The most primitive approach to the study of transporting capacities of weakly stratified estuaries is the evaluation of flushing time, either by means of the total river flow through the estuary, or by the volume of sea water entering the estuary during the tidal period. The flushing time, defined as the volume of the estuary divided by the water flux rate, is a measure of the residence time of the water in the estuary and is a useful figure for assessing load capacities. The application of the concept to the sea water entering the estuary with the tide rests on the assumption that the water in the estuary is being mixed completely during the tidal cycle.

b. Methods of Segmentation

The requirement of complete mixing in the estuary during a tidal cycle is only rarely satisfied. As a solution to this problem Ketchum[23] proposed a method in which the estuary was divided into a number of segments, each determined by the length of the tidal excursion. The assumption of complete mixing could then be applied to each segment more realistically and the corresponding budget equations for the transfer of fresh water, salt, or any pollutant be established. From these equations it was possible to compute the flushing time for any segment and for the estuary as a whole.

The actual rate of exchange between neighboring segments is determined by the relative amounts of water entering the segment during flood. These can easily be obtained from tidal and river discharge data applied to the geometry of the estuary. It is also possible to seek the corresponding exchange parameters as solutions to a longitudinal dispersion problem, for instance by application to the longitudinal distribution of salinity.[24,25] The derived exchange parameters can then be applied to problems of dispersion of pollutants. The method of segmentation has been applied to a number of estuaries with reasonable success. The reader is referred to papers by Ketchum and Kean[26] (Bay of Fundy), Preddy[27] (Thames estuary), and Neal[28] (Columbia River). It appears that satisfactory results depend on the number of segments being fairly high, as evident from the study of the Severn estuary by Stommel.[24]

Another approach also resting in principle on segmentation was taken by Gade[29] who considered exchanges between fjord basins connected by narrow sounds. In these cases the river flow during the tidal cycle was but a small fraction of that of the tidal currents during flood. A set of linear first-order equations expressing flux rates of any constituent could then be established. The set was easily reduced to one linear equation of order n, n being the number of segments or sections present. The method was applied to the three innermost basins of the Oslofjord with connection to a fourth, the Dramsfjord. For each of the adjoining basins, as for the system as a whole, residence times or, rather, half-life times for the concentration of any initial state were derived.

c. Methods Involving Simple Diffusion Models

An entirely different approach to the problem of assessing the transport capacity or dispersion in vertically well-mixed and weakly stratified estuaries was introduced by Arons and Stommel.[30] With the concept of a vertically well-mixed estuary in mind they considered the long-term steady state as a balance between advection of salt with the river flow and eddy transports caused by tidal currents. As with Ketchum's segmentation method the transporting effect of the estuarine circulation was ignored or assumed incorporated in the diffusivity term. In fact the same argument applies to the third term on the right hand side of Equation 14, whereas only the last term actually represents eddy fluxes. The method involves formation of cross-sectional averages over a tidal cycle so as to reduce the three-dimensional dispersion problem to the one-dimensional:

$$U \cdot S = K \frac{\partial S}{\partial x} \qquad (15)$$

Here, x is the longitudinal coordinate, U and S the respective sectional average velocity and salinity, or any other similarly distributed conservative property. K is the coefficient of longitudinal diffusion, later to be appropriately named "apparent", as it includes both the shear effect and that of the estuarine circulation. Its value can easily be determined from field observations at sufficiently dense intervals in time and space all along and across the estuary. It is more convenient to use a postulated form of K, which varies proportionally with the tidal velocity and the tidal excursion. These two are expected to be linearly related so that it should be satisfactory to derive the tidal excursion from the topography, knowing the tidal elevation. The problem then reduces to the determination of the constant β in the equation $K = \beta\xi^2_0$, where ξ_0 is the tidal excursion.

For simple one-dimensional diffusion from a point source in a tidal channel it can be shown that the coefficient β has the value $(6T)^{-1}$, where T is the tidal period (see Officer[2]). The relationship between the eddy diffusivity and the tidal excursion derived by Arons and Stommel[30] has also been confirmed theoretically by Bowden[13] for the shear effect associated with vertical shear. Fisher[20] points out that transverse shear effects may be as important as that of vertical shear and expands the theory to include circulation in the horizonal and vertical planes. Particular attention is given to the transverse circulation, otherwise much neglected in the literature on estuarine circulation.

Arons and Stommel[30] applied the theory of one-dimensional diffusion to a uniform channel with a rectangular cross section and derived analytical solutions. Comparison with field data showed reasonable agreement (see Figure 2), even in the case of Alberni Inlet which does not belong to the weakly stratified estuaries but indeed has a strong first order circulation mode.

The method outlined above can be applied to long-term stationary dispersion problems provided the pollutant is conservative. Depending upon the nature of the problem at hand, the apparent coefficient of diffusion can be determined either from measurements obtained in connection with test releases or from analysis of the distribution of salinity in the estuary. In the former case it is advised to check the approach by per-

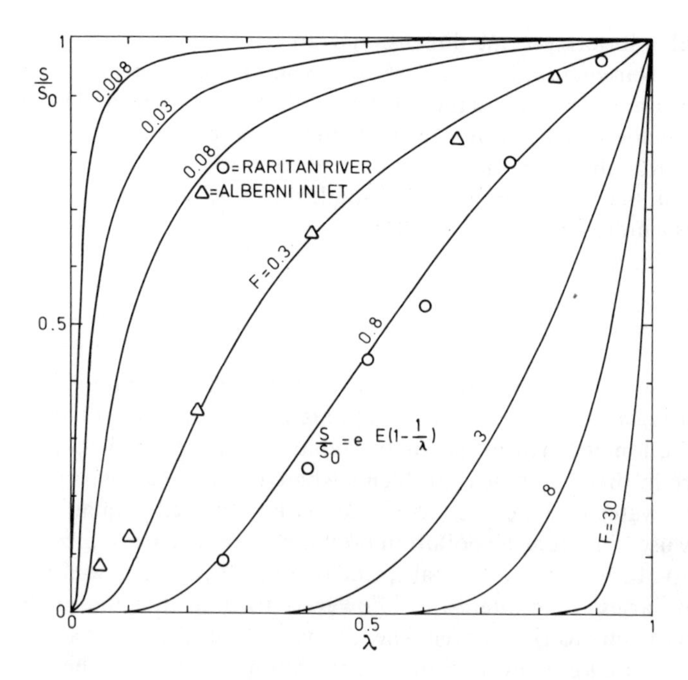

FIGURE 2. Theoretical solutions of the diffusion-advection equation (Equation 15) showing relative salinity vs. relative distance γ (= x/L) from the inner end of a uniform channel for selected values of a flushing number E, related to the river discharge, the height of the tidal prism, and the intensity of the mixing in the estuary. O, Raritan River; Δ, Alberni Inlet. (From Arons, A. B. and Stommel, H., *Trans. Am. Geophys. Un.*, 32, 421, 1951. With permission.)

forming integration of the corresponding salinity distribution problem. The coefficient of diffusion thus derived can then be applied to the appropriate equation for dispersion of a pollutant discharged into the river or at the head of the estuary:

$$Uq - K\frac{\partial q}{\partial x} = Q/A \tag{16}$$

Here, Q is the source strength and q the sectional concentration average of the pollutant in the estuary. A is the cross-sectional area of the transporting layer.

For problems involving nonconservative substances it is necessary to take into account the production or consumption. This is most conveniently handled by adding a decay term to the corresponding second-order equation. The topic is dealt with further in Section D of this chapter.

Application of Equation 16 is limited to steady-state problems. The more general case of time-dependent problems requires use of the corresponding second-order differential equation

$$\frac{\partial q}{\partial t} = -U\frac{\partial q}{\partial x} + \frac{\partial}{\partial x}\left(K\frac{\partial q}{\partial x}\right) \tag{17}$$

where in general, K = K(x,t) must be determined as before by application of observations. However, most time-dependent problems in estuaries are on time scales sufficiently long to allow the apparent eddy diffusivity to be assumed locally constant (K = K(x)). The validity of this assumption should in each case be checked in respect of the fortnightly tidal spring/neap variation.

Analytical solutions to one-dimensional diffusion from a point source have been discussed by Harleman[31] for the case of constant apparent diffusivity. Both steady (Equation 16) and finite (Equation 17) sources are considered. The discussion is extended to include problems of stratified flow, and reference is made to results from experiments in a laboratory flume.

With an estuary of varying width in mind, the cross-sectional averaging leads to the following as a more appropriate form:

$$A \frac{\partial q}{\partial t} = - \frac{\partial}{\partial x} \left(A\ Uq - A\ K \frac{\partial q}{\partial x} \right) \qquad (18)$$

This case has been dealt with by Kent,[32] who suggested that in the cases of spontaneous or transient releases the coefficient of apparent diffusion be reduced in a ratio determined by the longitudinal extent of the pollution in the estuary. This was clearly of consequence in time-dependent problems where complete similarity with the distribution of fresh water in the estuary could rarely be expected. Equations 17 and 18 have been widely used in estuarine pollution problems. For a general introduction the reader is referred to Harleman,[33] who deals principally with vertically well-mixed and slightly stratified estuaries. It is interesting, however, to note that the method can also be applied successfully to fjords even when, as is normal, they have a strong haline stratification and a predominant first-order circulation. Apart from the study of the distribution of salinity in the Alberni Inlet mentioned above, the method has been applied to the transport of phosphates in the brackish layer of the Oslofjord, predicting the longitudinal concentration distribution with a fair degree of accuracy.[34] The results of the computations are shown in Figure 3 and comparison is made with observations in January, during which the low ambient temperature and poor daylight conditions justify the assumption that phosphate was a conservative constituent.

d. Methods Involving Layered Circulation

The simplicity of the segmentation method introduced by Ketchum[23] and of the balance equation of Arons and Stommel[30] (or the corresponding second order equation employed by Kent[32]) has made them attractive for application to pollution problems. Nevertheless, wherever the estuarine circulation (first-order mode) is a predominant feature, it cannot easily be assumed incorporated, but should appear separately in the budget equations.

The topic has been dealt with by a number of authors since the time of Martin Knudsen,[35] who formulated overall budget equations for estuaries and basins. Probably the approach most relevant to pollution problems is that of Pritchard,[36] who established budget equations for a segmented two-layer estuary. From analysis of field data, Pritchard[37] found that in many examples the longitudinal eddy diffusion could be safely neglected and he proposed budget equations for each segment considering the advection in both layers, the vertical eddy exchange between them, and the flux of water from the lower into the upper layer. An element of Pritchard's box model estuary is depicted in Figure 4.

Basing his argument upon the assumption that the concentration of a pollutant is constant throughout the segment and equal to the average at the boundary between adjacent segments, Pritchard[37] established budget equations for the pollutant throughout the estuary. With the help of corresponding equations for salinity and application of appropriate boundary conditions at the inner and outer limits of the estuary, the concentration of the pollutant could be computed for any of the specified segments.

The case of a continuously stratified estuary with a pronounced first-order circulation is considerably more complicated. For a complete model of the mass budgets,

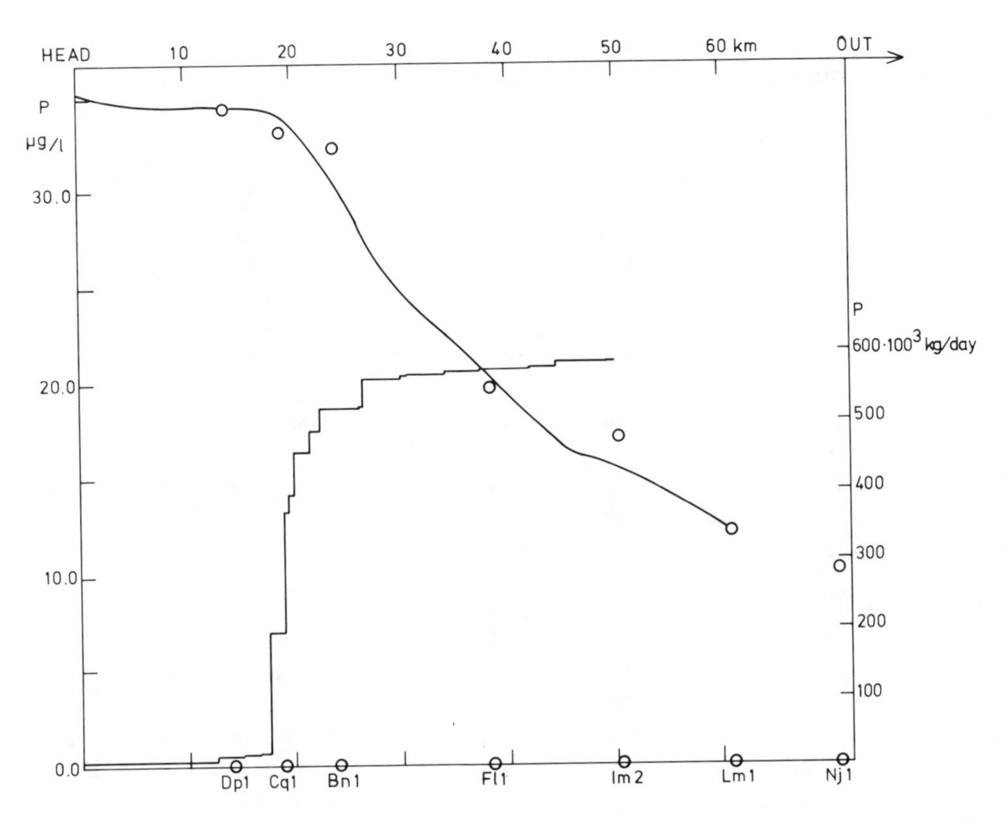

FIGURE 3. Computed (heavy line) and observed (dots) upper-layer phosphate concentrations (P) vs. distance along the Oslofjord. Also longitudinal distribution of discharge rates of phosphates (thin line). Note the impact of the city of Oslo at St. Cq 1.[34]

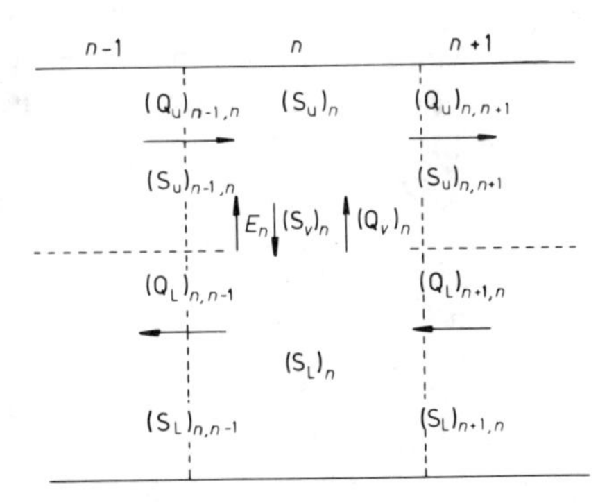

FIGURE 4. Flow diagram for the upper and lower box of number n in a two-dimensional model of estuarine circulation with vertical mixing. S, salinity; Q, volume flux rate; E, exchange coefficient for vertical diffusion. Indices u and l refer to upper and lower layer, respectively, v denotes the value pertaining to vertical fluxes. After Pritchard.[36] (Reproduced from Dyer, K. R., *Estuaries: A Physical Introduction,* John Wiley & Sons, London, 1973, 116. with permission.)

detailed knowledge of the current distribution is required. To obtain this by direct methods is usually quite costly and certainly not always feasible. The investigator may therefore choose to support his model by solutions of the dynamical circulation problems.

The dynamics of estuarine circulation is only adequately expressed by nonlinear equations. However, certain idealized situations can be approximated by linear equations and solutions can be sought by analytical or numerical methods. A fundamental problem in such approaches is the determination of the eddy diffusivity tensor and the corresponding eddy viscosity. This problem has no general solution.

A common approach is to assign certain values to the exchange coefficients, either constant or variable in space. This is also the case with the similarity model introduced by Rattray and Hansen[38] and later presented in more developed forms.[39,40] The equations established for two-dimensional flow and the corresponding density field were found to have semi-analytical solutions showing similarity along the longitudinal axis of the estuary, provided that certain boundary conditions were satisfied. By adjusting the exchange coefficients to fit observed distributions of density (salinity), the model is in principle applicable to problems of transport or dispersion of contaminants.

C. The Salt Wedge Estuary

The salt wedge estuary has many features in common with the fjord. Both are highly stratified and the dynamics of the flow is determined by entrainment of water from the lower into the upper layer. This situation is created whenever tidal mixing is weak or absent, allowing the sea water to flow up the estuary as far as to be nearly in hydrostatic balance with the fresh or brackish water above. The interface between the sea water and the brackish water is thus practically level making the sea water form a wedge against the generally sloping bottom of the estuary. In deep water, as in fjords, the flow of sea water compensating for the water entrained into the brackish layer meets very little resistance. Under these conditions the interface tends to rise toward the head of the estuary.[7,8]

In the more shallow estuaries the compensation current is confined in a relatively thin layer between the brackish water and the bottom. The higher resistance associated with higher velocities along the bottom and along the interface will in this case dominate the dynamic balance of the lower layer and cause the interface to slope downward toward the head of the estuary. A ''salt wedge'' will thus form even in a flat-bottomed estuary.

Distributions of contaminants in the brackish layer in salt wedge type estuaries follow much the same patterns as in fjords. Normally the pollutant is likely to follow the flow of the fresh water in the estuary with its dispersion mainly confined to the brackish layer. Some recirculation is bound to take place, the strength of which can be judged by the amount of dilution of the sea water in the salt wedge.

Properties of the salt wedge estuary have been discussed by Farmer and Morgan[41] and Ippen and Harleman.[9] A summary of salt wedge processes is given by Ippen.[42] For a more general introduction to the subject the reader is referred to Dyer[1] with a brief account of the Vellar estuary, and to Officer.[2] Insight into the mechanisms contributing to the eddy fluxes of mass and momentum in salt wedge estuaries has been gained by the extensive field measurements of Partch and Smith[43] and Gardner and Smith.[44]

D. Active and Nonconservative Constituents

With high concentration of a pollutant the density of the fluid may become sufficiently influenced to change the flow pattern in the estuary. This may particularly occur whenever effluents contain emulsified or suspended matter form such as is often released from milling industries. In more extreme cases local phenomena such as turbidity currents may be recognized. However, with the exception of thermal pollution, the overall behavior of the estuarine circulation, which depends upon the large-scale distribution of density, is rarely directly affected by effluents carrying matter introduced by man.

In highly stratified systems the flow of sea water up the estuary by the compensation current is often essential for the renewal of water in the estuary. It is therefore important to recognize that contamination of the deeper layers in the outer parts of the system may cause pollution in the entire estuary. The most common examples are seen in connection with submerged sewage outfalls, industrial or sewage dumping, and dredging. Dredging may be particularly harmful through the sudden release of large quantities of heavy metals deposited with recent sediments.

Pollutants which sink or float cannot adequately be described by the systems for conservative constituents presented in the preceding sections. Likewise, if a constituent is subject to conversion, reaction or dissociation, another term in the diffusion equation will be required. At the concentrations most commonly found in estuarine systems it is generally assumed that the decay rate is proportional to the concentration and thus appropriately described by a term of the form $-kq$, where k is the specific decay constant, and q the concentration.

The problem of dispersion of a nonconservative constituent in vertically well-mixed estuaries has also been dealt with by means of the modified tidal prism method. A brief discussion of nonconservative constituents is given by Officer.[2] The reader is furthermore referred to the papers by Ketchum et al.,[45] Ketchum,[46] Stommel,[24] O'Connor,[47,48] Thomann,[49] and Dyer.[1]

V. ESTUARIES WITH DYNAMIC CONTROL. OVERMIXED ESTUARIES

In estuaries having a constriction near the mouth or a sudden widening, for instance at the connection to the sea, stationary two-layered estuarine circulation is subject to certain restrictions known as dynamic control. Application of vertically integrated momentum equations to each layer in the constriction gives, under the assumption of stationary conditions, the critical flow condition

$$\frac{U_1^2}{gh_1\ \delta\rho/\rho_2} + \frac{U_2^2}{gh_2\ \delta\rho/\rho_2} = 1 \qquad (19)$$

where U_1 and U_2 are average velocities for each layer, h_1 and h_2 the thicknesses, and $\delta\varrho$ the density difference between them. The index 2 refers to the lower layer and also identifies its density ϱ_2.

The critical flow condition together with the equations of continuity of salt and mass give a relation of the form

$$f(\eta, Fe, P) = 0 \qquad (20)$$

where P is the sea water to fresh water admixture ratio, η is the thickness of the upper layer relative to the total depth of the constriction, and Fe is a Froude number measuring the rate of discharge of fresh water to the estuary. The problem was first approached by Stommel and Farmer,[50] who suggested that under certain conditions the flux of salt water in the estuarine circulation had an upper limit which could not be exceeded. When this limit was reached the estuary was said to be in an overmixed state or, simply, overmixed. Selected solutions to Equation 20 are shown in Figure 5.

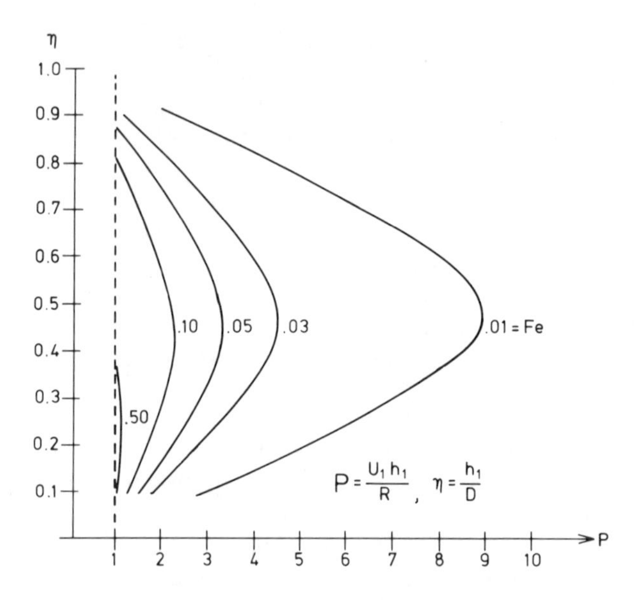

FIGURE 5. Relative thickness η of the upper layer in a constriction with critical flow as function of the admixture ratio P for selected values of the Froude number Fe measuring the fresh water flow through the estuary. (Reproduced with permission from Stigebrandt, A., *Stationar Tvalagerstromning i estuarier,* River and Harbour Laboratory, Trondheim 1975, 14. With permission.)

The significance of an estuary being overmixed can be studied by seeking the maximum of the salinity or of the volume flux, for example in terms of P, the admixture ratio, as suggested by Stigebrandt.[51] For estuaries with a relatively small fresh-water discharge rate the overmixed state is characterized by the two layers flowing through the constriction having approximately equal thicknesses.

In an overmixed estuary (or basin) the pressure forces driving the flow in the two layers are at their maximum. These pressure forces are related to the density differences outside and inside the estuary. That a maximum is reached implies that the water inside the estuary is essentially homogeneous down to a level below the sill of the constriction. No further vertical mixing will alter this condition.

According to Stommel and Farmer[50] a majority of estuaries around the world are overmixed. It is also true of a number of basins with shallow sills such as the Baltic, the Black Sea, and the Mediterranean Sea.

Tides entering estuaries with a pronounced first-order circulation may effect the transporting capacity of the two-way flow. Stigebrandt[52] has discussed this phenomenon with respect to overmixed estuaries and has shown that the transporting capacity can be expected to increase substantially if the amplitude of the barotropic tidal current velocity is significantly greater than the speed of the long internal wave.

The question of possible overmixing is relevant to problems of pollution when considering artificial means for increasing circulation in an estuary. Apart from bottom topography modification by dredging or excavation the most easily available means are submerged releases of fresh water or compressed air. For systems which are not overmixed, relatively small rates of energy input can often be sufficient to reduce the stratification and thereby enhance the estuarine circulation. Otherwise such effort is futile.

VI. THE FJORD-TYPE ESTUARY

Fjords are generally much deeper than coastal plain estuaries. This permits the existence of a relatively large body of semi-enclosed sea water below the brackish upper layer. Often one or several submarine ridges across the fjord form sills which hinder the free exchange of the deep water between the fjord basin and the adjacent coastal water. The deep water of a fjord therefore has very much a life of its own, but interacts with the upper layers in several ways which will be dealt with in the following. It is possible to group the mechanisms contributing to the exchange of heat and matter in a typical fjord as follows:

1. Estuarine circulation in fjords
2. Intermittent and alternating density currents
3. Frictionally driven flows
4. Eddy exchange processes

A. Estuarine Circulation in Fjords

In a fjord the fresh water mixes with sea water to form a brackish layer flowing slowly toward the sea. The type of stratification resulting from this process depends very much on the rate of fresh-water discharge and the general climatic conditions prevailing in the fjord. Mostly a fairly homogeneous brackish layer is observed with a transition to the sea water below. In the outer reaches of the fjord this transition often occupies an increasingly greater part of the upper layers, making the concept of a vertically homogeneous brackish layer less valid. A detailed classification of fjord stratification is given by Pickard.[53]

The forces responsible for maintaining the flow of brackish water toward the sea follow from the pressure field associated with the sloping free surface. The gradual rise of the free surface toward the head of the fjord is small, but measurable. An example of computed surface slope from the Sognefjord is given in Figure 6.

The vertical exchange between the brackish layer and the sea water below is very often dominated by entrainment of lower water into the brackish layer. This indicates that the brackish layer is turbulent, at least intermittently.

The dynamics of two-layered flow in fjord type circulation was first formulated by Stommel and Farmer,[54] and later developed by Long[8] who presented a complete model. The model is discussed by Gade and Svendsen[7] and McClimans and Mathisen.[55] Application to field observations and laboratory experiments indicates that beyond the entrainment and acceleration of sea water, interfacial boundary friction is not a predominant feature of estuarine dynamics in fjords. Inherent in the model is also the simple expression that $\delta\varrho h^2$ is nearly constant in the inner reaches of a fjord of constant width (see also Volume I, Chapter 1).

The ideal assumptions made in the Stommel/Long model are hardly fulfilled in nature. There is thus growing evidence[56-58] that the compensation current in many fjords is noticeably affected by brackish water from above, so that the undercurrent in the inner reaches of the fjord carries a significant portion of entrained brackish water. In this sense we are faced with recirculation of appreciable amounts of fresh water, a fact which has considerable significance in relation to the transport of dissolved and suspended matter.

Although the major part of the first-order circulation is contained within the zone of immediate influence of the brackish current, virtually all levels in a fjord are affected by eddy exchange with the upper layers. As discussed in Section D.1 the process is greatly enhanced by the existence of internal waves which generate turbulence by breaking, presumably against the sloping bottom. In this manner the estuarine circulation may be said to comprise the entire water mass in the fjord.

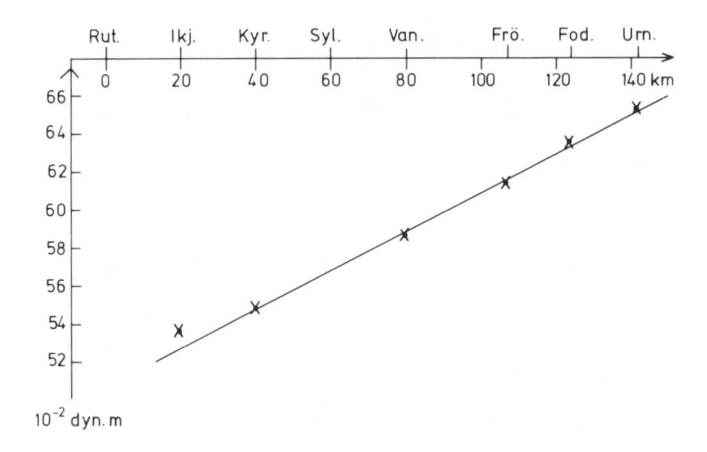

FIGURE 6. Longitudinal slope of the Sognefjord as it appears from the dynamic height of the surface. The horizontal axis covers nearly the full length of the fjord from the mouth (left) to the head (right of the diagram).[57]

With the very restrictive proviso that the density stratification of the coastal water adjacent to a fjord remain perfectly steady, sea water will enter the fjord and descend to the greatest depths at all times. The flow of sea water will then propagate at a speed determined by the vertical eddy diffusion in the basin water chiefly according to the equation

$$\nabla \cdot (\rho \underline{v}) = \frac{\partial}{\partial z}\left(K_z \frac{\partial \rho}{\partial z}\right) \tag{21}$$

where $\underline{v}$ is the time averaged velocity vector.[59] Equation 21 differs from the normal form of the continuity equation by including a diffusivity term, which in this case also incorporates eddy fluxes. Lateral diffusion is neglected here because of its minor role in the transport of mass in the interior of the stratified basin water (see also Section VI. D.2). Thus, the relation above may not be valid for the immediate vicinity of the bottom current of descending sea water where the lateral terms may be as important as the vertical.

B. Intermittent and Alternating Density Currents

Unless the constriction at the sill renders the fjord overmixed or the rate of vertical exchange is extremely low (which may be the case in the fjords with low tides) the deep water renewal will generally be sufficient to keep the basin water well aerated. Completely steady conditions of the adjacent coastal water are, however, hardly met in nature. Normally, seasonal and other fluctuations of the density field offshore interfere with the inflow of sea water to the fjord. With reduced density of the water available above sill depth the compensation current weakens or may for periods vanish altogether. In that case the basin water becomes stagnant. On the other hand, increased density of the coastal water may accelerate the intrusion of sea water to the fjord to the extent that the inflow reaches the maximum corresponding to overmixed conditions.

The alternation between circulating and stagnant conditions in basins of the type discussed above is not solely determined by the periodicity (seasonal or otherwise) of the fluctuations of the coastal water, but responds to the actual density of the intruding water. Thus, deep water renewal of particularly high density may cause extension of the stagnant interval until the density has been reduced to more "normal" levels by the vertical diffusion. In this way the extension may appear as multiples of the seasonal or yearly period. Such behavior is observed in the deeper water in the Baltic Sea and in many Norwegian fjords.

The cyclic behavior of the deep water in estuarine-type basins, alternating between active and stagnant states, has received considerable attention in the literature, presumably because of the often dramatic biological consequences. The reader is referred to Gade[34,60] and Gade and Edwards.[59] The first of the cited papers is a study of the Oslofjord where the regularity and separation of the two processes is remarkable, In the second paper the statistical behavior of such systems is analyzed theoretically with respect to frequency and density of complete deep-water renewals. The last of the mentioned papers is mainly a review of historical and current literature concerning deep-water renewals in fjords.

More rapid fluctuations of the density field in the coastal water are seen to propagate easily into fjords. This behavior is particularly noticeable within the sea-water phase where rise or fall of the isopycnals is followed by exchange of considerable quantities of water. The pycnocline separating the brackish water from the sea water below is also subject to disturbances of level, but because of the greater stability is is less influenced by external agents and more readily reassumes its equilibrium position.

The water exchange associated with the more rapid disturbances of the type described above has not received particular attention in the literature, but is usually considered as another form of horizontal eddy exchange. Among the more conspicuous seasonal events are the spring intrusions to the deep water in the fjords of southern Norway following the rise of Atlantic water along the coast. The opposite process of inflow of coastal water at higher levels with corresponding outflow below has also been recognized as a recurrent major event in the fall. Both processes are linked to the seasonal shift of longshore winds, being predominantly southerly in the winter and northerly in the summer. Similar phenomena are recognized in North American west coast fjords.[59]

C. Frictionally Driven Flows

It has long been recognized that in coastal waters local winds may generate surface currents of exceptional strength in the sense that the surface velocities by far exceed those of open waters under comparable winds. During the last decade these observations have been corroborated by measurements in various degrees of stratification.[34,61,62] The measurements indicate strongly that a frictionally balanced winddrift depends largely upon the static stability, as one would expect from the influence of the static stability on the eddy viscosity. However, the measurements exhibit a low degree of coherence, a fact consistent with the generally poor fulfillment of steady-state conditions.

The static stability of the stratification not only reduces the eddy viscosity, but inhibits free vertical circulation along bordering coastlines. Instead, an internal pressure field is established which tends to guide the currents along the coast. Also this favors higher velocities of the wind drift, as may be seen by comparison with unrestricted open ocean flow. It can be shown by application of simple Ekman theory that the horizontal stress balance in a steady wind-driven two-dimensional current is maintained by a constant velocity gradient of

$$\frac{\partial V}{\partial z} = \pi \sqrt{2}\ \frac{V_E}{d}$$

where V_E is the corresponding surface velocity of the Ekman drift and d is the Ekman depth with the assumption of identical eddy viscosity. With a linear velocity profile originating at the Ekman depth, two-dimensional wind drift would thus attain a surface velocity more than four times that of an Ekman current under similar conditions. It should be noted however, that longitudinal barotropic and baroclinic pressure gradients build up fairly quickly in an estuary or fjord and within a few hours modify the velocity field.

As a consequnce of the wind action the stratification will gradually break down so that the eddy viscosity in the affected layers will increase substantially. Nevertheless, during the initial stages the surface water is likely to attain appreciable velocities. Measured velocities in the range of 10% of the wind velocity are not uncommon.[62]

D. Diffusive Processes

The nonadvective transports within estuarine systems have traditionally been referred to as eddy or turbulent transports. These designations will be retained although it is generally recognized that in most forms of horizontal dispersion the effective mechanisms originate in interaction between turbulence and mesoscale advective processes.

1. Vertical Eddy Diffusion

Measurements of vertical fluxes of temperature and mass indicate that above the pycnocline in typical Norwegian fjords the eddy diffusivity is predominantly related to the action of the winds. The pronounced influence of the vertical stability, discussed by Gade,[34] Rye,[62] and Kullenberg,[16] should not be ignored. The topic is dealt with further in Volume I, Chapter 1. There is, however, a clear discrepancy between determinations based upon budget methods and direct measurement with artificial tracers released as point sources.[63] Stigebrandt[64] has explained this discrepancy for the deep water as a result of localized turbulence caused by breaking of internal waves of tidal origin. This explanation may not apply above the pycnocline, where the mixing appears to be mainly caused by wind action. Below the pycnocline the effect of the wind is greatly reduced. In some fjords the deep water therefore remains relatively inactive whereas in others it may not.

Investigations of the deep water of the inner Oslofjord have revealed fairly intense vertical mixing even during stagnant periods. This mixing is also remarkably steady throughout all seasons of the year, a fact which points toward the tides as the only possible source of energy for the process. In this case the mechanism is the generation of internal waves of tidal period at the sill and their subsequent conversion to turbulence.[34] Stigebrandt[65] presented a linear model for generation of such internal waves and also suggested that the conversion to turbulence is effectuated by breaking of the waves against the sloping bottom.

From heat and salt budgets of selected fjord basins, overall vertical eddy diffusion coefficients have been established during stagnant periods. Thus, Gilmartin[66] reported from Indian Arm, a British Columbia fjord, values ranging from 0.5×10^{-4} m^2 s^{-1} to 10^{-3}m^2 s^{-1}, but mostly between 10^{-4} and 3×10^{-4} m^2 s^{-1}. Corresponding values computed for Norwegian fjords cover mainly the same range, but with a tendency toward the lower end. The lowest known vertical diffusivities in fjords are of the order 10^{-6} m^2 s^{-1}, and have been computed from observations in Framvaren (southern Norway), a shallow silled fjord practically free of tides.

The computed eddy diffusivities are found to vary with the static stability. In both the innermost basins of the Oslo-fjord their dependency on the vertical stability is quite clear, being related to the Brunt-Väisälä frequency N according to the equation

$$K_z = a N^{-2\alpha} \tag{22}$$

where α is a nondimensional constant close to 0.8, and a is another constant which, however, differs widely from basin to basin.[34] Examples of computed diffusivities vs. N^2 are shown in Figure 7 in a double logarithmic representation. Determinations of vertical diffusivities from other fjords support the relationship, but with different values of a and α. Thus, Gøransson and Svensson[67] found $\alpha = 0.6$ for the Byfjord (Gothenburg), and values of α about 0.5 have been reported by Aure.[68]

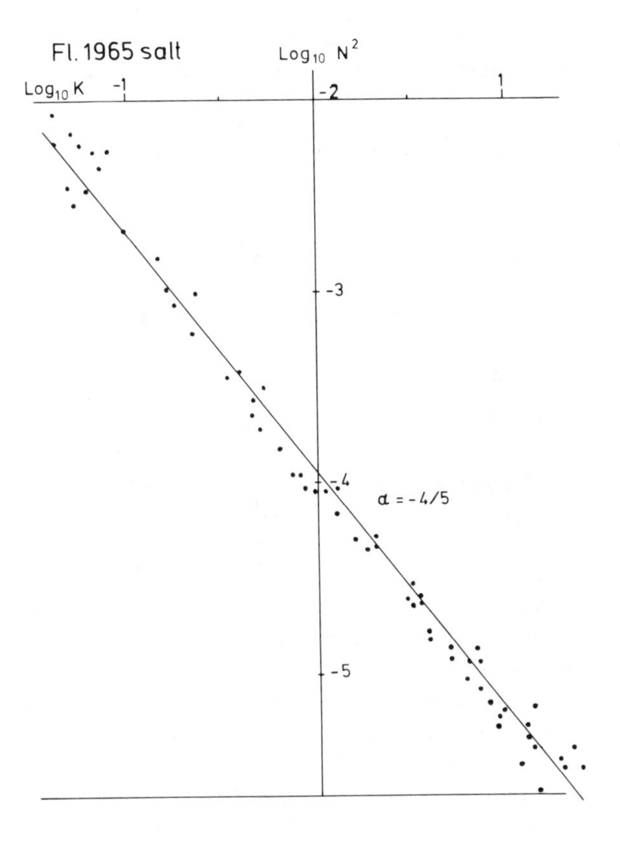

FIGURE 7. Overall vertical diffusivity, K, (cm^2s^{-1}) in the Vestfjord (Oslofjord) vs. static stability, N^2, s^{-2} in double logarithmic representation.[34]

2. *Horizontal Eddy Diffusion*

In the early models of salt and heat budgets for typical fjords,[29,38] it was assumed that horizontal eddy diffusion of salt was negligible in comparison with advection. This idea was based on the observation that the stratification in the upper layers of fjords was generally close to hydrostatic balance and that sustained lateral differences were diminutive. It was thus clearly implied that advective processes were overwhelmingly responsible for the lateral transports. This did not mean that horizontal eddy diffusion of other constituents was of no importance in fjords, merely that the horizontal eddy fluxes of density could in most cases be neglected.

Constituents distributed differently from the salinity must be expected to be subject to lateral dispersion to which eddy diffusion contributes significantly. It is, however, essential to distinguish between the local diffusivity and the apparent diffusivity, the latter usually referring to vertically integrated fluxes, including that of the estuarine circulation.

As in most estuarine systems the shear effect is important in fjords. In this respect the fjords differ from the more shallow estuaries in that the shear effect is not significantly related to the barotropic motion, except for areas of highly irregular shorelines or in constrictions. However, considerable effect must be attributed to the estuarine circulation and other baroclinic processes. Thus, internal waves can be expected to contribute significantly wherever present. The importance of the shear effect in fjords has been demonstrated by experiments in the Oslofjord[34] where purely horizontal eddies have been found to contribute less than 10% to the lateral dispersion of dye.

Table 1
SOME DISPERSION TESTS IN DISTORTED FROUDE MODELS

Basin	Distortion	Artificial roughness	Model disp: prototype disp.	Ref.
San Francisco Bay	10	Yes	0.5	78
San Diego Bay	5	Yes	0.75	79
Bergensfjorden	25	No	0.55	80
Seto Inland Sea	12	No	0.5	81
Brouwershavense Gat	3 37½	Yes Yes	<10% difference, except for the small scales	82

After McClimans and Gjerp.[69]

VII. MODELING OF DISPERSION AND TRANSPORTS IN ESTUARINE SYSTEMS

A. Scale Models

The concept of modeling applies to physical and mathematical models, both of which have found application in transport and dispersion problems of estuaries. The attraction to physical models has been obvious, particularly in the early stages of modeling before a deeper insight into estuarine processes was developed. However, because of the difficulties with scaling of the generally distorted physical models, the attempts have mostly been limited to Froude models of estuaries where tidal processes are predominant. McClimans and Gjerp[69] have compiled the following table listing the effectiveness of the dispersion relative to the prototype. Other Froude models have been reviewed by Sugimoto.[70]

In general the models give less dispersion than observed in nature, but the agreement is remarkably good considering that only the tidal currents are simulated.[69] It has been argued, however, that with the distortion the relative importance of lateral and vertical eddies is disturbed and therefore difficult to model adequately.[71,72] Similar criticism may also apply to a varying degree of stratification.

B. Mathematical Modeling

By far the majority of attempts to model estuarine processes has comprised mathematical methods ranging from simple box models and diffusion/advection models expressing conservation of mass (see Section IV.B.2) to more advanced dynamic models invoking the laws of physics. As the more complex problems rarely can be dealt with by purely analytical methods, the modelers have been obliged to seek solutions by analog computation or by numerical methods. The numerical modeling may imply that solutions are obtained by finite difference techniques of integration or that the field processes are simulated by a set of equations involving unknown coefficients expressing exchange rates of momentum and mass. Similarity with nature is then sought by adjustment of the coefficients (systems analysis) to give the best possible fit (optimization).

By far the majority of numerical models designed for simulation of estuaries is barotropic, dealing mostly with tidal currents and related dispersion phenomena. However, in recent years there seems to have a been a breakthrough of models dealing with the more complex baroclinic processes and the number of reports of successful modeling of specific cases is growing steadily.

Among the first complete numerical models to include baroclinic effects is that of Leendertse and Liu.[73] The presented scheme is, however, designed for (and applied to) processes on rather short time scales, ranging from hours to a few days. Computed examples show barotropic and baroclinic responses of various estuaries and coastal bays to tidal forcing. Attempts to deal with longer term processes, such as the salinity balance in estuaries and basins, have been presented by Svensson and Wilmot[74] who considered the two-way flow through Øresund (connecting the Baltic with Skagerak).

The particular considerations and techniques used in numerical modeling are outside the scope of the present chapter. The reader is referred to papers by Fischer,[75] Abbott et al.,[76] and Leendertse and Liu.[73] Modeling of branched flow such as is commonly found in delta areas has been discussed by Bennet.[77] The topic is also covered by several articles in *Hydrodynamics of Estuaries and Fjords*.[4]

ACKNOWLEDGMENT

The author is indebted to Dr. Anton Edwards for reading through the manuscript and offering valuable suggestions.

NOTATIONS

A	Area of vertical cross section
B	Flow ratio ($= U_f/U_t$)
$C, C_{1,2,3,4}$	Constants
D	Characteristic or total depth of estuary
d	Ekman depth
Fe	Froude number characterising fresh water flow in estuary
F_i	Interfacial (densimetric) Froude number
F_m	Densimetric Froude number according to Rattray and Hansen[6]
g	Acceleration of gravity
H	Tidal height (height of tidal prism)
h, h_1	Thickness of upper (brackish) layer
h_2	Thickness of lower layer
K	Coefficient of eddy diffusion, also apparent longitudinal eddy diffusivity
K_z	Vertical eddy diffusivity, also laterally or time averaged diffusivity
L	Total length of estuary
M	Stratification number
$m_{1,2,3}$	Quantities of mass
N	Brunt-Väisälä frequency
P	Admixture ratio ($= U_1h_1/R$)
Q	Source strength of pollutant (mass per unit time)
q	Concentration of pollutant
R	Fresh-water discharge or net flow per unit width
Re	Reynolds number of fresh-water (net) flow in estuary
S	Salinity
S_o	Salinity of lower layer, bottom, or source water
δS	Salinity difference between layers
δS_o	Vertical salinity difference at head of estuary
t	Time
U, U_f	Average velocity through vertical section in estuary
U_1, U_b	Average velocity of upper (brackish) layer
U_2	Average velocity of lower layer
U_{bo}	Average velocity of brackish layer at head of estuary
U_d	Speed of long internal wave, critical relative velocity of layers
U_t	Peak or rms velocity of tidal flow

U_s	Average surface velocity
V_E	Surface speed of Ekman spiral
u	Longitudinal velocity component
w	Vertical velocity component
x	Horizontal, longitudinal coordinate, sometimes measured from head of estuary
z	Vertical coordinate
ξ_o	Tidal excursion
η	Relative thickness of upper layer (h/D)
ζ	Surface elevation
α	Nondimensional coefficient
ϱ	Density
ϱ_s	Surface density
$\delta\varrho$	Density difference between layers
$\delta\varrho_o$	Density difference between layers at head of estuary
ν	Coefficient of kinematic viscosity
τ	Shear stress

REFERENCES

1. **Dyer, K. R.,** *Estuaries: A Physical Introduction,* John Wiley & Sons, London, 1973.
2. **Officer, C. B.,** *Physical Oceanography of Estuaries,* John Wiley & Sons, New York, 1976.
3. **Lauff, G. H.,** *Estuaries,* Publ. 83, American Association for the Advancement of Science, Washington, D.C., 1967.
4. **Nihoul, J. C. J.,** *Hydrodynamics of Estuaries and Fjords,* Elsevier, Amsterdam, 1978.
5. **Freeland, H. J., Farmer, D. M., and Levings, C. D.,** *Fjord Oceanography,* Plenum Press, New York, 1980.
6. **Hansen, D. V. and Rattray, M., Jr.,** New dimensions in estuary classification, *Limnol. Oceanog.,* 11, 319, 1966.
7. **Gade, H. G. and Svendsen, E.,** Properties of the Robert R. Long model of estuarine circulation in fjords, in *Hydrodynamics of Estuaries and Fjords,* Nihoul, J. C. J., Ed., Elsevier, Amsterdam, 1978, 423.
8. **Long, R. R.,** Circulations and density distributions in a deep, strongly stratified, two-layer estuary, *J. Fluid Mech.,* 71, 529, 1975.
9. **Ippen, A. T. and Harleman, D. R. F.,** One-Dimensional Analysis of Salinity Intrusion in Estuaries, Tech. Bull 5, Committee on Tidal Hydraulics, Corps of Engineers, Vicksburg, Miss., 1961.
10. **Simmons, H. B.,** Some effects of upland discharge on estuarine hydraulics, *Proc. Am. Soc. Civil Eng.,* 81, 742, 1955.
11. **Taylor, G. I.,** The dispersion of matter in turbulent flow through a tube, *Proc. R. Soc. London Ser. A* 223, 446, 1954.
12. **Bowles, P., Burns, R. H., Hudswell, E., and Whipple, R. T. P.,** Sea disposal of low activity effluent, Proc. 2nd Conf. Peaceful Uses of Atomic Energy, Vol. 18, U.N., Geneva, 1958, 376.
13. **Bowden, K. F.,** Horizontal mixing in the sea due to a shearing current, *J. Fluid Mech.,* 21, 83, 1965.
14. **Okubo, A. and Carter, H. H.,** An extremely simplified model of the "shear effect" on horizontal mixing in a bounded sea, *J. Geophys. Res.,* 71, 5267, 1966.
15. **Okubo, A.,** The effect of shear in an oscillatory current in horizontal diffusion from an instantaneous source, *Int. J. Oceanol. Limnol.,* 1, 194, 1967.
16. **Kullenberg, G.,** An Experimental and Theoretical Investigation of the Turbulent Diffusion in the Upper Layer of the Sea, Rept. 25, Institut for Fysisk Oceanografi, Copenhagen, 1974.
17. **Fischer, H. B.,** Mass transport mechanisms in partially stratified estuaries, *J. Fluid Mech.,* 53, 671, 1972.
18. **Dyer, K. R.,** The salt balance in stratified estuaries, *Estuarine Coastal Mar. Sci.,* 1, 411, 1974.
19. **Talbot, J. W. and Talbot, G. A.,** Diffusion in shallow seas and in English coastal and estuarine waters, *Rapp. P. V. Reun. Cons. Int. Explor. Mer,* 167, 93, 1974.

20. **Fischer, H. B.**, Mixing and dispersion in estuaries, *Annu. Rev. Fluid Mech.*, 8, 107, 1976.
21. **Smith, R.**, Longitudinal dispersion of a buoyant contaminant in a shallow channel, *J. Fluid Mech.*, 78, 677, 1976.
22. **Smith, R.**, Coriolis, curvature and buoyancy effects upon dispersion in a narrow channel, in *Hydrodynamics of Estuaries and Fjords*, Nihoul, J. C. J., Ed., Elsevier, Amsterdam, 1978, 217.
23. **Ketchum, B. H.**, The exchanges of fresh and salt water in tidal estuaries, *J. Mar. Res.*, 10, 18, 1951.
24. **Stommel, H.**, Computation of pollution in a vertically mixed estuary, *Sewage Ind. Wastes*, 25, 1065, 1953.
25. **Dorrestein, R.**, A method of computing the spreading of matter in the water of an estuary, in Scientific Conf. Disposal of Radioactive Wastes, Monaco 1959, IAEA, Vienna, 1960, 163.
26. **Ketchum, B. H. and Kean, D. J.**, The exchanges of fresh and salt waters in the Bay of Fundy and in Passamquoddy Bay, *J. Fish. Res. Board Can.*, 10, 97, 1953.
27. **Preddy, W. S.**, The calculation of pollution of the Thames estuary by the theory of quantized mixing, *Int. Conf. Water Pollution Research*, Vol. 42, Pergamon Press, Oxford, 1962.
28. **Neal, V. T.**, Predicted flushing times and pollution distribution in the Columbia River estuary, in Proc. 10th Conf. Coast. Eng., Am. Soc. Civil Eng., in 1966, 1463.
29. **Gade, H. G.**, Some hydrographic observations of the inner Oslofjord during 1959, *Hvalradets Skr.*, 46, 1, 1963.
30. **Arons, A. B. and Stommel, H.**, A mixing-length theory of tidal flushing, *Trans. Am. Geophys. Un.*, 32, 419, 1951.
31. **Harleman, D. R. F.**, Diffusion processes in stratified flow, in *Estuary and Coastline Hydrodynamics*, Ippen, A. T., Ed., McGraw-Hill, New York, 1966, 575.
32. **Kent, R. E.**, Diffusion in a sectionally homogeneous estuary, *Proc. Am. Soc. Civil Eng.*, 86, 15, 1960.
33. **Harleman, D. R. F.**, Pollution in estuaries, in *Estuary and Coastline Hydrodynamics*, Ippen, A. T., Ed., McGraw-Hill, New York, 1966, 630.
34. **Gade, H. G.**, Hydrographic Investigations of the Oslofjord, A Study of Water Circulation and Exchange Processes, Geophys. Inst. Rep. 24, Bergen, Norway, 1970.
35. **Knudsen, M.**, Ein hydrographischer Lehrsatz, *Ann. Hydrographie Maritimen Meteorol.*, 28, 316, 1900.
36. **Pritchard, D. W.**, Dispersion and flushing of pollutants in estuaries, *Proc. Am. Soc. Civil Eng.*, 95, HY1, 115, 1969.
37. **Pritchard, D. W.**, A study of the salt balance in a coastal plain estuary, *J. Mar. Res.*, 13, 133, 1954.
38. **Rattray, M., Jr. and Hansen, D. V.**, A similarity solution for circulation in an estuary, *J. Mar. Res.*, 20, 121, 1962.
39. **Hansen, D. V. and Rattray, M., Jr.**, Gravitational circulation in straits and estuaries, *J. Mar. Res.*, 23, 104, 1965.
40. **Rattray, M., Jr.**, Some aspects of the dynamics of circulation in fjords, in *Estuaries*, Lauff, G. H., Ed., Publ. 83, Am. Assoc. Adv. Sci., Washington, D.C., 1967, 52.
41. **Farmer, H. G. and Morgan, G. W.**, The salt wedge, Am. Soc. Civil Eng., Proc. Third Conf. Coast Eng., 54, 1953.
42. **Ippen, A. T.**, *Estuary and Coastline Hydrodynamics*, McGraw-Hill, New York, 1966.
43. **Partch, E. N. and Smith, J. D.**, Time dependent mixing in a salt wedge estuary, *Estuarine Coastal Mar. Sci.*, 6, 3, 1978.
44. **Gardner, G. B. and Smith, J. D.**, Turbulent mixing in a salt wedge estuary, in *Hydrodynamics of Estuaries and Fjords*, Nihoul, J. C. J., Ed., Elsevier, Amsterdam, 1978, 79.
45. **Ketchum, B. H., Ayers, J. C., and Vaccaro, R. F.**, Processes contributing to the decrease of coliform bacteria in a tidal estuary, *Ecology*, 33, 247, 1952.
46. **Ketchum, B. H.**, Distribution of coliform bacteria and other pollutants in tidal estuaries, *Sewage Ind. Wastes*, 27, 1288, 1955.
47. **O'Connor, D. J.**, Oxygen balance of an estuary, *Am. Soc. Civil Eng. Sanit. Eng. Div. J.*, 86, SA3, 35, 1960.
48. **O'Connor, D. J.**, An analysis of the dissolved oxygen distribution in the East River, *J. Water Pollut. Contr. Fed.*, 38, 1813, 1966.
49. **Thomann, R. V.**, Systems analysis and water quality management, Environmental Research and Applications, New York, 1972.
50. **Stommel, M. H. and Farmer, D.**, Control of salinity in an estuary by a transition, *J. Mar. Res.*, 12, 13, 1953.
51. **Stigebrandt, A.**, Stationär Tvålagerstrømning i Estuarier, River and Harbour Laboratory Rep., Trondheim, Norway, 1975.
52. **Stigebrandt, A.**, On the effect of barotropic current flucuations on the two-layer transport capacity of a constriction, *J. Phys. Oceanogr.*, 7, 118, 1977.

53. **Pickard, G. L.**, Oceanographic features of inlets in the British Columbia mainland coast, *J. Fish. Res. Board. Can.*, 18, 908, 1961.
54. **Stommel, H. and Farmer, H. G.**, On the nature of estuarine circulation I, Tech. Rept. 52288, Woods Hole Oceanographic Institution. Woods Hole, Mass., 1952.
55. **McClimans, T. A. and Mathisen, J. P.**, Stationary wind mixing in a deep-silled, narrow fjord: comparison of laboratory observations with a numerical model, *Mar. Sci. Commun.*, 5, 127, 1979.
56. **Gade, H. G.**, Note of the Reverse Current in Fjords and Estuaries, Geophys. Inst. Rep. 30, Bergen, Norway, 1972.
57. **Gade, H. G.**, Transport mechanisms in Fjords, in *Fresh Water on the Sea,* Skreslet, S., Leinebø, R., Matthews, J. B. L., and Sakshaug, E., Eds., The Association of Norwegian Oceanographers, Oslo, 1976, 51.
58. **Edwards, A. and Edelsten, D. J.**, Deep water renewal of Loch Etive: a three basin Scottish fjord, *Estuarine Coastal Mar. Sci.*, 5, 575, 1977.
59. **Gade, H. G. and Edwards, A.**, Deep-water renewal in fjords, in *Fjord Oceanography,* Freeland, H. J., Farmer, D. M., and Levings, C. D., Eds., Plenum Press, New York, 1980, 453.
60. **Gade, H. G.**, Deep water exchanges in a sill fjord: a stochastic process, *J. Phys. Oceanogr.*, 3, 213, 1973.
61. **Farmer, D. M.**, The Influence of Wind on the Surface Waters of Alberni Inlet, thesis, University of British Columbia, Vancouver, 1972.
62. **Rye, H.**, A Meteorological Study of the Winds in the Langfjord Area, Port and Ocean Engineering, University of Trondheim, Norway, 1973.
63. **Bjerkeng, B., Gøransson, C. G., and Magnusson, J.**, Investigations of Different Alternatives for Waste Water Disposal from Sentralanlegg Vest, Norwegian Institute of Water Research, Rep. 013276, Oslo, 1978 (in Norwegian).
64. **Stigebrandt, A.**, Observational evidence for vertical diffusion driven by internal waves of tidal origin in the Oslofjord, *J. Phys. Oceanogr.*, 9, 435, 1979.
65. **Stigebrandt, A.**, Vertical diffusion driven by internal waves in a sill fjord, *J. Phys. Oceanogr.*, 6, 486, 1976.
66. **Gilmartin, M.**, Annual cyclic changes in the physical oceanography of a British Columbia fjord, *J. Fish. Res. Board Can.*, 19, 921, 1962.
67. **Gøransson, C.-G. and Svenssson, T.**, The Byfjord: Studies of water exchange and mixing processes, Statens naturvårdsverk, Solna, 1975 (Sweden).
68. **Aure, J. N.**, Hydrography of the Lindåspollene, thesis, University of Bergen, Bergen, Norway, 1972.
69. **McClimans, T. A. and Gjerp, S. A.**, Numerical study of distortion in a Froude model, Am. Soc. Civil Eng. Proc. 16th Int. Conf. Coast. Eng., Vol. 3, 1978, 2887.
70. **Sugimoto, T.**, Similitude of the hydraulic model experiment for tidal mixing, *J. Oceanogr. Soc. Jpn.*, 30, 260, 1974.
71. **Fischer, H. B. and Holley, E. R.**, Analysis of the use of distorted hydraulic models for dispersion studies, *Water Resour. Res.*, 7, 46, 1971.
72. **Fischer, H. B.**, Mixing and dispersion in estuaries, *Annu. Rev. Fluid Mech.*, 8, 107, 1976.
73. **Leendertse, J. J. and Liu, S.-K.**, Modeling of three-dimensional flow in estuaries, *Am. Soc. Civil Eng. Proc. Symp. Modeling Techniques,* 635, 1, 1975.
74. **Svensson, J. and Wilmot, W.**, A numerical model of the circulation in Øresund. Hydrology and Oceanography Rep., 15, SMHI, Norrkøping, 1978.
75. **Fischer, H. B.**, Some remarks on computer modeling of coastal flows, *Proc. Am. Soc. Civil Eng. J. Waterways, Harbors, Coastal. Eng.*, 102, WW4, 395, 1976.
76. **Abbott, M. B., Bertelsen, J. Aa., Kej. A., and Warren I. R.**, Systems modelling of stratified fluids, *Am. Soc. Civil Eng., Proc. Symp. Modeling Techniques,* 1, 605, 1975.
77. **Bennet, J. P.**, General model to simulate flow in branched estuaries, *Am. Soc. Civil Eng. Proc. Symp. Modeling Techniques,* 1, 643, 1975.
78. **Bailey, T. E., McCullough, C. A., and Gunnerson, C. G.**, Mixing and Dispersion Studies in San Francisco Bay, Sanitary Eng. Div. ASCE Paper 4936, SA5, 1966, 23.
79. **Simmons, H. B. and Hermann, F. A.**, Effects of proposed second entrance on the flushing characteristics of San Diego Bay, California, in *Marine Pollution of Sea Life Fishing,* News Books Ltd., London, 1970. 460-464.
80. **McClimans, T. A.**, Physical oceanography of Borgenfjorden, *K. Norske Vidensk. Selsk. Skr.*, 2, 1, 1973.
81. **Higuchi, H. and Sugimoto, T.**, Experimental study of horizontal diffusion due to the tidal current, *Rapp. P-v. Reun. Cons. Int. Explor. Mer.*, 167, 177, 1974.
82. **Abraham, G.**, Hydraulic far-field modeling, European course on heat disposal from power generation in the water environment. Delft Hydraulics Laboratory, 1975, Ch. 10B.

Chapter 5

THE EFFECTS OF WEATHER SYSTEMS, CURRENTS, AND COASTAL PROCESSES ON MAJOR OIL SPILLS AT SEA

Stephen P. Murray

TABLE OF CONTENTS

I. INTRODUCTION

A wide variety of material is dumped into the oceans and coastal waters of the world both by plan and by accident. The accidental spills of petroleum products and crude oil from ship accidents and damaged oil drilling or production platforms are, of course, the best publicized examples. Additionally, man deliberately introduces into the ocean (in order of decreasing total volume) dredge spoil, industrial wastes, sewage sludge, refuse, radioactive wastes, construction debris, military explosives, and chemical wastes.[1] The movement of oil is generally confined (if not treated with surfactants) to the surface layer of the ocean and is controlled to a large extent by the physical processes that characterize the movement of water in the surface boundary layer. In some cases treated oils from spills in well-stratified waters near river mouths, such as off the Mississippi River delta, have been observed to form subsurface plumes on the pycnocline or density interface between the upper and lower layers. The remaining materials are generally negatively buoyant; they sink to the bottom, and their movement and eventual fate are determined by bottom boundary layer processes known to us mainly from research in the field of sediment transport.

The accidental introduction of oil into the ocean is clearly on the rise as oil exploration on continental shelves intensifies and supertanker traffic increases each year.[2] Oceanic disposal of other wastes has decreased in recent years in the industrialized nations. Accordingly, this chapter will emphasize the processes controlling the spread of oil at the sea surface.

Review of the effect of weather phenomena[3] on continental shelves and coastal waters has clearly demonstrated their dominant role in forcing the motion of these waters. As the response of oil to wind forcing will be largely controlled by the type of weather characteristic of different climatic regions throughout the world, the chapter will begin with a brief summary of the time and length scales of weather systems. Next the oil dispersal problem will be discussed in terms of the physical and chemical properties of oil on the sea surface, and then the geophysical processes (currents, tides, etc.) that experience has shown us to be important will be considered. Hydrodynamic details will not be emphasized, but rather case studies from major oil spills will next be called on to demonstrate the role of basic physical processes in determining oil motion. A brief review of the major types of models currently employed to aid in understanding or predicting oil spill movement will then be presented; it will be followed by a summary and evaluation of what I see to be the role physical oceanography will play in future oil spill incidents.

II. TIME AND LENGTH SCALES OF WEATHER SYSTEMS

While local winds are well recognized as the principal driving force of oil on the sea surface, the character of the wind field as an element of a coherent, predictable weather system is largely ignored. Moreover, as major oil spills have already occurred in all latitudes, from polar regions to the tropics, an appreciation of the types of weather systems typical of the various regions of the world will greatly aid engineers and administrators responsible for planning for such emergencies.

Table 1
VOLUME-WEIGHT CONVERSIONS FOR OIL OF SPECIFIC GRAVITY 0.90 (API 30)

	Gallons	Meters^3	Barrels	Tons (long)
1 gallon =	1	0.00379	0.0238	0.00334
1 meter^3 =	264.3	1	6.29	0.885
1 barrel =	42	0.159	1	0.141
1 ton (long) =	299.8	1.13	7.14	1

The relationship between the length scale (size) of the oil spill and the length scale of the weather system acting on it is important. For example, a 1000-bbl/day (Table 1) oil spill will have a length of about 50 km. Convective cells (thunderstorms) of a 5- to 10-km scale, being smaller than the slick, will act to disperse or spread it, while larger scale systems such as mid-latitude cyclones ($\sim$1000 km) will tend to move or advect the slick intact. Weather systems smaller than the slick will act to disperse it, whereas weather systems larger than the slick will act to advect it. Thus we see that dispersive processes increase in importance with increasing slick size.

Meteorologists have long recognized concentrations of energy in the atmosphere at various time scales. Figure 1 shows a schematic distribution of familiar meteorological phenomena, most of which strongly affect coastal and shelf waters, the primary area of interest with respect to pollutant transport. At the shortest time and length scales of atmospheric motion are turbulence and gustiness (or unsteadiness), which have been extensively studied for their role in air-sea transfer processes. For present purposes turbulence can be thought of as rapid variations in wind speed and direction, commonly studied at time scales of seconds and length scales of centimeters. It is this phenomenon through the Reynolds or frictional stresses that transfers energy from the wind to the sea surface, providing the principal mechanism of oil slick movement at sea.

Owing to its rarity over water, the next larger ambient weather phenomenon, the tornado, can be safely ignored with respect to marine pollutant transport. On a slightly larger time and length scale (Figure 1), thunderstorms can, however, exert significant influence on surface water movement and hence oil slick development, in particular those small slicks on the order of 1000 bbl. The "Thunderstorm Project",[4] from innumerable field observations, has synthesized a typical thunderstorm as a multicelled system 10 to 20 km in width, each cell of which is characterized by an intense core of downdraft adjacent to a weaker region of updraft. The intense downdraft, reaching speeds of 10 m/sec just above the surface, impacts the surface and then spreads out radially at speeds of 5 to 10 m/sec. Because of the multicellular nature, there are several strong convergences and divergences present in the surface wind field. The movement of such a complex surface wind field over an oil slick would quickly tear the slick into several parts, acting as a very efficient dispersal agent. In oceanography we are accustomed to think of thunderstorm effects as being in the "noise" level, but there are vast stretches of water in the humid-tropical areas around the globe where such processes may well be dominant with respect to oil slicks. Hsu,[5] for example, describes such an intense convective activity as a dominant factor in the local weather off the Nicaraguan coast.

Differential heating of adjacent land and water in coastal areas leads to low-level atmospheric pressure gradients that produce the sea breeze-land breeze dynamical system of daily time scale (Figure 1) well known to mariners. Studies by Johnson and O'Brien[6] and Hsu[7] show the mechanics of the flow to be well understood, and Sonu et al.[8] have documented the surprisingly coherent response of coastal currents, waves,

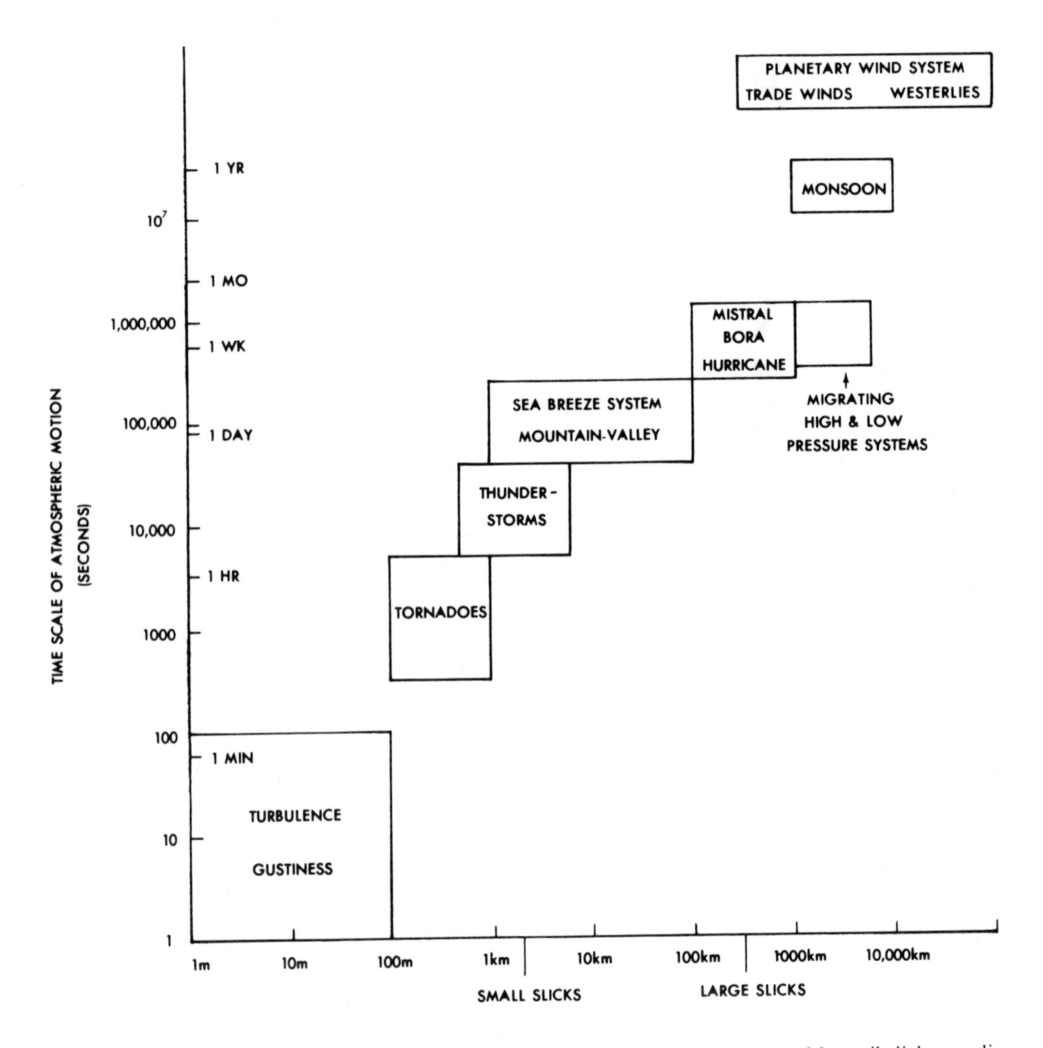

FIGURE 1. Length and time scales of atmospheric motions and weather systems. Most oil slicks are dispersed by processes of smaller length scale than the slick size and advected by processes of larger length scale than the slick size.

and beach processes to sea breeze forcing. As shown in Figure 2, onshore surface wind speeds routinely reach 5 to 7 m/sec by late afternoon, die off, and then blow offshore in the early morning hours. A lower speed, more diffuse return flow is centered about 2 km above the surface. Hsu[7] has emphasized the three-dimensional nature of the system, showing it to extend 50 km offshore and the wind velocity to rotate clockwise throughout the day under the influence of the Coriolis force. His model indicates that a daytime divergence and nighttime convergence develop in the surface layer. Thus oil spilled within 50 km of a coast subject to sea breeze forcing will eventually be driven ashore by the sea breeze (as the morning return flow is considerably weaker and short lived). A slick seaward of the surface wind divergence has a much greater chance to drift offshore, provided of course no other current mechanism dominates the flow pattern. In many areas of the world sea breeze conditions reach storm levels and probably are the dominant force driving coastal waters. Off Chile, for example, afternoon sea breezes routinely reach 15 m/sec.[9]

Diurnal heating of upper slopes of hills and mountains and subsequent rapid cooling in the evening produce pressure gradients very similar to those of the sea breeze system. This new system rises from topographic differences (steep slopes) rather than air-sea

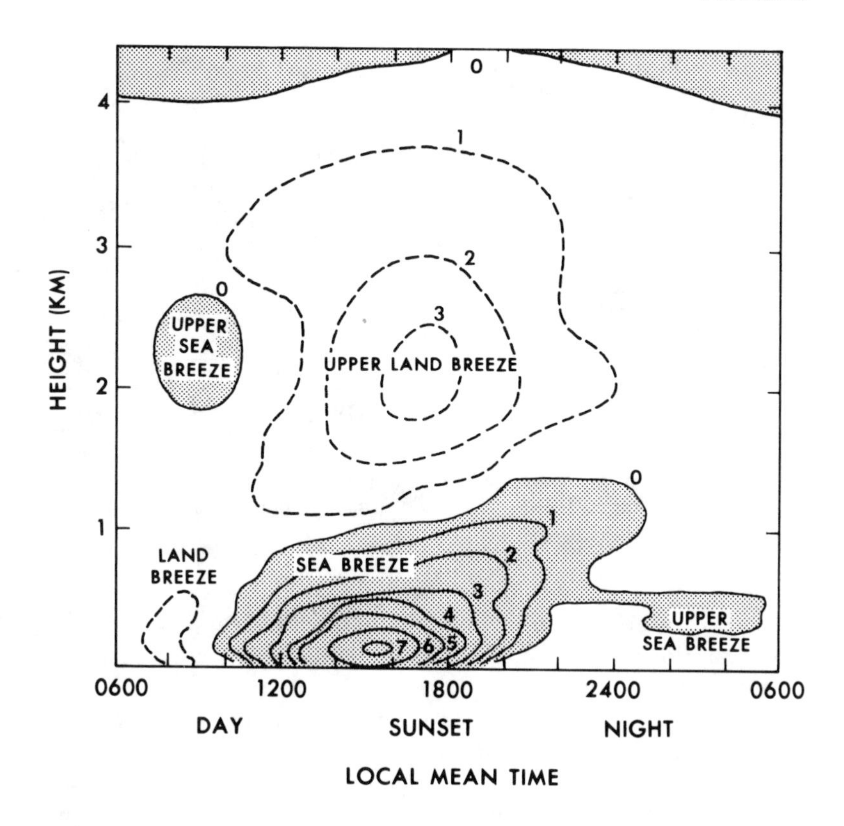

FIGURE 2. Vertical distribution of wind speeds (m/sec) on the coast of Indonesia through a daily cycle of a sea-breeze system. Strong onshore winds (shaded area) predominate during the day, and they are followed by a weak offshore wind at the surface in the evening. (After Atkinson.[9])

temperature differences and is referred to as a mountain valley wind system. Pronounced local daily circulations occur where land-sea breeze and upslope-downslope wind regimes reinforce each other. An example is the Red Sea-Gulf of Aden system illustrated in Figure 3, one of the major oil tanker routes of the world. Steep escarpments rise abruptly from the coastline in this region, and generally fair skies allow a maximum of surface heating. During the morning, downslope land breezes prevail, causing convergence over the water; by afternoon upslope sea breezes prevail, causing divergence over the water and convergence at selected locations over the land. Rainfall and vegetation patterns are known to be controlled by this phenomenon, and it is quite clear that an oil spill in the summer would tend to migrate in a southerly direction in the Red Sea and in a northeasterly direction in the Gulf of Aden, but at the same time would exhibit daily pulses of onshore and offshore movement caused by this local wind system. The Pacific coast of Guatemala and the Mediterranean coast of Asia Minor also have strong wind systems of this type and time scale.[9]

In contrast to our ignorance of the smaller scale systems, tropical cyclones (hurricanes and typhoons) are well recognized as discrete forcing agents capable of modifying the water mass characteristics and current and wave fields over distances of hundreds of kilometers.[10] Forristall,[11] using the numerical model approach, has studied the currents produced on the shelf by hurricane winds, and he predicts speeds in excess of 1.5 m/sec. Murray[12,13] has measured hurricane-driven currents in shallow coastal waters of 60 to 80 cm/sec, four times the normal value, at a distance of 200 km from the eye of the storm. The impact of the intense cyclonic winds, in excess of 125 km/hr, associated with one of these systems could undoubtedly destroy an oil slick through surface mixing and turbulence caused by whitecapping and breaking.

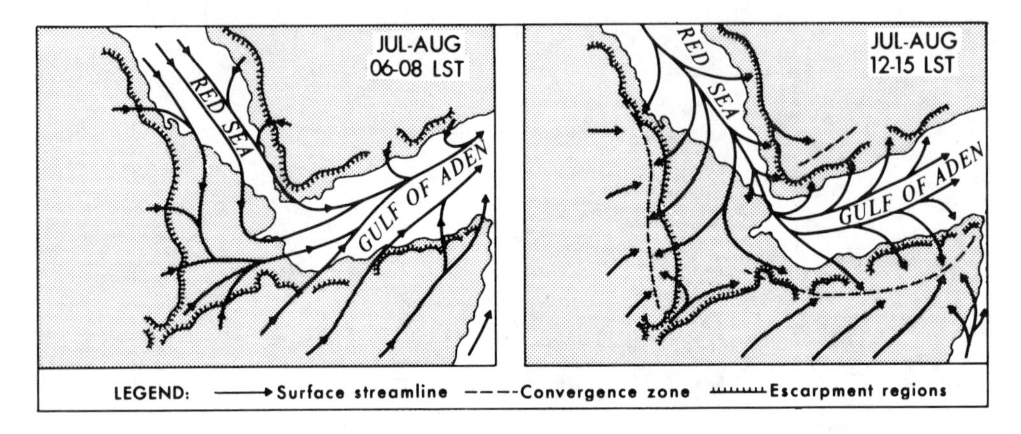

FIGURE 3. Streamlines of the surface wind flow in the mountain-valley wind system prevalent in the southern Red Sea — Gulf of Aden area. (After Atkinson.[9]) Note the onshore-offshore reversal of wind direction as a function of time of day.

Studies of large-scale local wind effects of a 5- to 10-day time scale such as the mistral have shown how these intense winds, generated by gravity driving of high-altitude cold air blowing down the Rhone and adjacent valleys, affect the deep waters of the Mediterranean over a scale of hundreds of kilometers.[14] Coastal waters must also be strongly influenced by the mistral, both directly and in response to the temperature transformation of the offshore water. Along the coast of the Adriatic Sea, the bora, a large-scale local wind effect very similar to the mistral, is well recognized as a controlling agent of local and nearshore weather. Similar gravity-driven winds, draining off the glaciers and ice caps in polar regions, are well known for nearly instantaneous initiation of severe weather conditions.

Another extremely interesting meteorological influence on nearshore waters disrupting the generally onshore northerly winds that predominate along the Egyptian coast is the khamsin, known in other areas as the sirocco. This is a dry, dusty, and generally quite hot southerly wind that blows out of the desert and is apparently related to moving low-pressure cells, as its approach is generally signaled by a sharp drop in the barometer. Khamsin conditions generally last from 1 to 3 days and as winds increase from the south, severe dust storms, with visibility sometimes reducing to only 50 m on the land, are not uncommon. Thus an oil spill in a specific area may come under strong coherent mesoscale local wind effects that will tend to push it either onshore, in the case of the sea breeze and mountain valley wind systems, or offshore, if under the influence of the bora, mistral, local cold gravity winds, or khamsin.

Operating over time scales of a day to a week but coherent over greater distances — thousands of kilometers — are the familiar high- and low-pressure systems (anticyclones and cyclones) that control the weather in much of the middle and higher latitudes, especially in the winter half of the year.

A situation fairly common in winter along the North American shores of the Atlantic and the northern Gulf of Mexico is illustrated by the weather maps for February 25 and 26, 1979 (Figure 4). The low-pressure center in the middle of Figure 4A has moved northeasterly 870 km in the previous 18 hr, as shown by the chain of arrows. Locations of the center at 6, 12, and 18 hr preceding map time are indicated by heavy dots. Winds circulate counterclockwise around the low, blowing roughly parallel to the isobars following the laws of geostrophic adjustment. In fact, they blow 20 to 30° to the left of the isobars, spiralling into the low center. Note the cold front with its accompanying sharp wind shift from southwesterly to northwesterly flow. A weak high-pressure cell is seen to the west in the upper left corner of Figure 4A , following behind

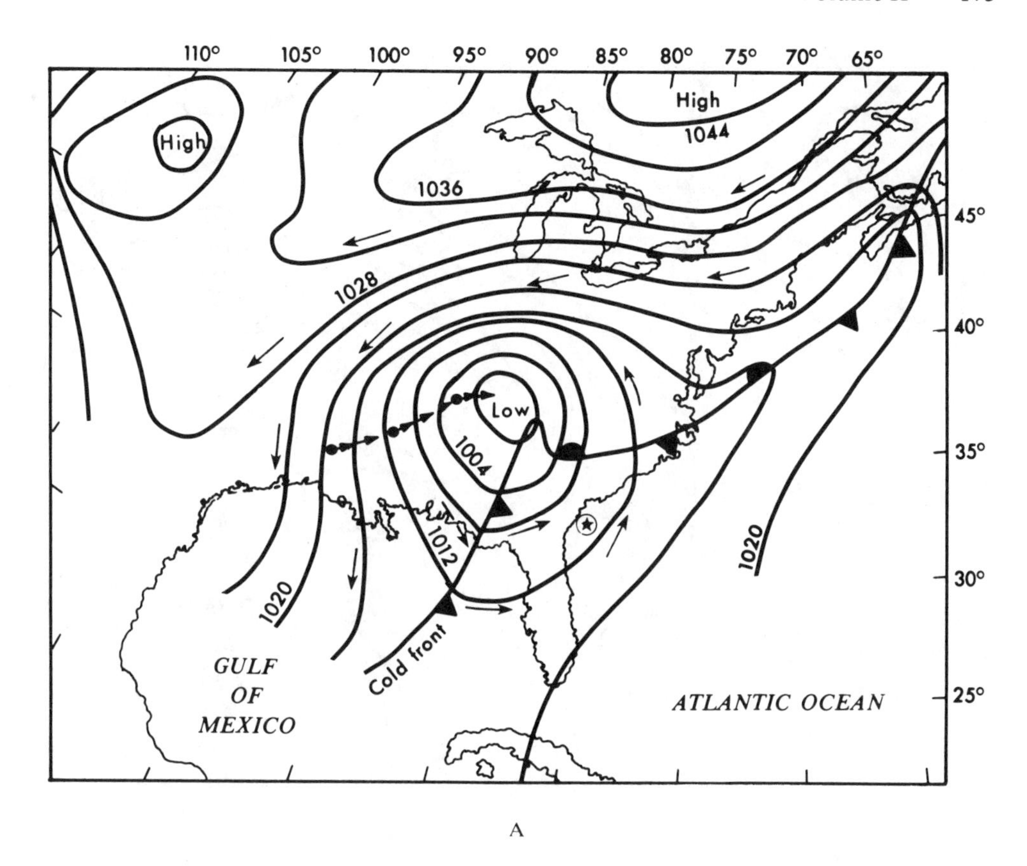

A

FIGURE 4. (A) Weather map (0700 hr) along the Atlantic and Gulf of Mexico coasts of North America on February 25, 1979. The locations of the low-pressure center 6, 12, and 18 hr before map time are shown as heavy black dots linked by arrows. The star symbol located off the Atlantic coast north of Florida is the site of a hypothetical oil spill discussed in the text and illustrated in Figure 5. Contours in millibars. (B) Weather map (0700 hr) along the Atlantic and Gulf of Mexico coasts of North America on February 26, 1979. The locations of the low-pressure center 6, 12, and 18 hr before map time are shown as heavy black dots linked by arrows. The star symbol located off the Atlantic coast north of Florida is the site of a hypothetical oil spill discussed in the text and illustrated in Figure 5. Contours in millibars.

the moving low, and a stronger one lies to the north. Twenty-four hours later (February 26, 1979) the low center has moved 600 km farther to the northeast and the trailing high has formed a broad ridge from Canada to the Gulf of Mexico, as seen in Figure 4B. On February 27 the high has moved onto the coastal belt and by February 28 dominates the Gulf Coast and entire eastern seaboard. A hypothetical oil spill in the Atlantic off the North American coast on February 24 (at the position marked with a star on Figure 4A) would be influenced by the systematic wind shifts shown in Figure 5 as the tandem high- and low-pressure systems passed up the coast. Initial southeasterly and southerly winds would veer to southwesterly and then due westerly as the center of the low passed directly to the north. On the back side of the low, northwesterly winds would rapidly change to northerly and northeasterly and then easterly as the following high was pushed off the weather map by another fast-moving low developing to the west. Thus in the course of 5 days wind-driven oil slicks would be moving off in all directions of the compass but in a systematic and predictable fashion, exhibiting a complete clockwise revolution in 5 to 7 days. If the line of centers of the pressure cells were south of the spill instead of north as in the example (which occasionally happens over North America after a severe outbreak of arctic air), then the rotation of the wind and the oil slicks would be similar to that in Figure 5, except that they would be counterclockwise.

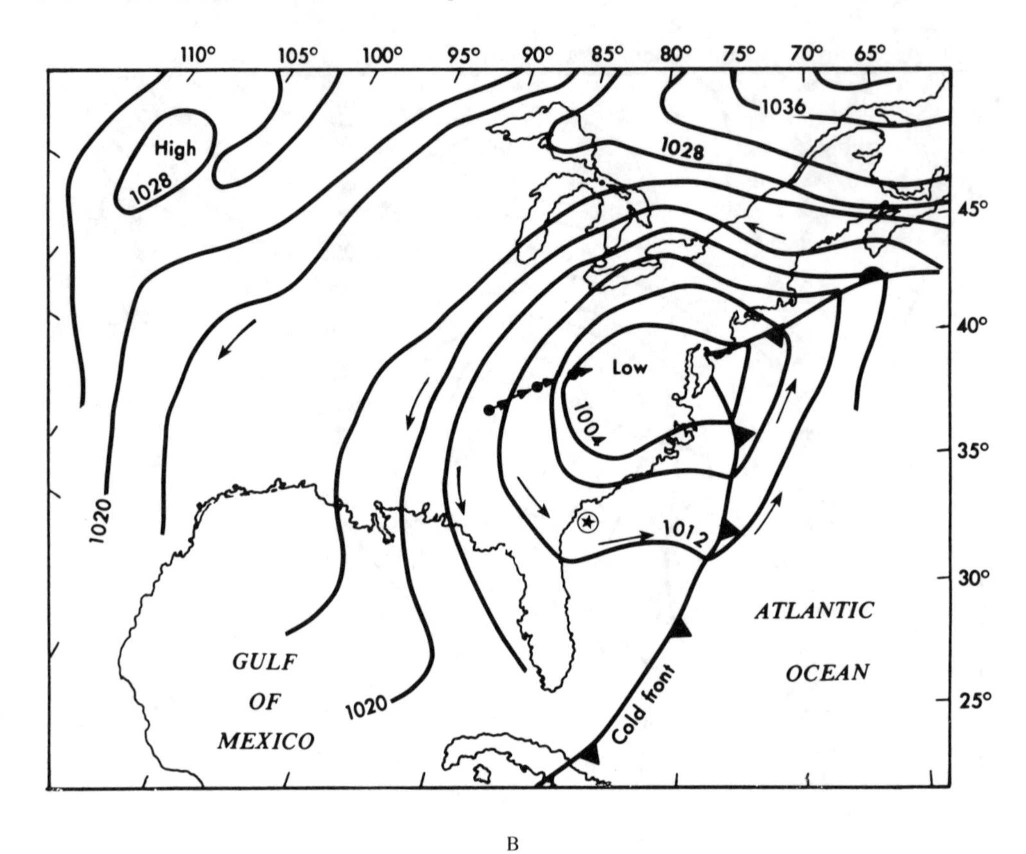

B

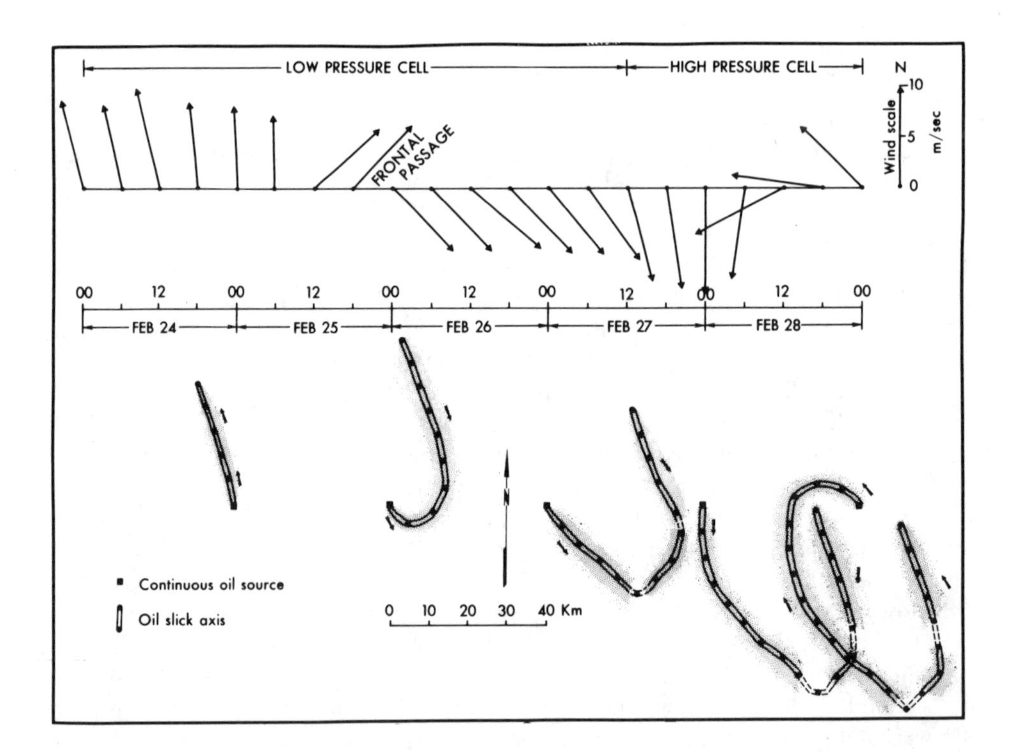

FIGURE 5. Wind velocities off the American coast during passage of a tandem high- and low-pressure system of February 25 to 28, 1979. Oil slicks produced by these wind shifts are shown in the lower part of the figure.

Oil spilling continuously from a fixed source would reflect these wind shifts with arcuate patterns, as shown in the lower part of Figure 5. In this figure parcels of oil (the black dots on the slick axis) released at 6-hr intervals are advected by surface currents moving in the wind direction at 3% of the wind speed. The first parcel is released at 0000 hr on February 24, and slick maps are calculated at the beginning of each of the following 5 days by simply tracking the parcels. The slicks form a smooth arc under the slow rotation of winds under the low, but a tight hook shape arises after the wind shift. Oil emitted after the wind shift will stream out in the new wind direction, while old oil in the prewind-shift slick will be translated en masse in the new wind direction. In practice the slick will probably break up at the tight bends, leading to a series of oil patches.

The semiannual time scale is often illustrated by references to Arabian Sea monsoons. Winds blow 6 months from the northeast and 6 months from the southwest, bringing monsoonal rains to the Indian subcontinent. The East Asian and Tropical South Asian monsoons are other well-known examples. In recent years the early simplistic model of a seasonal wind reversal of thermal origin arising from differential heating of land and water surfaces has been all but abandoned. Certainly with respect to surface transport of pollutants it is the great number of moving cyclonic storms and their associated wind fields at lower time scales (whose long-term effect makes up the monsoonal low) that is important.

At the longest time scale the planetary circulation of polar easterlies, mid-latitude westerlies, and tropical easterlies or trades will at times, despite their long-term statistical nature, become important to pollutant transport. Byers[15] notes that the steadiness of the trade wind belts is an exception in the planetary surface wind system, and it is perhaps here and in the prevailing westerlies in the southern hemisphere, with constant winds for weeks at a time, that oil slicks could grow to mammoth size, the very effective dispersive action of variable winds being absent.

III. PHYSICAL AND CHEMICAL CHARACTERISTICS OF OIL ON THE SEA SURFACE

When petroleum products leak or spill onto the sea surface they are subject to a variety of processes that alter their makeup and exert strong control on how they will move about on the sea surface or sink below it. Among these processes are evaporation, emulsification, turbulence and surface tension effects, sedimentation, and microbial activity.

A. Evaporation, Emulsification, Sedimentation, Microbial Activity

Evaporation, the process by which low to medium molecular weight components of low boiling point are volatized into the atmosphere, is now well recognized as a significant factor in the removal of oil from the sea surface.[16] Milgram[17] notes that if a relatively volatile oil is spilled under warm conditions half of the oil can evaporate in a day or two. Observational work by Harrison et al.,[18] who performed multiple releases of 1.04 m^3 of crude oil spiked with cumene on a natural water surface, indicated that the cumene and all lower boiling point aromatics had disappeared within only 90 min. Similarly, Sivadier and Mikolaj[19] measured evaporation rates over natural oil seeps and found that most of the volatile components were lost in 1 to 2 hr. They noted that wind speed and sea surface roughness exerted significant control on the rate of evaporation. In general, evaporation will also be a function of slick area and thickness, temperature, the concentrations of the various components of the oil, and their vapor pressure. Galt,[20] for example, notes that the loss resulting from evaporation ranges from essentially none in the *Argo Merchant* spill (No. 6 oil) to over 50% in the spill of light crude oil at the *Ecofisk* platform. Harrison et al.[18] point out that, as many of the volatile components of spills are toxic, this rapid evaporation is biologically advantageous. Caution is advised on this point, however, as evaporation in thick spills may form a surface crust that traps volatile toxics underneath until mechanical forces such as wave action break the crust and restart the evaporation cycle.

Emulsification is the formation of a colloidal suspension of one liquid in another. Oil can form two types of emulsions. The first is an oil-in-water emulsion, in which the water is the continuous phase. This is considered a "good" emulsion in that the microscopic size of the oil droplets provides much more surface area for microbial degradation than would be true if they were to remain in the underside of an oil slick. The second type is water in oil emulsion, containing up to 80% water. This is a different problem, as this type of emulsion floats on water and agglomerates into large masses bearing resemblance to a chocolate mousse, hence the nickname "mousse". The incorporation of water into the slick through this process considerably enlarges its size and impact. Smith,[21] for example, found that the volume of slick impacting one beach after the *Torrey Canyon* spill was greater than the original spill values. Such emulsions resist weathering and take on physical properties quite different from those of surface oil. Berridge et al.[22] found that they were stable for periods of up to 100 days and formed naturally on an agitated sea surface. After their importance in the *Argo Merchant* and *Amoco Cadiz* spills, Galt[20] emphasized that the development of algorithms to predict mousse formation is of major importance in predicting overall impact of many oil spills. An in-depth review of the roles of evaporation and emulsification is given in Stolzenbach et al.[23]

Sedimentation processes can also greatly alter the character of spilled oil. Rough seas apparently increase the chances that oil droplets dispersed into the water column will contact and be absorbed onto particulate matter such as silts, clays, and shell material. When calm seas return, the increased effective specific gravity of the mixture sinks the oil. In the Santa Barbara channel spill suspended sediment from the Ventura River served to sink large amounts of oil on contact, while in the *Argo Merchant* case very little oil sank to the bottom.[20] Once the oil is trapped on the bottom in the marine sediment, microbial degradation seems to be extremely slow; for example, Blumer and Sass[24] report that the area of polluted bottom sediment was much greater 10 months after an oil spill on Cape Cod than it had been initially.

After a sufficiently long residence time in the water, biological processes rapidly gain in significance.[16] Over 90 species of bacteria and fungi have been identified that are capable of degrading oil by biological oxidation. These processes operate on a time scale of several months and, while greatly affecting the ultimate fate of the oil, they will play no role in the shorter-term physical processes that are the focus of this paper.

B. Spreading and Dispersion

The spreading of oil on a water surface has been extensively studied over the past 10 years; however, the early work of Fay[25] and Hoult[26] remains, despite its limitations pointed out by Milgram[17] and Stolzenbach et al.,[23] a favorite choice for application purposes because of its simplicity and utility. Basically this approach assumes a constant volume spill of a single component petroleum which spreads out radially with a monotonically decreasing thickness. Initially the oil spreads in a buoyant lens with gravitational spreading balanced by inertial forces, i.e., an accelerated mode that lasts only for the time

$$t = \sqrt{\frac{h_0 \rho_0}{g(\rho_W - \rho_0)}} \qquad (1)$$

To estimate a time scale, let h_o, the initial height of the oil, equal 100 cm and the oil-water density difference equal 0.01 g/cm^3. The duration is only a few seconds for this regime. A second nonaccelerated phase follows, in which gravity forces are balanced by viscous retardation set up by the boundary layer flow in the water underneath the oil slick. Rapidly the slick thins to a point where surface tension becomes the dominant spreading agent, with viscous drag again the retarding force. The slick diameter D in this final and by far most significant regime is given by

$$D = 2\,k_3 \left[\frac{\sigma^2\, t^3}{\rho_w^{\;2}\, \nu_w} \right]^{1/4} \tag{2}$$

where k_3 is a constant $= 1.6$, σ is the air-water surface tension, t is time since inception of spill, and ϱ_w and ν_w are the density and viscosity of water, respectively. Milgram[17] emphasizes that oil does not spread uniformly, as assumed in this theory, but characteristically separates into thick (>0.5 mm) and thin (<0.1 mm) regions with varying levels of spreading pressure. Fractionation of a real oil slick into its various components is cited as the cause of this phenomenon.

Dispersion and diffusion theory has also been applied to oil slick spreading. Basically this approach assumes that the physical spreading related to oil slick properties such as surface tension and buoyancy effects are secondary to the effects of the shear on the oil slick exerted by turbulent motions of the underlying water. Murray[27] found that many of the geometric properties of oil slicks leaked from a damaged platform could be explained by continuous point source Fickian diffusion theory. Such an approach is covered in detail elsewhere in this volume, and the interested reader is directed there. After considering most of the available data, Stolzenbach et al.[23] concluded that both spreading and dispersion processes are likely to be important in the growth of real oil slicks in nature. No doubt this is the case, and the importance of these two processes will vary tremendously depending on the site, the local physical oceanography, and the type of oil.

IV. ADVECTION OF OIL BY WAVES AND CURRENTS

The stress exerted by wind blowing over a water surface has been extensively studied, as outlined in most texts on physical oceanography (see, for example, Neumann and Pierson[28]). The usual formula that arises from the study of forces on submerged objects in hydrodynamics takes the form

$$\vec{\tau}_0 = \rho_a\, C_d\, |W|\, \vec{W} \tag{3}$$

where τ_o is the surface stress of the wind on the water, ϱ_a is air density, C_d is a drag coefficient, and W is wind velocity. The absolute value operator keeps the stress vector oriented in the wind direction. The drag coefficient is itself the subject of a great number of investigations to establish its dependence on wave characteristics[29] and wind speed.[30] $C_d = 2.5 \times 10^{-3}$ is a nominal value for scaling purposes. The action of this wind stress on the surface generates waves and currents, and perhaps has some effect directly on the oil itself. Our analytical understanding of the combined motion of waves and currents is quite primitive, and so each is treated separately. Let us briefly examine the effects of waves first.

A. Waves

First-order or linear theory for monochromatic waves (i.e., a wave train composed of a single repetitive wave form) predicts an orbital water motion with closed orbits.

Therefore, there is no net movement of the water particle after each successive wave period. When this periodic surface wave theory is carried out to second or higher order, terms arise that lead to a net movement of a particle over each wave period. This net translation (or current due to the wave field alone) is referred to as the "Stokes drift," which is a function of the period, length, and amplitude of the wave and the water depth. As the real sea surface is highly irregular, modern wave research emphasizes of amplitudes of varying wave period. For a field of random waves Chang[31] found the net surface drift $\overline{U}_s$ to be given by

$$\overline{U}_s = 4\int_0^\infty \frac{\omega^3}{g} S(\omega)\, d\omega \tag{4}$$

where ω is the wave frequency and S is the amplitude spectra.

Following Stolzenbach et al.[23] we can represent the amplitude spectra with the expression

$$S(\omega) = \alpha \frac{g^2}{\omega^5} \exp - \beta \left(\frac{g}{W\omega}\right)^4 \tag{5}$$

where α and β are empirical constants. Substituting Equation 5 into Equation 4 indicates that the net wave-induced surface drift in random waves in deep water is 1.5 to 2.5% of the wind speed, i.e.,

$$\overline{U}_s = \lambda\, 10^{-2}\, W \tag{6}$$

where the constant λ takes a value between 1.6 and 2.7, depending on the value of α and β used in Equation 5. In the shallow-water case of interest for pollution transport, Stolzenbach et al.[23] derive an expression suggesting that the Stokes drift at the water surface is reduced by a factor W/gh, i.e.,

$$\overline{U}_s = 1.33 \times 10^{-2} \frac{W}{\sqrt{gh}} W \tag{7}$$

Nominal values in coastal water at a depth of 30 m indicate that the wave-induced drift varies between 0.4% of a wind speed at 5 m/sec to 0.8% of a wind at 10 m/sec.

Using the tenuous assumption that a no-slip condition exists between the water surface and a thin viscous layer of oil floating on top of it would allow us to estimate the wave-induced speed of the oil as a function of wind velocity from Equation 7. Milgram[17] has in fact recently presented an analytical model of wave-induced mass transport that includes the presence of a viscous oil slick. His results indicate that the oil slick speed is essentially identical to that given by the Stokes drift at the water surface, but it includes an additional nonlinear term resulting from the diffusion of vorticity through the oil. This additional term indicates that the oil will move slightly faster (on the order of 1% of the wind speed) than the surrounding water. Munday et al.[32] found that in their experiments with small oil slicks in the field the oil did indeed move faster than the surrounding wind-driven current, with a threshold at a wind speed of 4 m/sec and most values centered at 1% of the wind speed.

During the *Argo Merchant* spill in December 1976 the oil formed thick pancake-shaped bodies that could be tracked for measurements of relative velocity between oil and water. After tagging the water with dye pills, aerial photography was used to measure the time and distance required for the pancakes to overtake the dye pills.[2]

Four values, 1.1, 0.7, 1.1, and 0.8%, were determined in a range of wind speeds between 5 and 15 m/sec. While it is unclear whether Milgram's process is responsible, the clustering of observations around 1% does indicate that the excess speed of the oil is indeed a real phenomenon and the advection by waves is a factor that must be included in modeling the trajectories of oil slicks.

B. Wind-Driven Currents

Incidents in the field to date have clearly shown the direct stress of the wind on the water surface to be a principal (but not exclusive) force driving oil on the water surface. Knowledge of the dynamics of the surface layer of water will, then, give us considerable insight into the oil motion problem. The momentum equations governing the horizontal movement of water at sea can be written

$$\frac{\partial u}{\partial t} + u\frac{\partial u}{\partial x} + v\frac{\partial u}{\partial y} = -g\frac{\partial \eta}{\partial x} + \frac{g}{\rho}\int_0^z \frac{\partial \rho}{\partial x}\,dz + fv - \frac{1}{\rho}\frac{\partial \tau_x}{\partial z} \tag{8A}$$

$$\frac{\partial v}{\partial t} + u\frac{\partial v}{\partial x} + v\frac{\partial v}{\partial y} = -g\frac{\partial \eta}{\partial y} + \frac{g}{\rho}\int_0^z \frac{\partial \rho}{\partial y}\,dz - fu - \frac{1}{\rho}\frac{\partial \tau_y}{\partial z} \tag{8B}$$

Term 1 2 3 4 5 6 7

where u and v are the current speed components in the x and y directions, respectively, t is time, g is gravity, η is the height of the free surface above an undisturbed mean water level (positive up), ϱ is water density, f is the Coriolis parameter $= 2\Omega \sin \phi$, Ω is the angular velocity of the earth, ϕ is latitude, and τ is the frictional or eddy stress arising from turbulent fluctuations in the speed components. The z axis is taken as positive up. The upper equation governs the motion and forces acting along the x axis, usually taken as the east-west direction (east taken as positive), while the lower equation governs the motion and forces acting in the y or north-south direction, with north taken as positive.

The terms on the left-hand side of Equation 8 are due to temporal (t) (term 1) and spatial (x, y) (terms 2 and 3) changes in current speed. Term 4 is the component of the pressure gradient force resulting from a slope in the water surface, while term 5 is the component of that force due to horizontal density gradients. Term 6, the Coriolis force, reflects the effect of the rotation of the earth. The final term, 7, is the force of internal friction arising from turbulent motions in the water. The eddy stress τ in the last term is frequently parameterized as the product of an eddy coefficient, A_x, A_y, and the local velocity gradient, i.e., $\tau_x = A_x\, \delta u/\delta z$, $\tau_y = A_y\, \delta v/\delta z$. As a result, the frictional force, term 7 of Equation 8, is frequently written $1/\varrho\ \delta/\delta z\ (A\ \delta u/\delta z)$, $1/\varrho\ \delta/\delta z\ (A\ \delta v/\delta z)$.

Our quantitative knowledge of wind-driven surface currents at sea stems from Ekman's[33] classic work in which his theoretical treatment predicted the familiar 45° deflection of the surface current to the right of the wind. He assumed, however, the restrictive conditions of a steady, uniform flow, no pressure gradients, and an eddy viscosity constant with depth, i.e., he solved Equation 8 keeping only the last two terms on the right. Attempts to confirm Ekman's predictions of the current velocity as a function of wind speed have led to inconclusive results, as outlined by Neumann and Pierson,[28] except that the current deflection to the right of the wind, decreasing with decreasing total water depth, is well substantiated. In Ekman's analysis the surface current speed V_o is given by

$$V_0 = \frac{\vec{\tau}_0}{(Af\rho)^{1/2}} \tag{9}$$

Many empirical expressions relate A to the square of the wind speed, e.g., $\varrho A = 4.3 \times 10^{-4} W^2$.[34] Substituting this expression for A and Equation 3 for τ_o, Ekman's expression (Equation 9) becomes

$$V_0 = k\,W \tag{10}$$

where k, the so-called wind factor, $\simeq 1.2 \times 10^{-2}/(\sin\phi)^{1/2}$. The weak inverse dependence of the surface current speed on the latitude suggested by Ekman's theory has not been established, but the form of Equation 10 has led to a great number of investigations to determine the wind factor k. Studies by Hughes[35] and Neumann,[36] among others, indicate that k is between 0.033 and 0.043.

Thus it is common practice in the practical prediction of surface currents (e.g., in search and rescue missions and oil slick trajectories) to use the empirical rule that the current speed is 3 to 5% of the wind speed and to assign a deflection angle of 10 to 15°, representative of the Ekman solution for finite depth.

In the presence of a coast an added boundary condition of no net flux perpendicular to the coast,

$$\int_0^h v\,dz = 0 \tag{11}$$

where h is water depth and v is the component of the current speed normal to the coast, allowed Jeffreys[37] to solve for the wind-driven current velocity vector as a function of depth. Jeffreys retained an eddy viscosity constant with depth but, unlike Ekman, he included the effect of a third term, the water surface slope ($g\,\partial\eta/\partial y$) normal to the coast. Analysis of current trajectory data by Murray[38] confirmed Jeffreys' theory within the first kilometer or so of the coastline, where the "no net flux" boundary condition is likely to be valid. Under such conditions the primary current flow is parallel to the coast independent of wind angle. The angle of the wind to the coast is as important as the wind speed in determining the surface current speed, as shown in Figure 6. At low wind speeds when the wind is nearly parallel to the coast the theory indicates that current speed is as much as 6% of the wind speed. This wind factor k decreases with both increasing wind speeds and increasing angle to the coastline down to values of 1 to 2% of the wind speed. Over a broad range of values of wind speed and wind angle common to marine conditions, both theoretical and observational values of k are seen to fall between 3 and 5%, consistent with earlier results.

A subtle vertical structure in Murray's data, i.e., a slight onshore component of flow in the surface layer (usually less than 10°), a slight offshore component of flow in a middepth layer, and a reversal back to a strong onshore component of flow in the lowest 1 to 2 m, is not explained by the constant eddy viscosity theory of Jeffreys. Murray[38] has shown this three-layer structure to be associated with a depth-dependent eddy viscosity by comparing field data (Figure 7B) with a numerical solution to the same differential equations of motion used by Jeffreys, but assuming a power law dependence of eddy viscosity on water depth (Figure 7A). In the final stage of an oil slick working up on a beach, positively buoyant oil will migrate ashore in the surface layer, while negatively buoyant subsurface residues (e.g., weathered tar balls) may work ashore in the thin layer hugging the bottom.

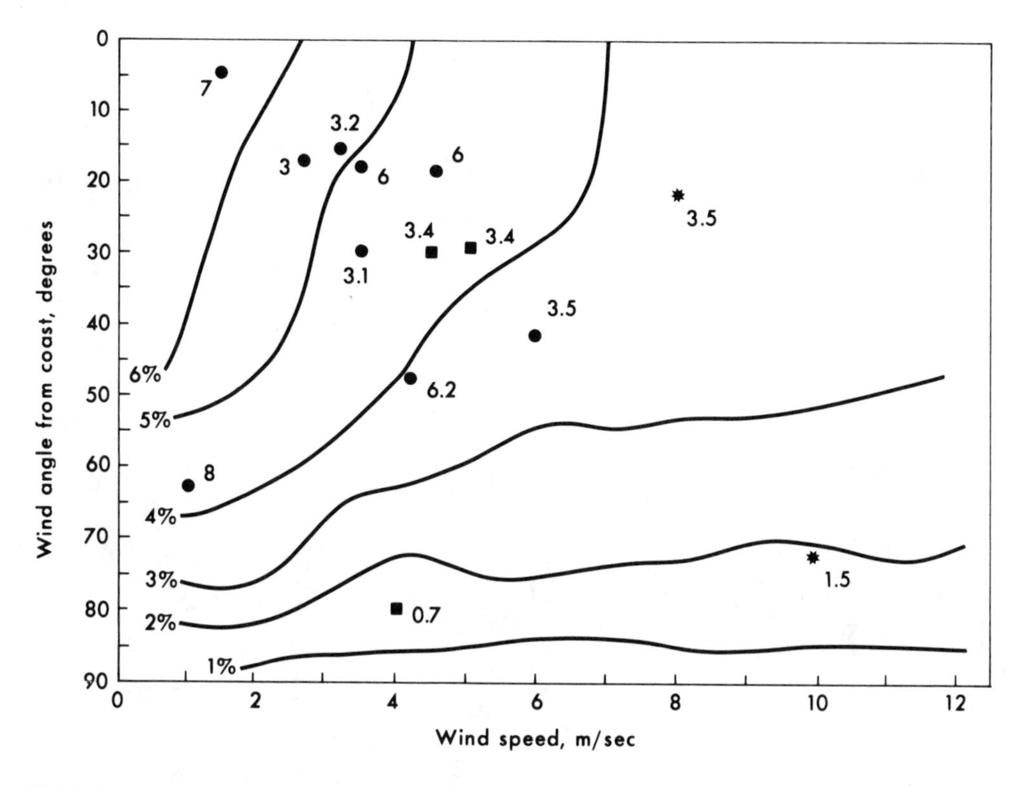

FIGURE 6. Longshore surface current speeds plotted as a percentage of wind speed; the solid lines are predictions from a constant eddy viscosity theory, the squares are data from the coast of the Gulf of Mexico, the dots are data from the Arctic Alaskan coast, and the asterisks are data from Lake Michigan. (Modified from Murray.[38])

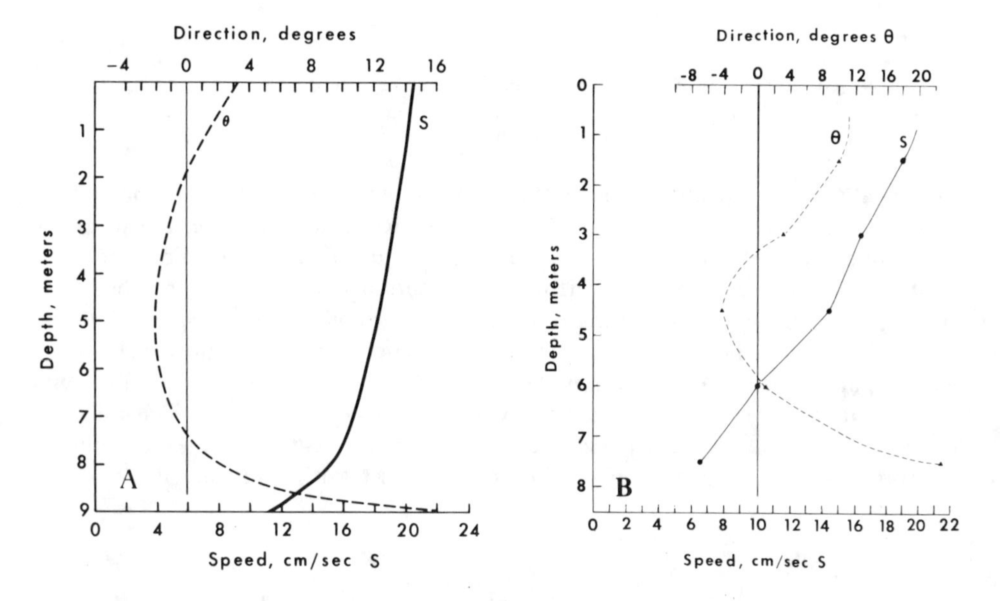

FIGURE 7. (A) Alongshore current speed S and direction θ (measured positive onshore from parallel to the coast) predicted by a numerical solution to the equations of motion including the effect of no net flow normal to the coast and a depth-dependent eddy viscosity. (B) Data from moderately stratified waters off the Florida coast. (Modified from Murray.[38])

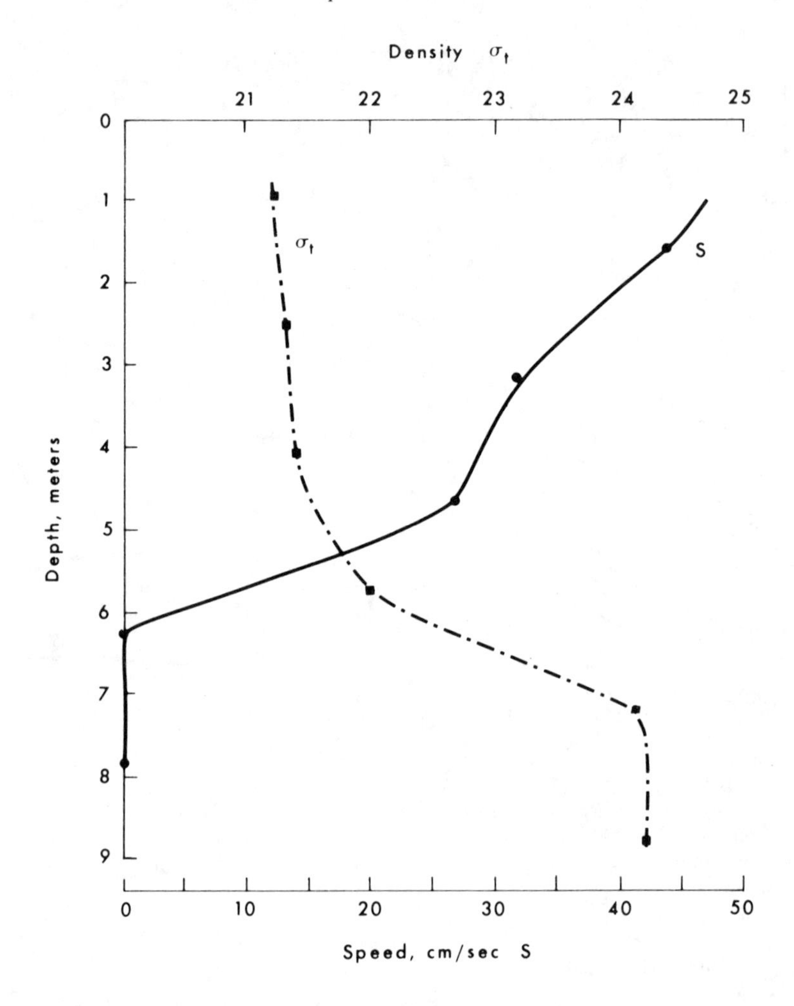

FIGURE 8. The effect of a density gradient on the vertical profile of a wind-generated current. Winds had been blowing onshore for several hours at 8 m/sec at angles of 55 to 65° from the shoreline.

In coastal waters vertically varying eddy viscosities will frequently be associated with density gradients which, if strong enough, force the Richardson number to exceed the critical value necessary to maintain turbulence. With turbulence damped out the vertical eddy viscosity or mixing coefficient of momentum tends to zero and the wind-generated current speed propagating down from the surface layer is cut off. Internal waves, however, may be effective in transferring momentum through the density interface. An example of this effect is shown in Figure 8. A wind blowing for several hours at 8 m/sec fails to generate a current below the 6-m level owing to the sharp density gradient in that region. The local Richardson number, a measure of the relative importance of mechanical and density effects in promoting or retarding mixing

$$\mathrm{Ri} = \frac{-g\partial\rho/\partial z}{\rho\,(\partial u/\partial z)^2} \tag{12}$$

in the pycnocline is calculated at Ri = 0.6, which satisfies the criterion for a stable, stratified shear flow that Ri must be greater than ¼. In this case the density difference from surface to bottom is caused by a temperature drop from 27 to 25°C and a salinity increase from 33 to 36°/oo. Thus, in such stratified conditions near river mouths and estuaries, since energy is effectively trapped in the upper layer, we may expect higher surface current speeds than the same wind velocity would produce in the well-mixed water farther offshore.

Madsen[39] has recently proposed an analytical solution to the wind-driven current problem with an eddy viscosity that increases linearly with depth and predicts a deflection angle between wind and surface current of approximately 10°. While he warns against oversimplification, his results also tend to support the use of a 3% rule until the details of the air-oil interface problem are resolved.

1. Inertial Oscillations

Another type of motion of great importance to the movement of oil in the surface layer is inertial oscillations. If we assume a frictionless flow, neglect the spatial accelerations and any pressure gradients, the horizontal equations of motion reduce to a balance between the temporal acceleration of the speed and the Coriolis force, terms 1 and 6 in Equations 8A and B. The solutions to these equations then are

$$u = u_0 \cos ft + v_0 \sin ft \qquad (13A)$$

$$v = v_0 \cos ft - u_0 \sin ft \qquad (13B)$$

which are recognized as a clockwise-rotating circular motion with a constant tangential speed $V_T = (u_o^2 + v_o^2)^{1/2}$, where u_o and v_o are the initial components of the speed. Thus particles in inertial motion will describe a circular orbit with frequency f, where f is the Coriolis parameter. It is easily shown (for example, Neumann and Pierson[28]) that this oscillation has a period T_P, which is a function of latitude only; i.e., $T_P = \pi / \Omega \sin \phi$ and has a radius $r = v_T/f$. A sudden increase or decrease in wind stress or an abrupt shift in wind direction is especially effective in their production.[40,41] While inertial oscillations have been observed in many parts of the ocean, they are believed to be common in well-stratified coastal waters at distances greater than about 15 km from the shore. Oil spills in shipping lanes near major river mouths should be especially prone to this type of motion.

As part of a study for an offshore oil port near the Mississippi Delta, biplanar drogues 1.3 m square were buoyed at a 1-m depth and tracked for 2 days. Figure 9 shows three such tracks, A and B on March 13 to 15 and C on January 29, 1974. The two tracks of March 13 to 15 clearly executed clockwise loops while traversing north entrained in the Mississippi River plume. The inertial balance differential equations can be combined with a steady current and integrated to obtain an expression for the complex particle trajectory Z. In complex notation

$$Z = Z_0 + \overline{W}t - \frac{W_0}{if}\left(1 - e^{-ift}\right) \qquad (14)$$

where Z is the complex displacement from the initial position Z_o, $\overline{W}$ is the steady current speed, W_o is the strength of the rotating inertial current, t is time, and f is the inertial frequency. The inertial period of 22.94 hr at this location is quite close to the diurnal tidal period. Five estimates of $\overline{W}$ can be obtained from fixes separated by 25 hr on track A, giving $\overline{W} = 0.12 \pm 0.01$ m/sec. The speed from positions A5 to A6, a portion of the trajectory nearly opposite to the mean drift, can be taken as $\overline{W} - W_o$, which gives a value of $W_o = 0.27$ m/sec. These two speeds can be used to predict a trajectory using Equation 14. The good agreement between the predicted trajectory (inset in Figure 9) using these two speeds and the actual drogue track strongly suggests that the drogues are undergoing inertial oscillations induced by a wind shift associated with a frontal passage on March 12 while at the same time translating shoreward in the freshwater plume from Southwest Pass.

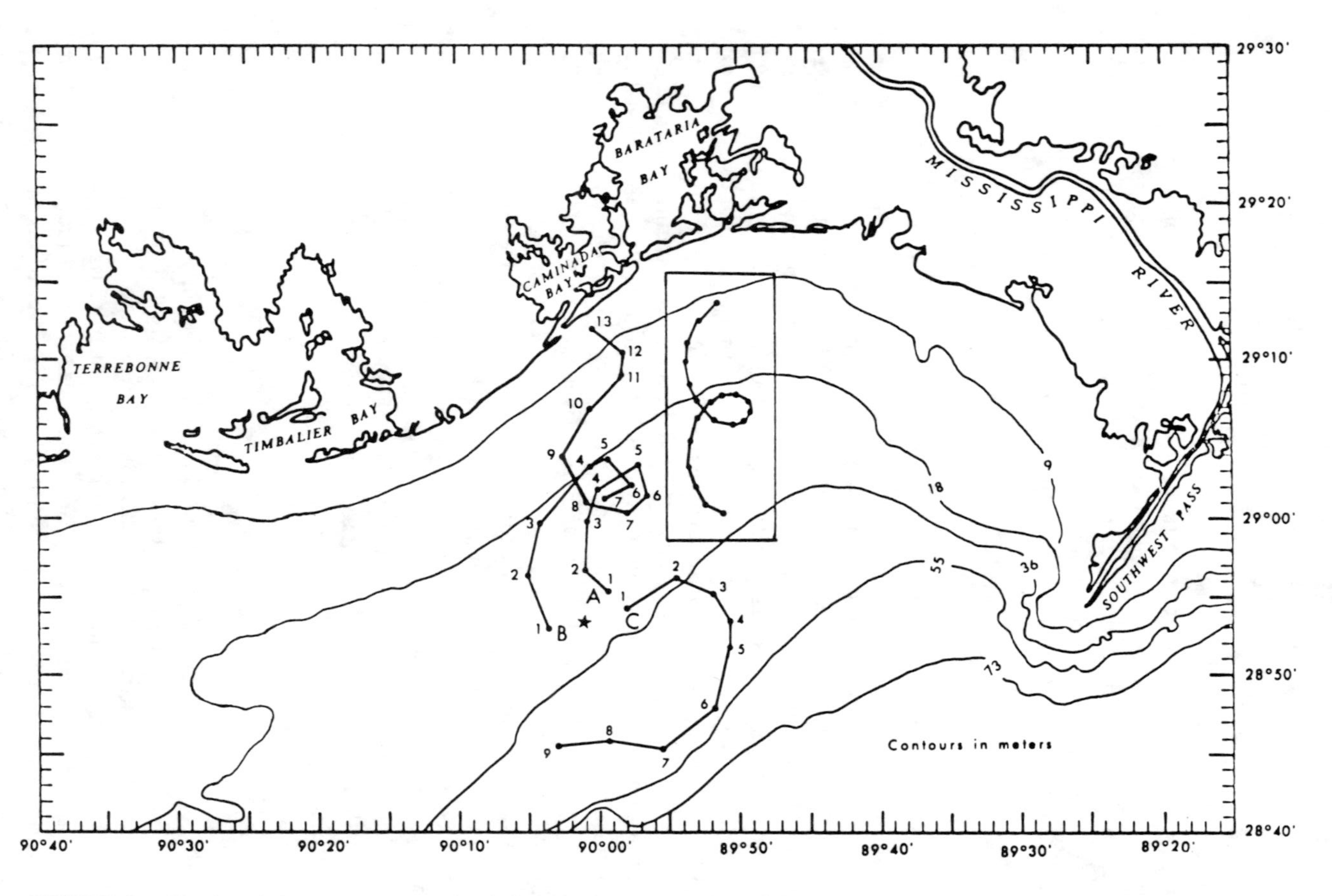

FIGURE 9. Tracks of drogue movement in the surface layer west of the Mississippi delta. The numerals on the track are fix numbers, and the star locates a current meter mooring discussed in the text. The inset is a surface current trajectory predicted by inertial current theory for comparisons to the observed tracks.

The drogue track C of January 29 to 30, although only 21 hr long, can be interpreted similarly in terms of an inertial oscillation. The direction of the mean drift is southwesterly. The distribution of current speeds suggests $\overline{W} \simeq 0.13$ m/sec, $\mathcal{W}_o = 0.64$ m/sec. The diameter of the inertial oscillation is given by $d = \mathcal{W}_o\Omega \sin \Omega$, which with $\mathcal{W}_o = 0.64$ m/sec gives a prediction of 18 km compared to a distance measured from position 1 to position 6 of 16 km. This oscillation is associated with a frontal passage on January 27.

Analysis of current meter records made over several months in this same area (see location on Figure 9) has similarly shown the prevalence of inertial oscillations. Daddio et al.[42] predicted u and v current speed components in the surface layer from the inertial equations, modified by wind stress driving at the surface and linear frictional damping, i.e., their model includes terms 1, 6, and 7 of Equations 8A and 8B. The model currents are in good agreement in both magnitude and frequent clockwise directional rotation with the currents observed by a current meter moored in the surface layer. There is little doubt that oil spilled in well-stratified coastal waters such as around the mouth of the Mississippi River will, under the impulse of high winds or abrupt wind shifts, describe clockwise rotational movements which, if understood, can be of great assistance in understanding oil spill movement and designing for its cleanup.

2. Cross-Shelf Wind Effects

The movement of oil on the sea surface is measured by its Lagrangian velocity, that is, its spatial displacement over a given length of time. This is not necessarily the same as the Eulerian velocity, which is measured at a fixed point such as by a moored current meter. Lagrangian velocities obtained by tracking drogues or drifters will thus give a better idea of what course drifting oil will follow and how fast it will move. In an interesting experiment off the East Coast of the U.S. designed to evaluate the trajectory of water motions on the shelf during the passage of a severe weather system, three drogues similar in geometry to those described earlier were placed in the water at 1-, 3-, and 5-m depths below the sea surface. Constant depth was maintained by tethering to a surface-piercing staff carrying a miniature high-frequency RF transmitter described by Murray et al.[43] Direction-finding bearings from shore stations allowed determination of the drogue path during severe weather when ship operations were impossible. As the 1-m drogue was affected by the strong wave action, the track of the 3-m drogue, shown in Figure 10, best represents the movement of the surface layer. After insertion at 1740 local time on January 21 the drogue moved directly to the southwest in line with the winds, which had been blowing in that direction for the previous 29 hr at speeds up to 20 m/sec under the influence of the eastern half of a high-pressure cell centered over Nova Scotia, to the north. The corresponding wind track in Figure 11 is from Ocean Weather Station Hotel, at 38°N, 71°W, or 335 km to the east, which was felt to be more representative of shelf wind conditions than any of the shore stations. Weak northerly winds then blew for 4 days as the diffuse high-pressure cell passed over the area to the drogue path, which was quite irregular, probably readjusting to the earlier intense winds of January 20 to 21.

A powerful new system extending over 45° of longitude dominated the weather maps beginning about noon on January 25. Low pressures to the north and high pressures to the south brought intense east-going winds to the shelf area. Winds of 15 to 25 m/sec blew all day on the 26th and 27th. The 3-m drogue moved (Figure 10) across the shelf to the shelf break, a traverse of over 100 km, in 24 hr; surface-layer speeds were apparently a little greater than 100 cm/sec. By the morning of January 27 the water had begun moving back toward the coast. This onshore movement of the 3-m drogue occurred during the period of weak winds, January 28 to 29. Thus it appears to result

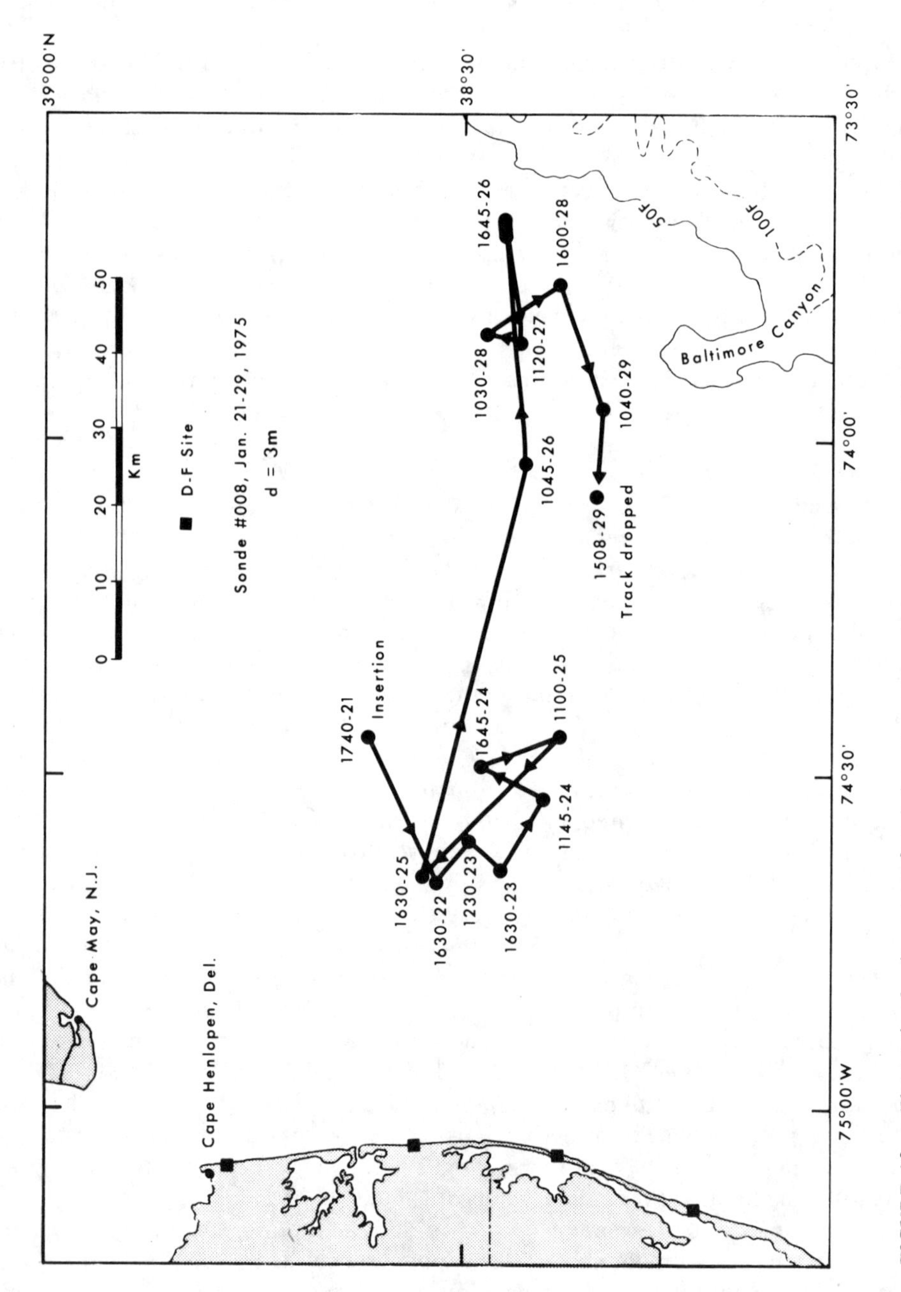

FIGURE 10. The track of a drogue set 3 m below the surface during a severe storm off the Atlantic coast of the U.S. (After Wiseman et al.[44])

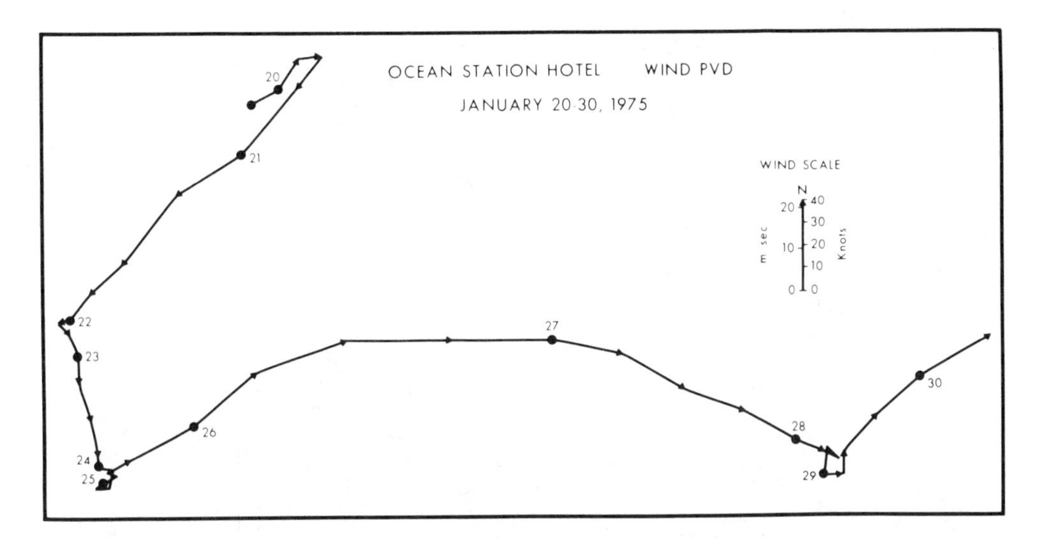

FIGURE 11. A progressive vector diagram of offshore winds during the drogue experiment in Figure 10. (After Wiseman et al.[44])

from the relaxation of the shoreward-directed pressure gradient built up by the intense offshore winds of the previous 2 days. On the basis of this data we may expect strong offshore winds to advect oil all the way across the shelf, but it may return shoreward with the drop out of the wind.

C. Tidal Currents

The results of the gravitational attractions of the moon and sun on the oceans and smaller water bodies of the earth are well known. The vertical rise and fall of the water surface owing to the cyclic nature of these forces is called the tide, while the accompanying horizontal movement of these waters, demanded by the laws of conservation of mass, are referred to as tidal currents. In the oceans beyond the continental shelves both the tides and tidal currents are barely perceptible because of the great water depths involved. Nearer the coast shoaling effects cause the tidal wave to reach easily measurable values, ranging from an order of 0.1 m in marginal seas such as the Mediterranean and Gulf of Mexico to formidable values in excess of 8 m at Darwin, Australia, and Inchon, Korea. Thus it is near the coast, where interaction between the tide and the topography causes amplification and other special effects to occur, that oil spills will be directly affected by this phenomenon.

When the coast is cut by a river, estuarine channel, or embayment whose width is on the order of a few kilometers, the rising tide will generally force a flood current up the channel and the falling tide will produce the reversing ebb current. The tidal currents will be constrained by the relatively narrow channel to flow parallel to its walls, merely reversing in direction and varying in speed harmonically (the so-called rectilinear tidal currents). In general, the dynamics are governed by the temporal acceleration, the pressure gradient due to the water surface slope, and the force of friction, terms 1, 4, and 7 in Equation 8.

With increasing channel width the Coriolis force (term 6) due to the rotation of the earth assumes an important role in the dynamics and quite a different flow arises in the embayment. Tides can become one half cycle out of phase from one side of the channel to the other. Flood currents and ebb currents then occur at the same time on opposite sides of the estuary, and an amphidromic point or position of zero tide range may develop in the center of the channel. The Rio Plata estuary, Argentina, is an example of this type of tidal response. This process can become a very efficient dispersal mechanism for oil spilled in an estuary or embayment.

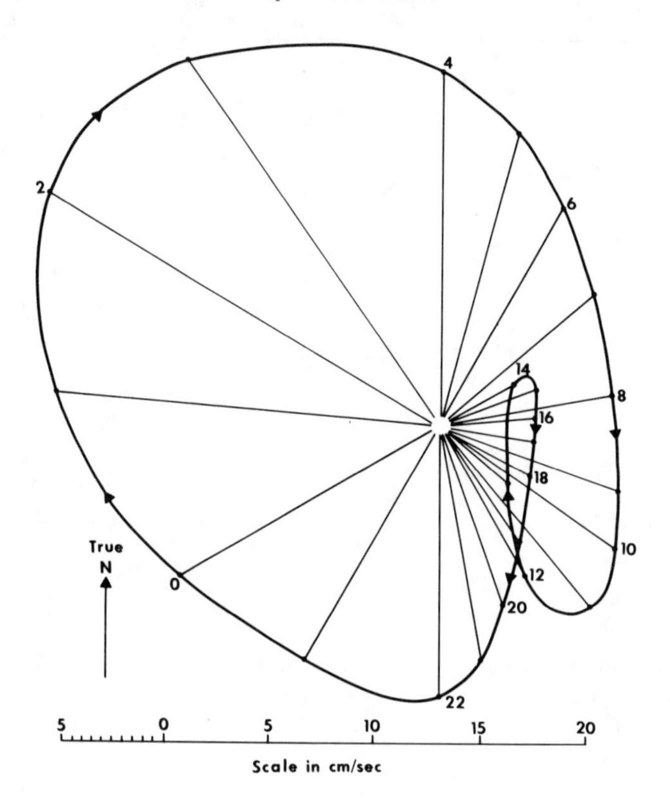

FIGURE 12. Rotary tidal currents at the *San Francisco* lightship.

Another interesting effect of the Coriolis force due to the rotation of the earth is Kelvin waves. Tidal waves that occur in large straits or channels open at both ends such as the English Channel (La Manche) are of this type. The tidal wave in the English Channel progresses south to north. Flood currents under the crest (high tide) tend to be deflected to the right by the Coriolis force, but the French coast acts as a barrier, forcing the deflected water to set up against it. The end result at high tide is that the water level increases in height laterally across the channel toward the French coast. The westward-directed pressure gradient due to the slope in the water surface at high tide balances out the eastward-directed Coriolis force. At low tide ebb currents flow southerly in the channel, and for the same reasons water will pile up against the English coast, producing a higher level of low tide on the English side than on the French side. It can be shown[45] that in this so-called Kelvin wave the difference in tidal range across the channel should be equal to $2\overline{U}_T f b/g$, where $\overline{U}_T$ is an average value of the tidal current amplitude, f is the Coriolis parameter, b is the channel width, and g is gravity. Near the southern end of the English Channel $\overline{U}_T$ is about 2 m/sec and b is 90 km, predicting a difference in range of 3 m across the channel, which is in good agreement with values seen in the tide tables. Thus the unusually high tidal range on the French coasts of Normandy and Brittany can be explained with rather simple tidal dynamics. We shall see that such effects can play important roles in the ultimate spread of spilled oil along a coast.

While tidal currents are essentially rectilinear in most channels and estuaries owing to side-wall constraints, off the open coast the situation is quite different. The Coriolis force due to the rotation of the earth is free to continually deflect the moving tidal current to the right, developing what are referred to as rotary tidal currents. Figure 12, from the lightship *San Francisco,* is a well-known example. The curved line connects the end points of the tidal current vectors, which are drawn from the origin to end point at each hour. The currents are seen to rotate clockwise continually through the 24-hr lunar day except for a 2-hr interval when the interaction of diurnal and semidiurnal effects causes a brief reversal. It is this type of tidal current that is most common off the coasts and on the continental shelves where oil spills are likely to occur.

D. Density Currents

In estuaries and along certain stretches of open coast the density gradient force (term 5 in Equations 8A and 8B) can dominate the momentum balance and determine the net movement of surface waters and surface-bound pollutants such as oil. The surface layer in an estuary, however, is extremely sensitive to wind. Hansen and Rattray[46] have shown that at very small values of τ_o the wind stress can reverse the current direction in a typical estuary driven by gravitational convection. These effects are covered in detail in a separate chapter of this book. Such are the major physical processes that control the movement of oil on the sea surface. Now we shall examine their effect on some major oil spills in the field.

V. THE EFFECT OF PHYSICAL PROCESSES ON SOME MAJOR OIL SPILLS

A. *Torrey Canyon*

An early example of the dominant effects of winds on the long-term drift of oil was provided by the *Torrey Canyon* spill in 1967, shown in Figure 13. At about 0900 hr on March 18 of that year the *Torrey Canyon,* carrying 120,000 tons (Table 1) of Kuwait crude oil, ran hard aground on Seven Stones Reef, off the southwest coast of England. For the first 6 days winds were from the north and the 30,000 tons (Table 1) of oil that had leaked out of the hold of the ship had moved south and southeast, away from the English coast. On Friday, March 24, a southwest wind began to blow that moved about 20,000 tons of oil, newly escaped from the tanker over the previous 3 days, directly onto the English coast. These southwesterly winds pushed the old oil, which earlier had spread far to the south, into the English Channel toward Alderney and the Cherbourg peninsula (Figure 13). These general meteorological conditions held until April 9, when the winds shifted again and 14 to 17 m/sec winds pushed the oil south, past Guernsey and Jersey toward the Brittany coast, where major pollution affected the coastline. Thus despite the strong reversing tidal currents, which exceed 1.0 m/sec in this part of the channel,[48] the long-term path of the oil (as determined by several days of movement) was controlled directly by the wind velocity. The effect of tidal currents on the surface drift of the oil apparently acted only as dispersive perturbations.

B. The Chevron Spill

At distances offshore greater than about 10 km the blocking action of the coast on the current direction is greatly diminished and the surface water and entrained oil are free to move in the wind direction. The presence of the coast can still be important, however, through pressure gradients set up by water piling up against the coast, as seen in the following case.

Offshore production platform MP-41C, off the east flank of the Mississippi Delta in 15 m of water, suffered an explosion and fire on February 10, 1970. While escaping oil burned unchecked for the following 4 weeks, small, light slicks formed around the platform as a result of the pumping of water onto the flame. On March 10 the fire was extinguished, and large slicks immediately began to form from the steady discharge of at least 1000 bbl (Table 1) of oil per day over the next 20 days. The location

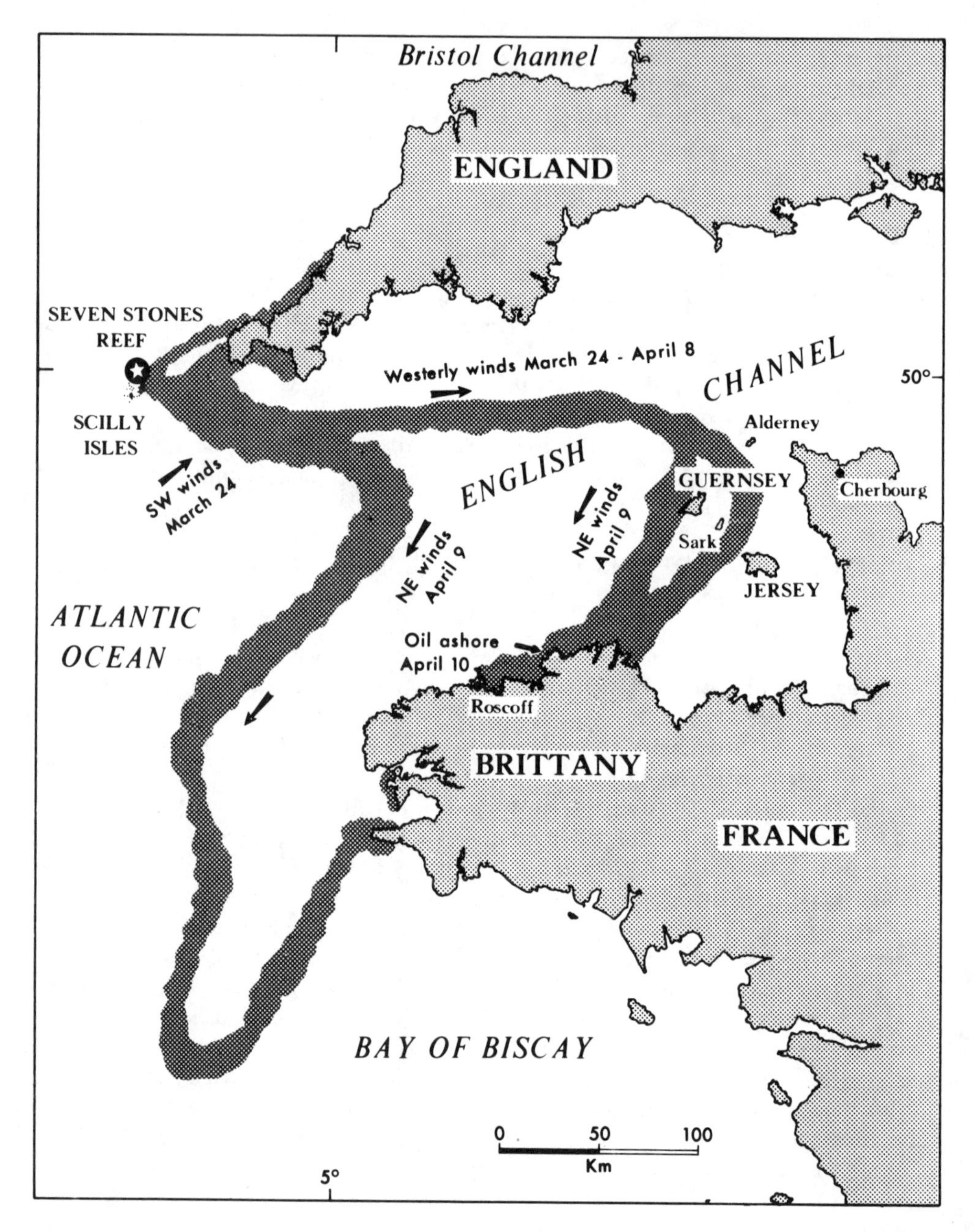

FIGURE 13. Composite oil slick map from the wreck of the *Torrey Canyon* on Seven Stones Reef. (Modified from Petrow.[47])

of the damaged platform, 41C, the anchor stations of the on-scene command post U.S. Coast Guard cutter *Dependable,* and the lightship *New Orleans,* where winds were routinely observed every 6 hr for the U.S. Weather Bureau, are shown in Figure 14.

In March 1970 the Mississippi River was at a normally high river stage and discharge for flood season. The river effluent is highly turbid and cool (~10°C) and contrasts sharply with the clear, warmer (~20°C), saline Gulf waters as it flows out in the form of buoyant plumes.[49] Sharp, quasi-permanent density fronts tend to form throughout the area at the leading edge of the buoyant plumes, usually near the major outlets

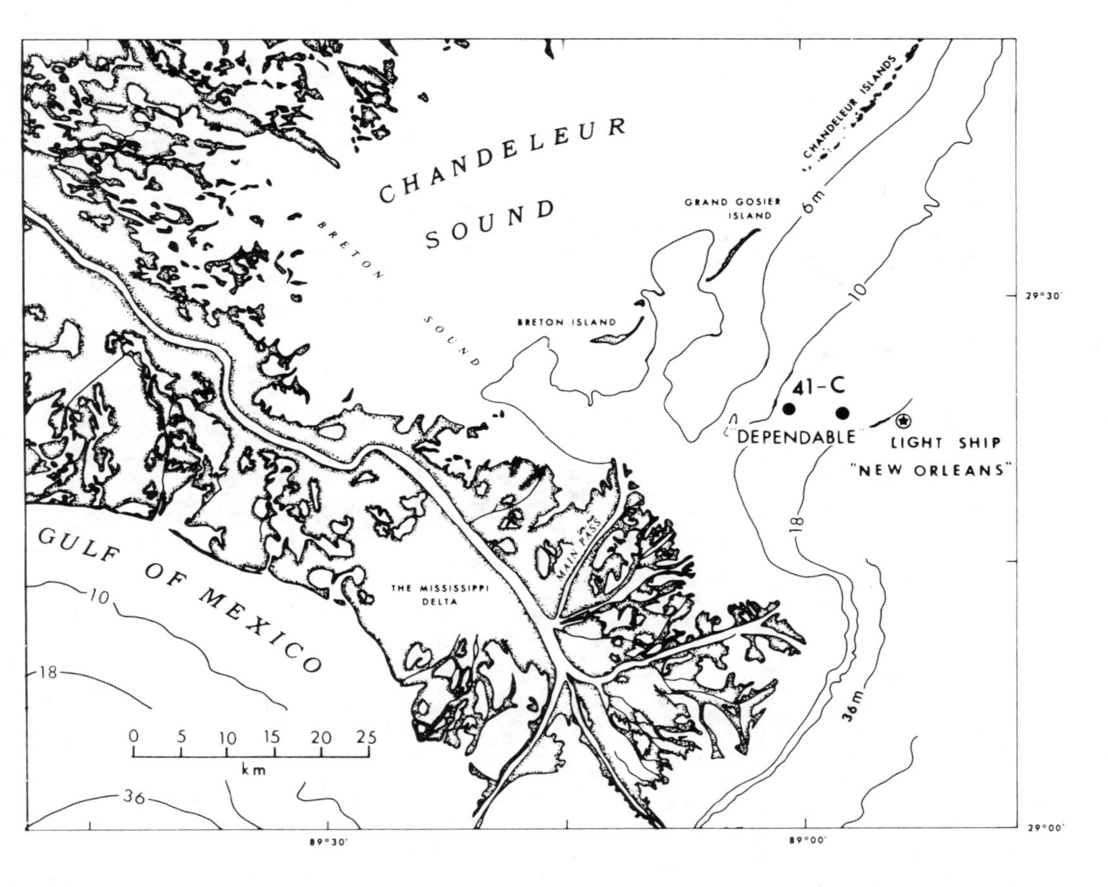

FIGURE 14. The locations of the damaged Chevron oil well MP41C, the Coast Guard cutter *Dependable*, and the lightship *New Orleans*.

from the river; their stability and movement are controlled by wind stress and tidal pumping. Tides in the area are diurnal, with a range of 30 cm. Tidal currents measured from the *Dependable* rotate clockwise and have an amplitude of 10 cm/sec.[50]

1. Effects of Weather Systems

Winds in March are strongly controlled by migrating high- and low-pressure cells and wind shifts associated with frontal passages, as discussed earlier in the section on weather systems. A typical current cycle during a frontal passage near the *Dependable* station is characterized by a strong southeasterly wind raising sea level against the coast by driving near-surface water toward the north-northwest, i.e., in the line of wind but deflected clockwise by the blocking action of the north-trending coastline (cf. Figure 14). Deeper waters show a southwesterly flow, providing a return mechanism for the onshore-moving surface waters. With the removal of the onshore-directed wind stress, either by a drop in the wind speed or an abrupt shift in the wind direction, setdown ensues and the current pattern completely reverses itself, i.e., flow is directed offshore in the surface layers and onshore in the deeper layers. This effect is sometimes referred to as current relaxation or rebound. A south to southwesterly current drift normally persists several days after such a storm pattern, during the ensuing period of weak, variable winds.[50]

The Chevron Oil Company, owner of MP-41C, mapped well-defined slick boundaries on a twice-daily basis from an S-55 helicopter equipped with a Raydist navigation system, giving a position accuracy of less than 10 m. The geometry and movement of 37 such slicks were analyzed in terms of the on-site winds, currents, and tides by Murray.[51] Figures 15 and 16 show a few examples of the time behavior of the oil slicks in comparison to progressive vector diagrams (wind tracks) of wind velocities observed at the lightship *New Orleans.* The arrows coming from the platform in these figures indicate the direction of movement of newly spilled oil at the time of the mapping.

For some 30 hr prior to the slick shown in Figure 15A a large high-pressure cell centered to the northeast over northern Florida produced a wind toward the west-northwest at 8 to 10 m/sec. These onshore wind components built up a pressure gradient directed seaward, and when the wind shifted suddenly prior to 1800 hr on March 11 (see Figure 15I) the surface current undoubtedly began to rebound seaward. The slick at 1115 hr the next day (Figure 15B), although lined up with the wind direction, is nearly 30 km long, a fact that indicates 20 to 24 hr of southeasterly current flow (at 0.35 to 0.45 m/sec). Inasmuch as the wind had turned toward the southeast no more than 6 hr earlier, the oil slick in Figure 15B must have been entrained in a fast-moving current setting southeast as a rebound or release motion after the onshore winds of March 10 and 11.

The following morning (March 13, Figure 15C) a slick initially extended southerly for about 10 km until it impinged upon the quasi-permanent Pass A Loutre density front, causing a clockwise detour. Upon reaching the South Pass area, this unusually large slick was deflected seaward along the density front associated with the northeastern boundary of the South Pass effluent, which frequently extends far offshore this time of year. The vertical thickness along the edges of these low-density effluent lenses is normally only 20 to 40 cm,[49] and therefore the slick movement is intimately related to the dynamics of the upper few decimeters of water.

The afternoon slick of March 13 and the morning slick of March 14 (Figures 15D and 15E) both followed the wind direction (from the northwest toward the southeast), a situation that the weather maps clearly show is produced by the fact that the spill area was in the northeastern quadrant of a high-pressure cell centered over the South Texas-Mexico coast. The northward migration of the Pass A Loutre front severely deformed the morning slick of March 14 (Figure 15E).

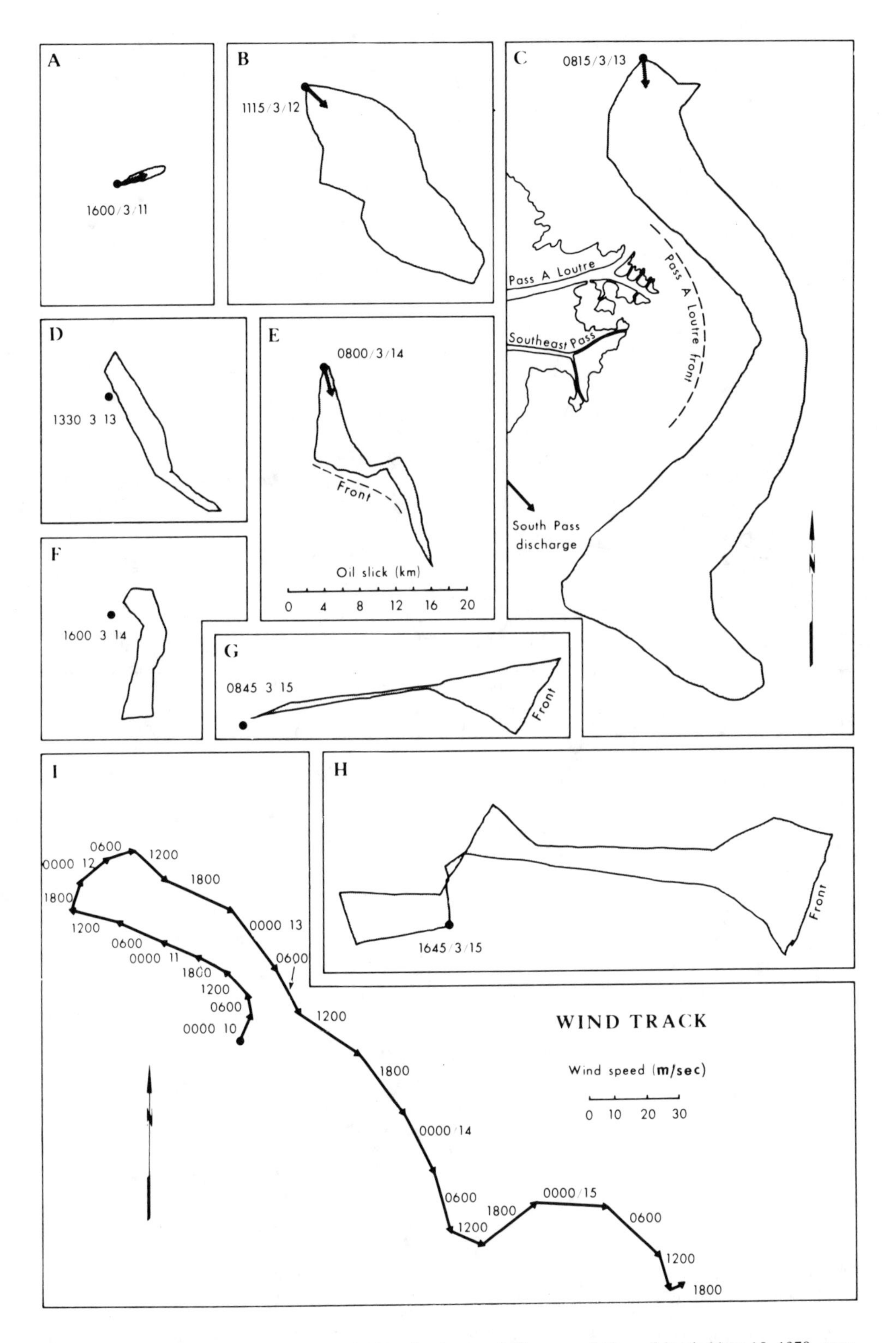

FIGURE 15. Maps of oil slicks emitted from the damaged Chevron platform March 11 to 15, 1970, compared to the wind velocities recorded at the lightship *New Orleans*. (After Murray.[51])

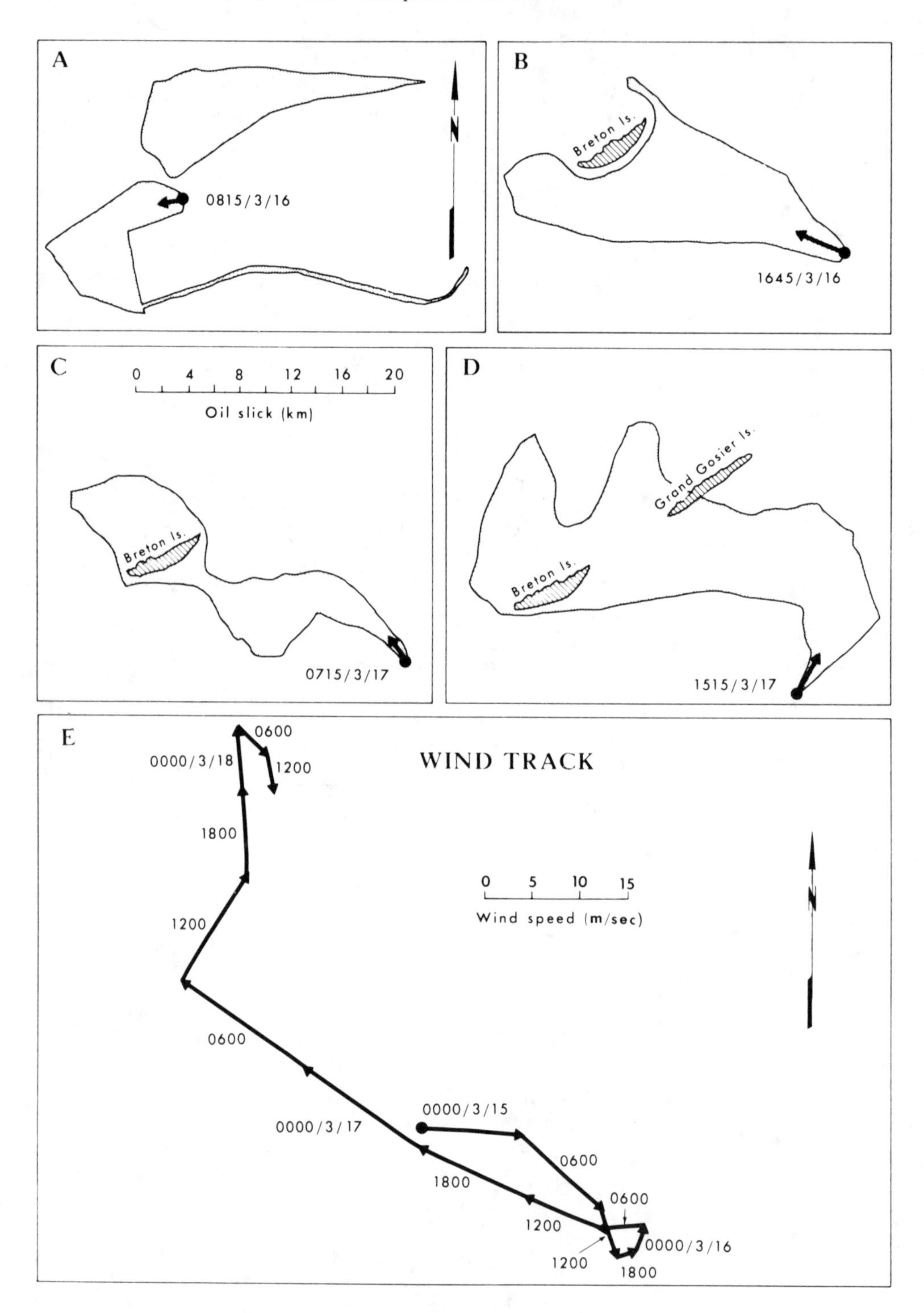

FIGURE 16. Maps of oil slicks emitted from the damaged Chevron platform March 16 to 17, 1970, compared to the wind velocities recorded at the lightship *New Orleans*. (After Murray.[51])

Approximately at noon on March 14 the northwesterly wind slackened abruptly and shifted westerly as the spill area passed under the north-central sector of the moving high-pressure cell. This wind shift is reflected in the motion of the slicks by a direct displacement of the old slick toward the east that afternoon (Figure 15F); the shift was followed by dissipation (degradation resulting from mechanical mixing of wave action) of the old oil south of 41C. Winds in excess of 10 m/sec from 1800 hr on March 14 to 1200 hr on March 15 resulted in the formation of a long (~40 km) new slick to the east-northeast (Figures 15G and 15H). This implies surface current speeds of 55 cm/sec, which is a reasonable value of 4 to 5% of the wind speed. The sudden drop in offshore wind speed (Figure 15I) as the high-pressure cell diffused rapidly on the afternoon of March 15 likely induced an onshore (westerly setting) rebound current, and the quadrilaterally shaped slick west of 41C in Figure 15H was formed.

On the morning of March 16 a slick (see Figure 16A) recently cut off from MP 41C lay to the north and east of the spilling well. The south and west winds of early that morning and late the previous evening (Figure 16E) had appropriate speeds and durations to account for this old slick, whose smooth outline suggests active dissipation. The new slick in Figure 16A initially trended off to the west-southwest, in agreement with the recently shifted 0600-hr wind vector. This new slick, however, was abruptly truncated about 10 km to the southwest by an old density front. Still older oil trapped along the front shows the interface to extend 25 km toward the east. The east and southeasterly winds of March 16 and 17 were the result of the same high-pressure cell that first appeared on the weather maps of March 13. The spill area was then located in the southwestern quadrant of this large, well-developed high, which extended from Canada to the Gulf of Mexico. The most serious pollution of this entire spill was the oil slick impingement on Breton Island (Figures 16B and 16C) which occurred during this episode of southeasterly winds. A protective barrier of a density front, usually located just south of Breton Island, was absent at this time, leaving the island vulnerable to oil slick encroachment.

Along the west side of this high-pressure cell the isobars were nearly north-south, and the wind shifted sharply as the spill area came under their influence during the afternoon of March 17. The afternoon slick of March 17 (Figure 16D) promptly followed the wind and flowed northeastward, and the old oil wrapped around Breton Island was also translated northerly by the new wind direction. A similar story of migrating pressure cells, wind shifts, and rebound of surface currents generally explains the slick trajectories of the remainder of this incident.

Current measurements on the *Dependable* were taken coincident with oil slick mapping only during the period March 16 to 20. Comparison of the near-surface (1.5-m level, Figure 17) currents with the winds and slicks on Figure 16 indicate that the slicks follow the wind direction very well but that the current 1.5 m below the surface has a more complex relationship with the wind. For example, after the sharp wind shift at 1200 hr on March 17 the currents do turn to the northeast in response, but continue turning clockwise until winds and currents are 180° out of phase by midnight of the same night. This priming of clockwise rotation of currents appears several other times in the records after wind shifts or sudden deceleration of wind speed, and suggests inertial oscillations are playing a stronger role in the current dynamics than in the movement of the surface constrained oil slicks.

Following Bowden and Hughes'[52] approach relating wind and current, a multivariate regression technique was run between the oil slick vector and the wind vector such that

$$\vec{S} = B\vec{W} + \vec{C} \tag{15}$$

where $\vec{S}$ is the oil slick vector with an orientation in the slick direction and a magnitude equal to the area of the oil slick, B is a 2 × 2 coefficient matrix dependent on geographic position, $\vec{W}$ is the wind vector with components W_x and W_y, and $\vec{C}$ is a constant vector representing all nonwind effects on the slicks as steady drift currents.

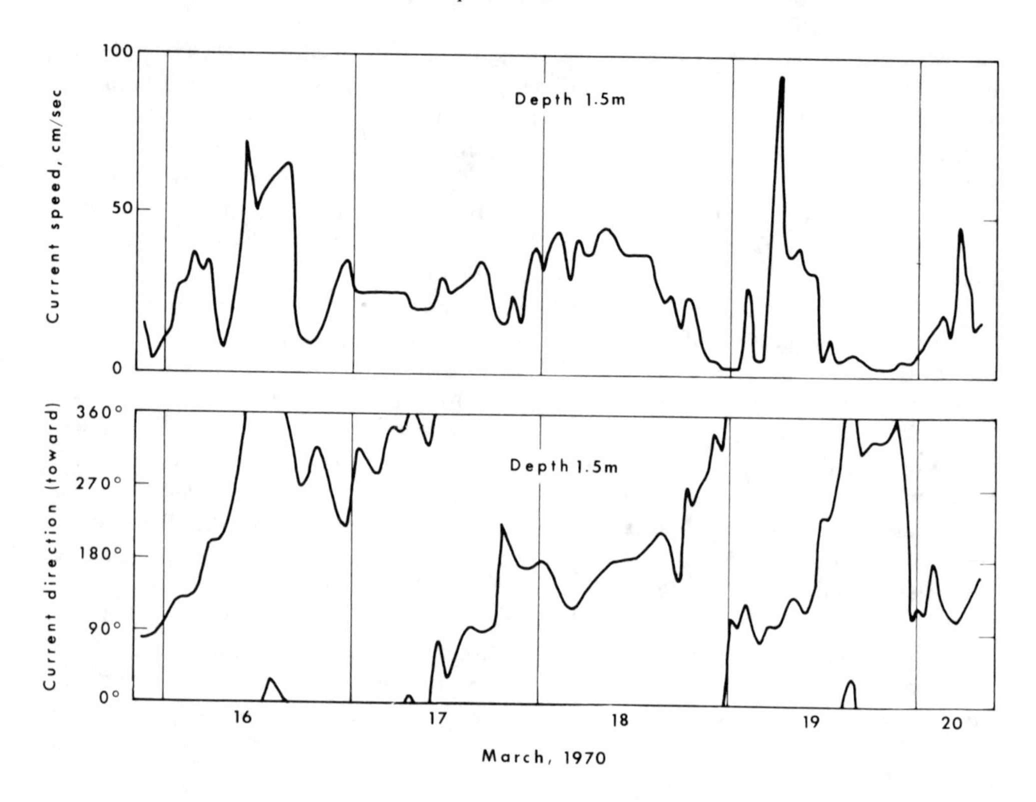

FIGURE 17. Near-surface current velocities March 15 to 20 at the site of the Chevron oil spill.

In component notation the preceding equation becomes

$$S_x = b_{11} W_x + b_{12} W_y + C_x \tag{16A}$$

$$S_y = b_{21} W_x + b_{22} W_y + C_y \tag{16B}$$

where the x component is east-west (positive to the east) and the y component is north-south (positive to the north). The multiple regression was performed on all slicks with wind speeds greater than 4 m/sec (26 cases), yielding the values of the coefficients and the vector constant. Small values of C_x and C_y reflect the dominant role of the wind in determining the size and orientation of the slick. The correlation coefficients of the regression are not impressive (0.45 to 0.55), but do indicate the presence of distinct trends. With the coefficients and constants derived, S_x and S_y can now be calculated for any wind speed and direction. The resulting slick area and orientation model is presented in polar coordinates as a function of wind speed in Figure 18. For a given wind speed and direction, plot the wind vector in the appropriate direction and read the slick area (dashed curves) and orientation (radiating solid lines).

Some interesting effects can be seen in the slick model (Figure 18). For example, slick area is predicted to be a strong function of wind direction, independent of wind speed; the largest slicks occur in the northeastern quadrant. Oil slicks also are consistently 20° or more to the right of the wind in the northeastern quadrant but only 10 to 15° to the right in the southeastern quadrant. The southwestern quadrant shows oil slicks only barely to the right of the wind, and even to the left in some areas, a situation that is due to the blocking effects of Pass A Loutre. In the northwestern quadrant, where both Ekman effects and coastal blocking effects tend to deflect the slick to the

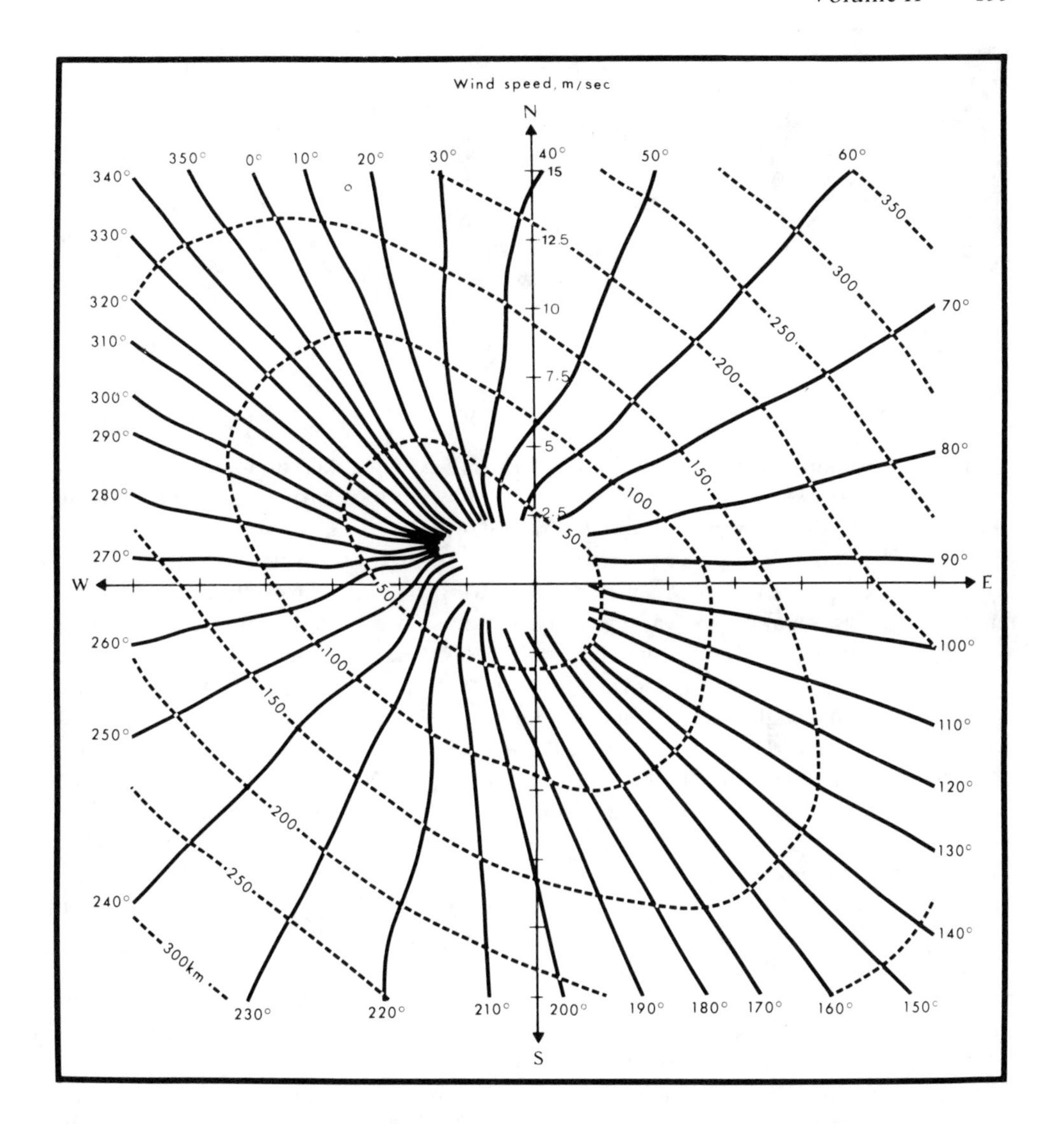

FIGURE 18. A multivariate regression model of oil slick area and orientation as a function of wind speed and direction. Plot the wind velocity in polar coordinates and read the slick area in square kilometers (dashed lines) and the slick orientation in degrees true (solid radiating lines). For example, a 10-m/sec wind toward the northeast produces a 200-km^2 oil slick oriented at 067°. (After Murray.[51])

right of the wind, such deflections reach 45°. It is clear from this data set that the large-scale forcing of oil-slick movement by atmospheric wind systems is strongly constrained by the topography, bathymetry, hydrology, and hydrography typical of this particular area. Further experience may show us that models of spills in most coastal areas may have to be treated on a case by case basis because the general driving forces are so strongly modified by the complex boundary conditions.

2. Fronts

The close control of the oil slicks during the Chevron spill by the winds associated with large easterly-migrating pressure cells is clear in the data. Also of major significance was the role of density fronts in blocking slick movement and realigning it to directions quite different from that of the local wind, as seen, for example, in Figures 15G and 15H. A thermal scanner image (Figure 19) of the Chevron slick by the U.S.

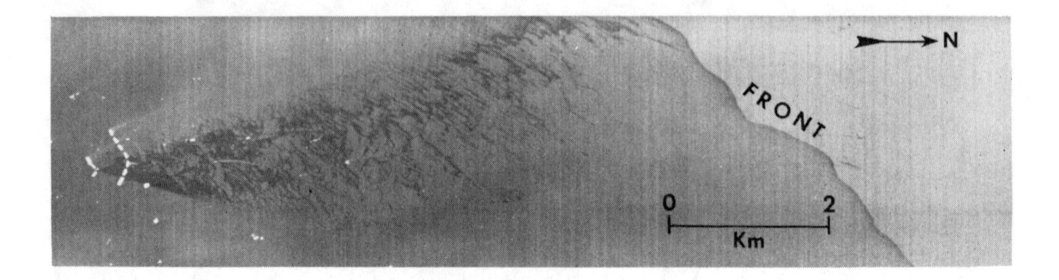

FIGURE 19. Thermal scanner imagery of Chevron slick on March 16, 1970, illustrating slick cutoff by a density front. The seven white rectangles in a curved line to the left of the figure are barges from which oil containment booms were suspended, obviously ineffectively (NASA imagery).

National Aeronautic and Space Administration (NASA), taken at 1500 hr on March 16 in the time between Figure 15A and 15B, shows an extensive front truncating the slick and a concentration of oil along the frontal line.

While it has long been general knowledge that such frontal zones are characterized by strong surface convergences, as evidenced by the usual collection of flotsom, an important series of papers by Garvine and associates[53] has brought considerable insight into their kinematics and dynamics. Figure 20 shows Garvine's data on the density and velocity structure across a front formed by the Connecticut river plume in Long Island Sound at seasonal high river discharge. The brackish water plume is coming in from the right of the figure, and a 10-m wide density front is seen between −10 and −20 m. Normal to the front in the surface layer (Figure 20B) offshore water approaches the front at speeds of 40 to 50 cm/sec, while inside the plume brackish water also approaches the front at speeds of 20 to 30 cm/sec. Garvine points out the obvious efficiency of this 80-cm/sec velocity difference in trapping and concentrating material at the front. The velocity component parallel to the front (Figure 20C) is "... rich in variation and seemingly uncorrelated to the density field...," emphasizing the three-dimensional complexity of the problem.

Several other types of fronts that could conceivably affect oil movement at the surface are discussed in a proceedings volume from a recent workshop on coastal fronts. Shelf break and upwelling fronts are discussed by Horne[57] and Mooers et al.[58] Of particular interest are the so-called "shallow sea fronts" produced by tidal mixing,[59] since they are often observed in the shallow seas of the European continental shelf, the location of many of the major tanker routes of the world. The ratio of the rate of production of potential energy necessary to maintain well-mixed conditions to rate of tidal energy dissipation is shown to be proportional to $h/|U_T|^3$, referred to as the stratification parameter, where h is water depth and U_T is tidal current amplitude. Given the same tendency to stratification in two adjacent volumes and intense tidal mixing in only one of them, we may expect a front to develop between them. Analysis of infrared (IR) images and ship data indicate that frontal regions in the Celtic Sea, English Channel, and Irish Sea between well-mixed and stratified regions do indeed reflect the shape of contours of the stratification parameter, thus suggesting mixing induced by tidal currents as the origin of these fronts.

Another type of front, referred to as a "headland front", is discussed briefly.[60] Such fronts are thought to form as a result of strong acceleration or jets around coastal headlands and accompanying flow separation, mainly at the tidal time scale. Although only poorly known, such features will certainly affect a spill moving along a coast displaying large promontories or irregularities. This process was clearly important in the *Amoco Cadiz* spill discussed subsequently.

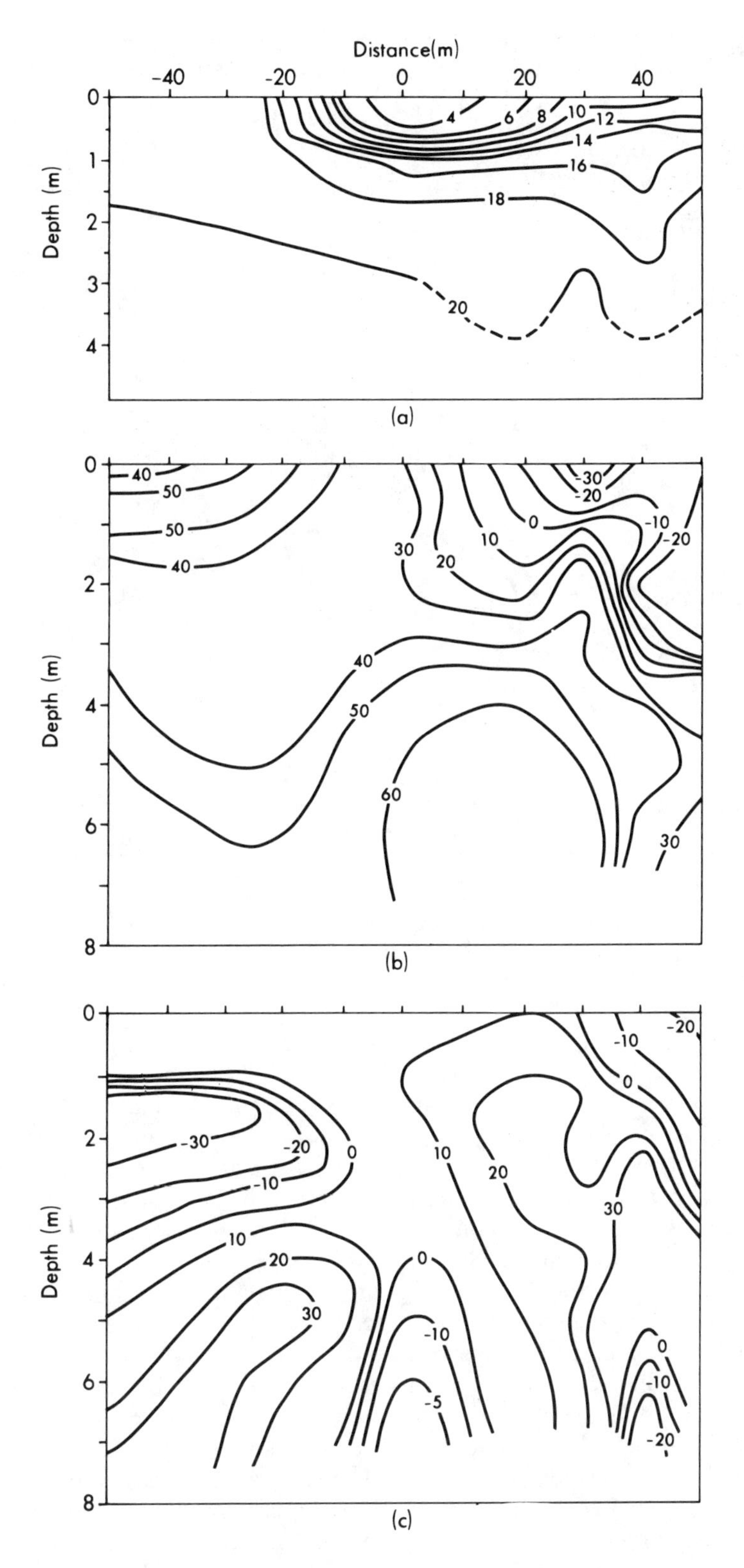

FIGURE 20. (A) Density section in σ_t units normal to the boundary of the Connecticut River plume. Note the surface front between −10 and −20 m. (B) Velocity component normal to the front (cm/sec). (C) Velocity component parallel to the front (cm/sec). (From Garvine, R. W. and Monk, J. D., *J. Geophys. Res.*, 79, 2251, 1974. With permission.)

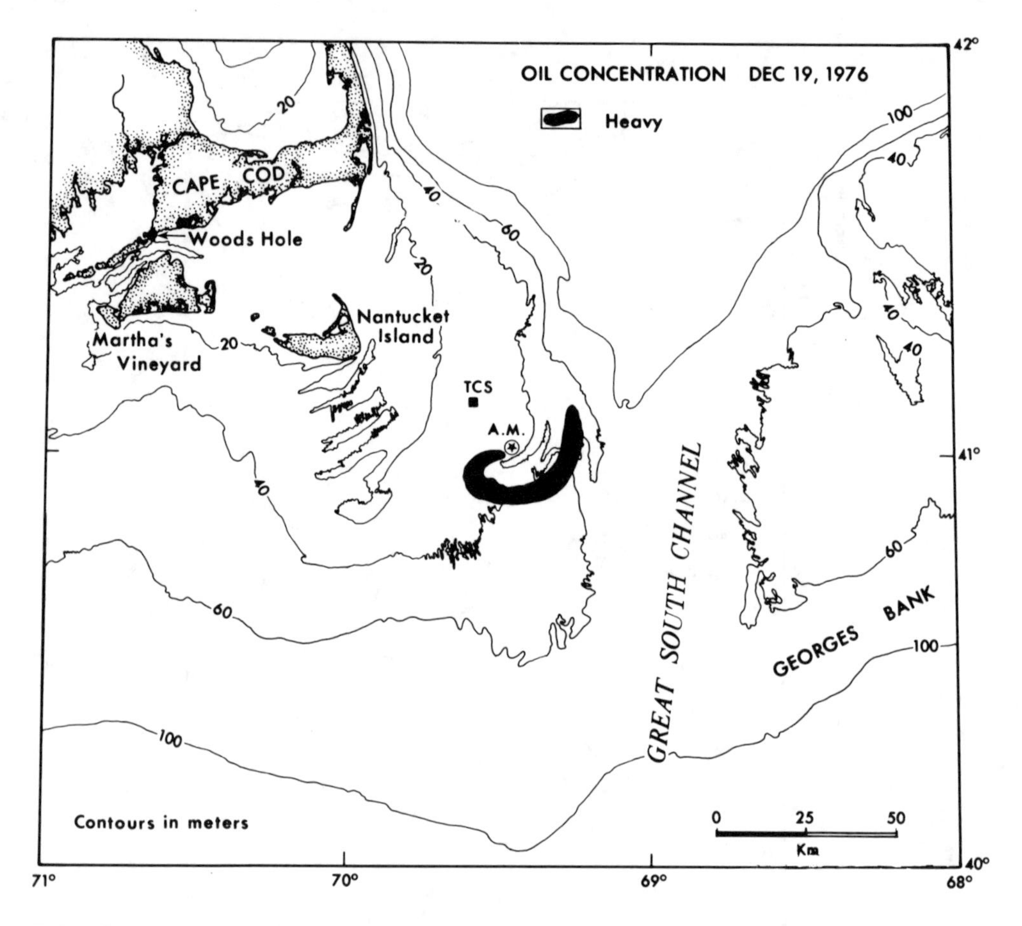

FIGURE 21. Location of the wreck of the *Argo Merchant* and an adjacent National Ocean Survey tidal current station (TCS). The hook-shaped oil spill of December 19 is caused by the combined effect of a southwesterly wind and strong tidal currents.

C. The *Argo Merchant* Spill

At about 0600 hr on December 15, 1976, the tanker *Argo Merchant*, carrying 7.7 million gallons (see Table 1) of No. 6 fuel oil, ran aground on Nantucket Shoals, about 50 km southeast of Nantucket Island, off Massachusetts (Figure 21). The eventual breakup and release of the oil resulted in the largest and most intensely studied oil spill to occur up to that time. It is a good example of the combined effects of wind and tidal currents. An excellent compilation of data taken on the site, with some preliminary analyses, is presented in Grose and Mattson.[2]

Tidal currents on the Shoals are quite strong, with a weakly rotary semidiurnal cycle being channelized by the complex north-northeast/south-southwest trending bottom topography in the area of the wreck. Monthly average tidal currents at a National Ocean Survey station 15 km to the northwest (TCS on Figure 21) are shown in Figure 22A. Maximum flood and ebb currents exceed 100 cm/sec. Winds from December 15 to December 31 (plotted from data in Grose and Mattson[2]) are shown as Figure 23. On the second day (December 16) they were from the east-northeast or onshore because the site of the grounding was in the northwestern quadrant of a well-defined low-pressure cell. Oil was observed to be moving shoreward toward Nantucket Island.

Fortunately, by the morning of December 17 the low center had moved well north of Nantucket, putting the site in its southwestern quadrant with accompanying northwesterly or offshore winds, a condition that persisted through December 18 and

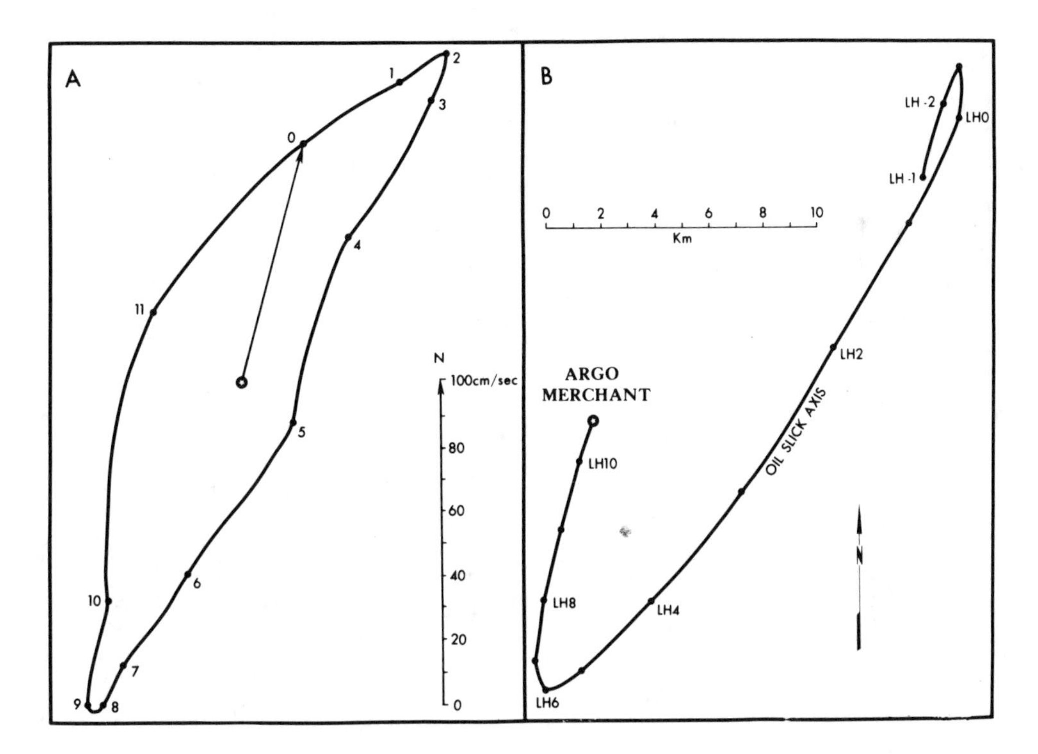

FIGURE 22. (A) Tidal current diagram at station TCS on Nantucket Shoal, located in Figure 21. (B) The axis of a slick computed hourly from the addition of a wind-driven current from southwesterly winds and the tidal currents of Figure 22A. Compare to observed slick in Figure 21.

until midday on December 19. The ship was still intact but leaking heavily when the oil slick shown in Figure 21 was mapped during the afternoon of December 19. The wind shift from northwest to southwest near midday of December 19 (Figure 23) was caused by the southeastern quadrant of a second massive low-pressure cell moving in over the spill area directly behind the earlier low. Both of these pressure systems are shown on the 7 a.m. weather map of December 19 (Figure 24); the first low is seen exiting the map in the top right of the figure, while the second low, moving easterly at 32 km/hr, is seen west of the Great Lakes. That the hook-shaped oil slick of Figure 21 is caused by the superposition of these southwesterly winds and the strong local tidal current is clearly shown by reference to Figure 22B. On this figure the axis of the oil slick is calculated by combining a wind-driven current from the southwest (moving at 4% of the 10 m/sec wind) and the tidal current of Figure 22B. The locations of oil parcels released at lunar hour 10.5 of the tidal cycle and preceding hours as shown are tracked and plotted as the axis of the slick. The agreement is quite good. Note especially that the observed 20-km northeastward excursion of the oil from the *Argo Merchant* (Figure 21) matches closely the calculated value (Figure 22B) of 19 km. The turning of the tidal current back to the southwest overpowers the northeast-setting wind-driven current and limits the northward migration of the oil until the next tidal cycle. Murray et al.[61] noted this same type of slick even in the weak tidal currents of the Gulf of Mexico during rare periods of calm winds and referred to it as a "tidal advection slick".

On December 20, 21, and 22 winds remained out of the southwest, west, and northwest as this second low-pressure cell traversed easterly with its center to the north of the wreck site. At 0835 on December 21, when winds were from due west and over 20 m/sec, the *Argo Merchant* broke in two, releasing an estimated 1.5 million gallons

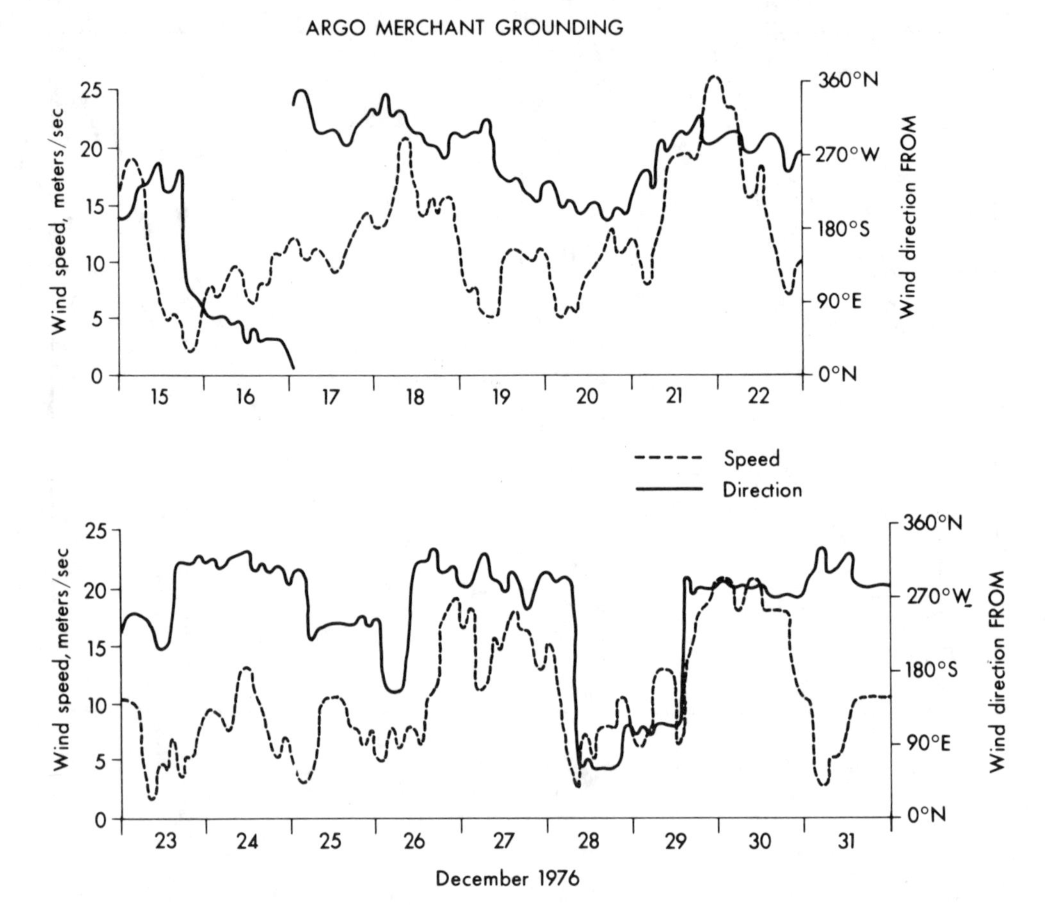

FIGURE 23. Winds at the site of the *Argo Merchant* grounding during the spill (plotted from data in Grose and Mattson[2]).

(Table 1) of oil. Fortunately, these high winds, generating about an 80 cm/sec current offshore (3% rule), moved the oil directly seaward, as shown in the slick map of Figure 25. As the wreck site is at the edge of Nantucket Shoal, water depth increases rapidly to the east with a corresponding decrease of tidal current velocity. The oil, then, was soon out of the strong tidal current regime and moving with the wind. On the following day, December 22, the ship split again, releasing another 1.5 million gallons (Table 1) of oil, which moved out easterly with the still strong westerly winds (Figure 23).

The second low was followed by a third strong low, also with its center tracking easterly to the north of the wreck site, bringing the southwesterly and northwesterly winds of December 23 and 24. A high-pressure cell briefly dominated the weather map on December 25, but the location of its center to the south of the wreck site also brought southwesterly winds, keeping the huge slick mapped in Figure 26 trailing out over 200 km to the east. Two more lows followed in succession to the end of December. The east winds of December 28 and 29 resulted from the fact that one of these lows was centered to the south of the wreck site, but fortunately the low wind speeds (~5 to 8 m/sec) could not move the oil far enough onshore to cause damage before the low center passed to the north and its southern quadrants brought the return of westerly offshore winds. On December 31 the leading edge of the spill was mapped 260 km southeast of the wreck, and it was clear that the bulk of the oil was lost to large-scale circulation of the Atlantic. Leaking continued from the ship for several more

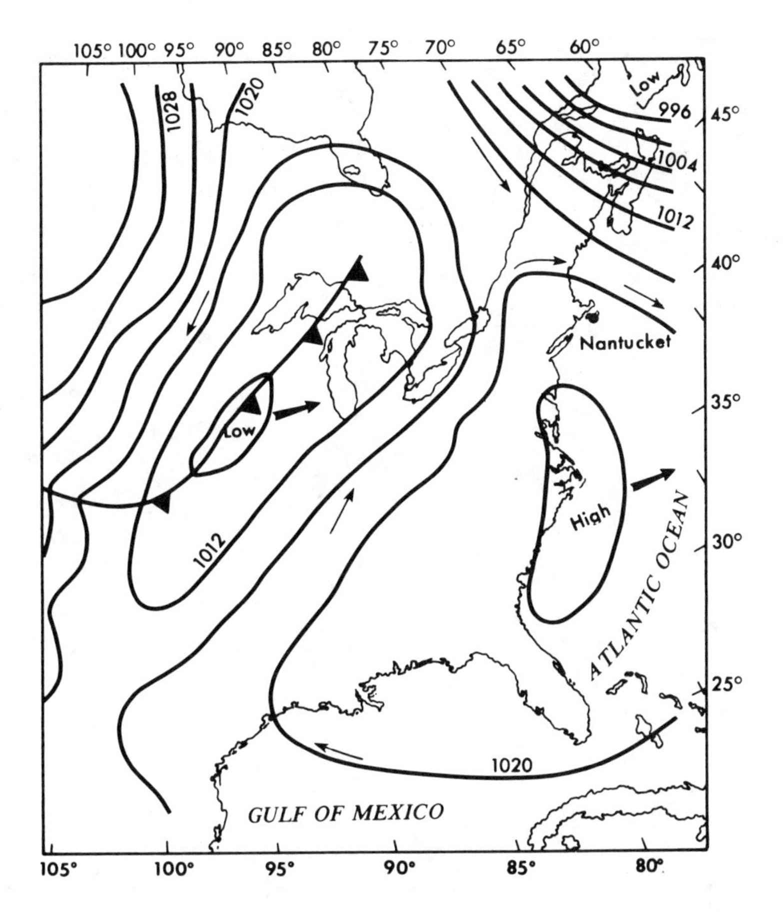

FIGURE 24. Surface map of atmospheric pressure on December 19, 1976, during the *Argo Merchant* oil spill. Heavy arrows indicate direction of movement of the pressure centers, lighter arrows are wind directions (contours in millibars).

weeks, but in retrospect the most critical time was the period of December 20 to 25, when the ship repeatedly broke up and the cyclonic weather patterns continually brought strong southwesterly winds to drive the oil offshore.

Among the first and most interesting results of the mapping of *Argo Merchant* slicks was the description of the formation and movement of large, thick oil slicks known as "pancakes", which maintained their cohesiveness and integrity for weeks at a time. On close inspection a typical area that appeared to be completely covered with oil was estimated roughly as 40% clear water, 60% sheen (a thin iridescent film of oil displaying the colors of the rainbow), and less than 1% covered with pancakes. A great variety of sizes were reported, from 1 m in diameter up to the giant Pancake I located on December 25, measuring 215 × 90 m. It was estimated to contain nearly ½ million gallons (Table 1) of oil.

The pancakes played such a prominent role in the spill that a Navy diving team was called to inspect one in the water on December 23. Their observations, summarized in Grose,[62] indicated the bottom of the pancake was flat and smooth and did not have a keel or ridges extending into the underlying water. This was an important point, suggesting that the oil pancake was somewhat uncoupled from the water column and that momentum from the wind would tend to concentrate in the oil. Frictional retardation might best be modeled by a skin drag rather than an eddy diffusivity. The edges of the pancake were sharp and relatively square in shape and did not taper at the edges

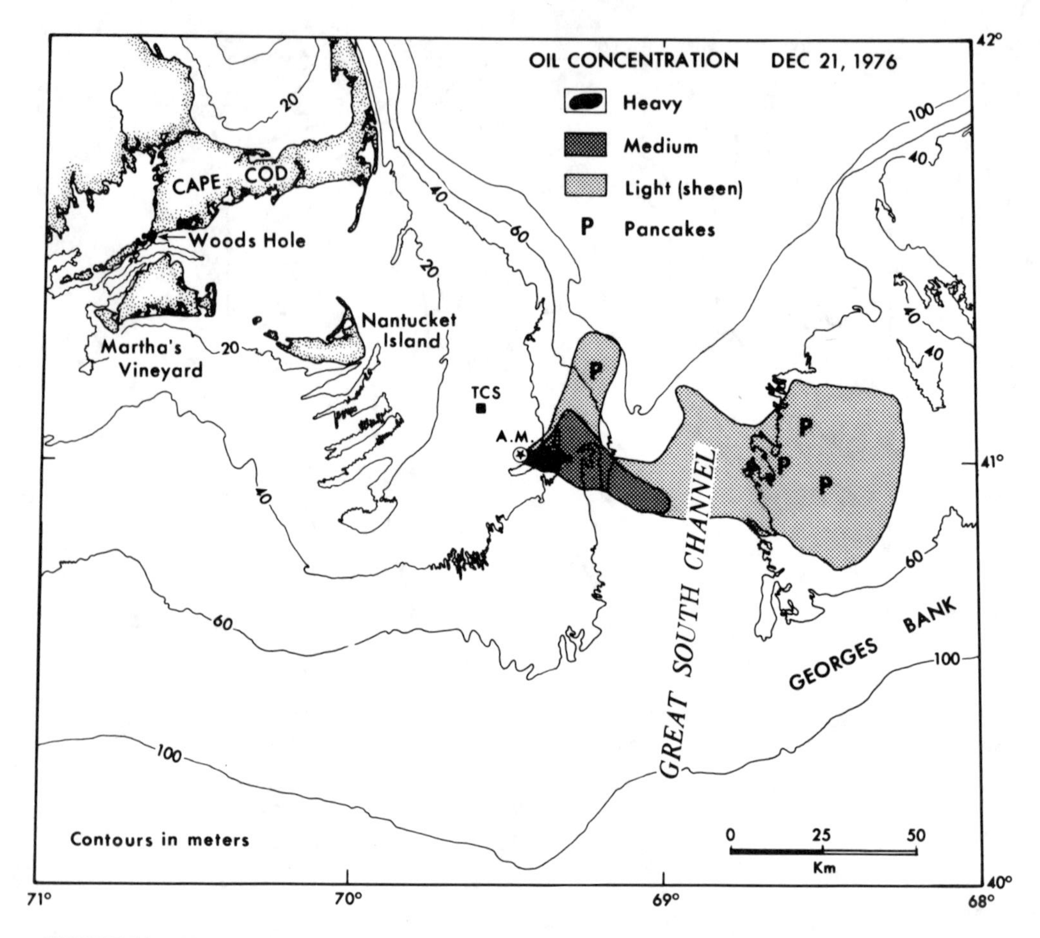

FIGURE 25. Map of oil slick from the *Argo Merchant* on December 21, 1976, the day the ship broke into two sections, releasing an estimated 1.5 million gallons of oil.

as in a buoyant lens. Bubbles of air from the divers would clear the oil in a small circle, but it soon coalesced again, "... indicating that there were forces acting on or within the pancake to maintain it as a distinct entity."

As the slick grew to a size unmappable from shore-based helicopters, a NOAA data buoy that located itself via the Nimbus F satellite was deployed on December 31 into what was thought to be Pancake I. This was a very successful move, as the buoy was used for the next 2 weeks to locate and map the oil slick. Speeds computed from daily fixes beginning January 1, 1978, generally range from 0.5 to 1.0 m/sec but sporadically reach high values such as the 2 m/sec speed on January 13. The buoy track indicated that the *Argo Merchant* oil continued moving easterly and eventually became incorporated into the general circulation of the North Atlantic Ocean.

D. The *Amoco Cadiz* Spill

While the spread of oil from the *Argo Merchant* was dominated by oceanic processes and unaffected by shelf or coastal current fields, the grounding of the *Amoco Cadiz* on March 16, 1978, gave rise to an oil spill quite different in nature and subject to quite different physical processes from those discussed up to this point. Carrying 216,000 tons (Table 1) of crude oil and 4000 tons of bunker fuel, the *Amoco Cadiz* ran aground only 1.5 km offshore of Portsall, on the northwest coast of France (Figure 27), and after breaking apart on March 17 released all of her cargo to the sea over the

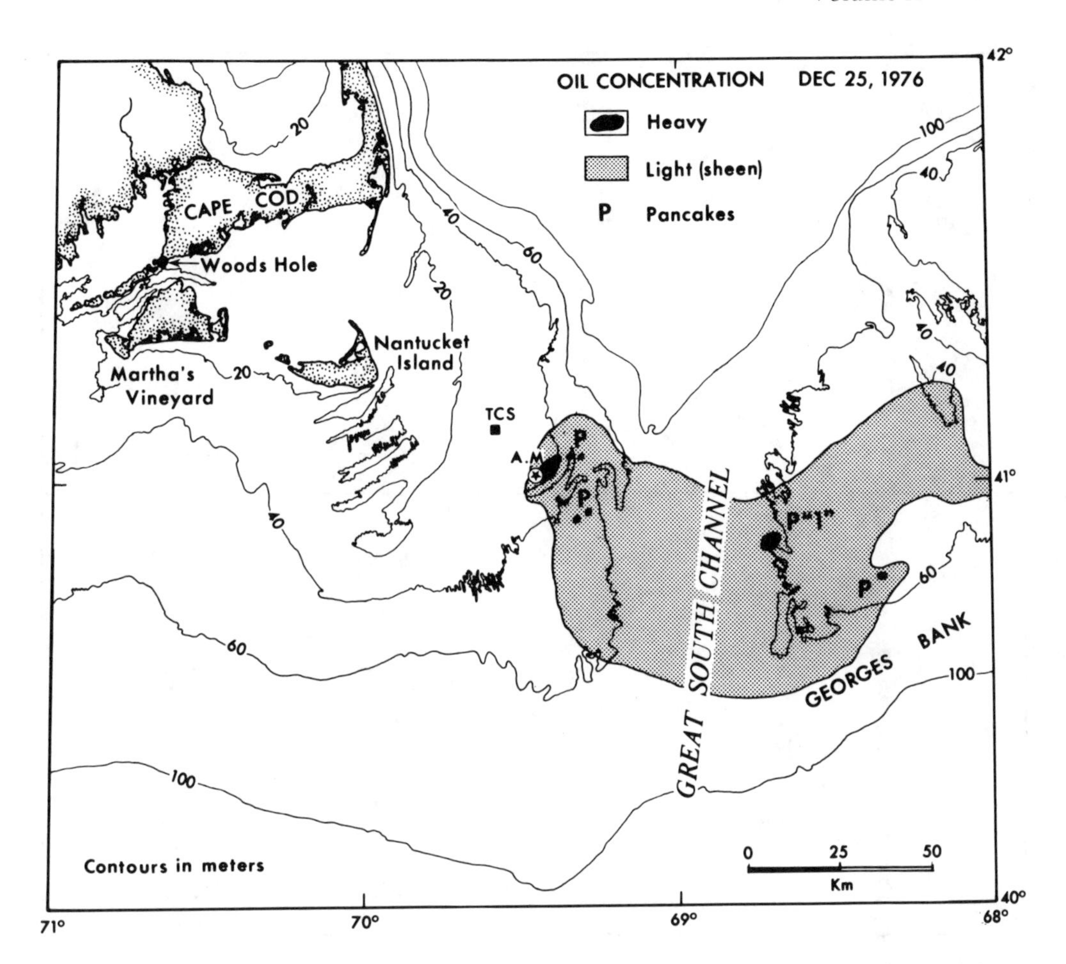

FIGURE 26. Map of oil slick from the *Argo Merchant* on December 25, 1976. Note the location of the large, thick, coherent oil slick referred to as Pancake I, which was estimated to contain nearly ½ million gallons of oil.

next 14 days. Due to the proximity to the coast and the persistent onshore winds (quite unlike the *Argo Merchant* situation), nearly half of the spilled oil came directly onshore within 100 km of the wreck site within the first 3 weeks of the spill,[63] and at one time or another oil was observed along 393 km of coastline. Unlike earlier spills discussed here, nearshore and inner shelf processes played a dominant role in the spreading of the oil.

1. Effects of Winds and Tides

An excellent series of maps utilizing an airborne thermal scanner to trace the *Amoco Cadiz* oil slicks has been published by a French scientific consortium.[64] Figure 27A shows the extent of the oil slick and its severity at about 1200 hr (GMT + 1) on March 18, 1978, or about 36 hr after the grounding. During the two and a half tidal cycles between the breakup of the ship at 0600 hr on March 17 and 0900 hr on March 18 (3 hr prior to Figure 27A), winds were from the north at an average speed of 7.6 m/sec, which could move the oil (4% rule) 33 km to the south, comparing favorably with the 27-km displacement seen on the figure. Heavy oil concentrations obviously just released from the tanker are seen, however, to extend out 6 km to the east-northeast.

As the winds had been blowing from the west at 4 to 5 m/sec for only about 3 hr, they were capable of extending the slick only 2 km in the appropriate direction. Tidal

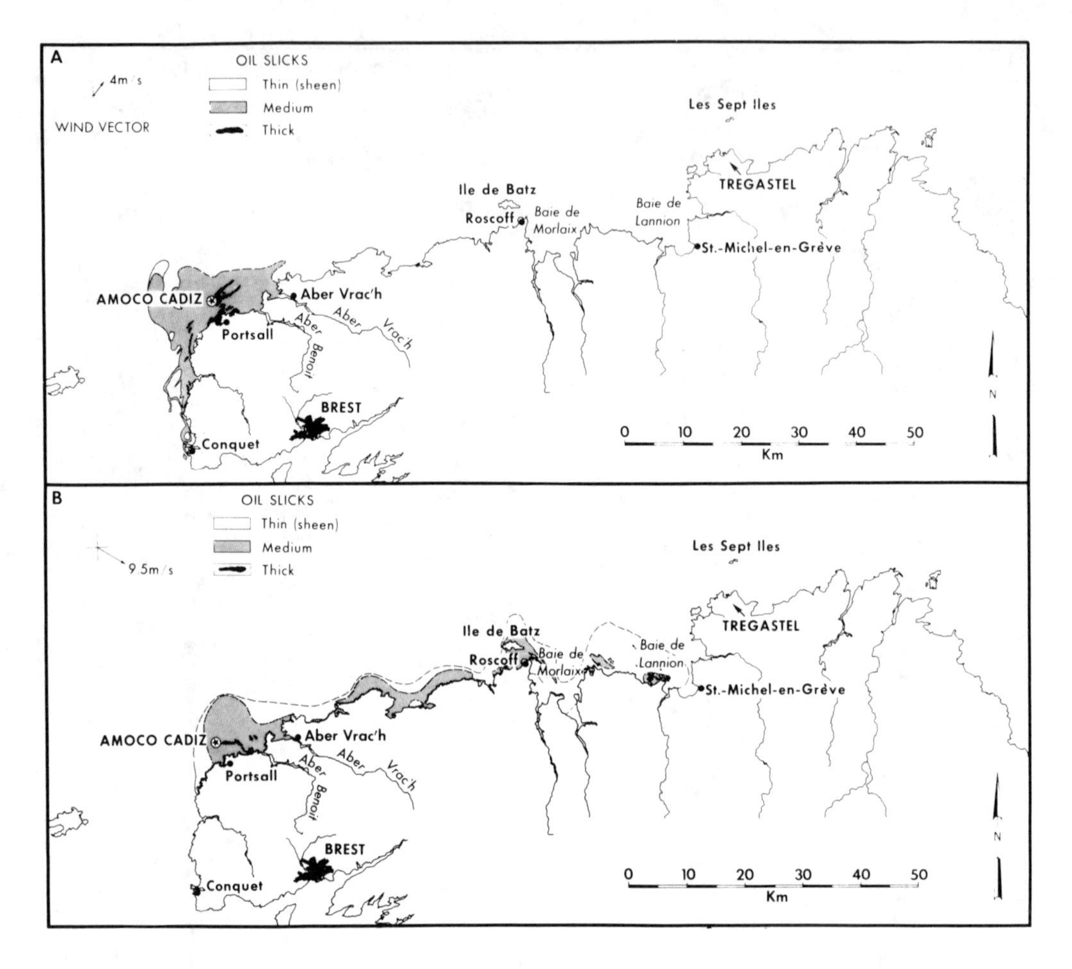

FIGURE 27. (A) The location of the wreck of the *Amoco Cadiz* off the Brittany coast of France. The oil slick is shown at 1200 hr on March 18, 1978, 36 hr after the grounding. The town of Roscoff, 50 km east of the wreck site, is also located on Figure 13 for geographic reference. (B) The *Amoco Cadiz* oil slick on March 21.

currents, on the other hand,[48] had been running to the east-northeast for 3 hr at an average speed of 55 cm/sec, and were easily capable of carrying the heavy oil concentrations 6 km in that direction. Here again, as in the *Argo Merchant* case, in an area of strong tidal currents the near-field distribution of oil is dominated by the tidal currents and the far-field by the long-term effects of the wind. Tidal currents[48] at the wreck site generally run parallel to the shore, making the near-90° turn in the coast at Portsall in a smooth curving motion with current amplitude of 75 cm/sec at neap tide and 125 cm/sec at spring tide. However, a distinct convergence in the tidal current field at the bend in the coast for a 2-hr period near low tide forces the flow to turn offshore in a northwesterly direction for 20 km. This local nonlinear effect perhaps explains the large amount of oil extending 15 km northwest of the wreck site in Figure 27A, even though the winds (Figure 29) had no component in that direction up to the time of this mapping.

Five days after the grounding of the tanker, oil had spread as far east as St. Michel-en-Greve, as seen in Figure 27B. Covering this distance in the 75 hr since the wind turned to blow from the west indicates an oil speed of 33 cm/sec, or 4.2% of the 7.8 m/sec average wind speed. Note how the oil closely follows the contours of the coast, illustrated especially by the salient of pollution-free water in the Bay of Morlaix.

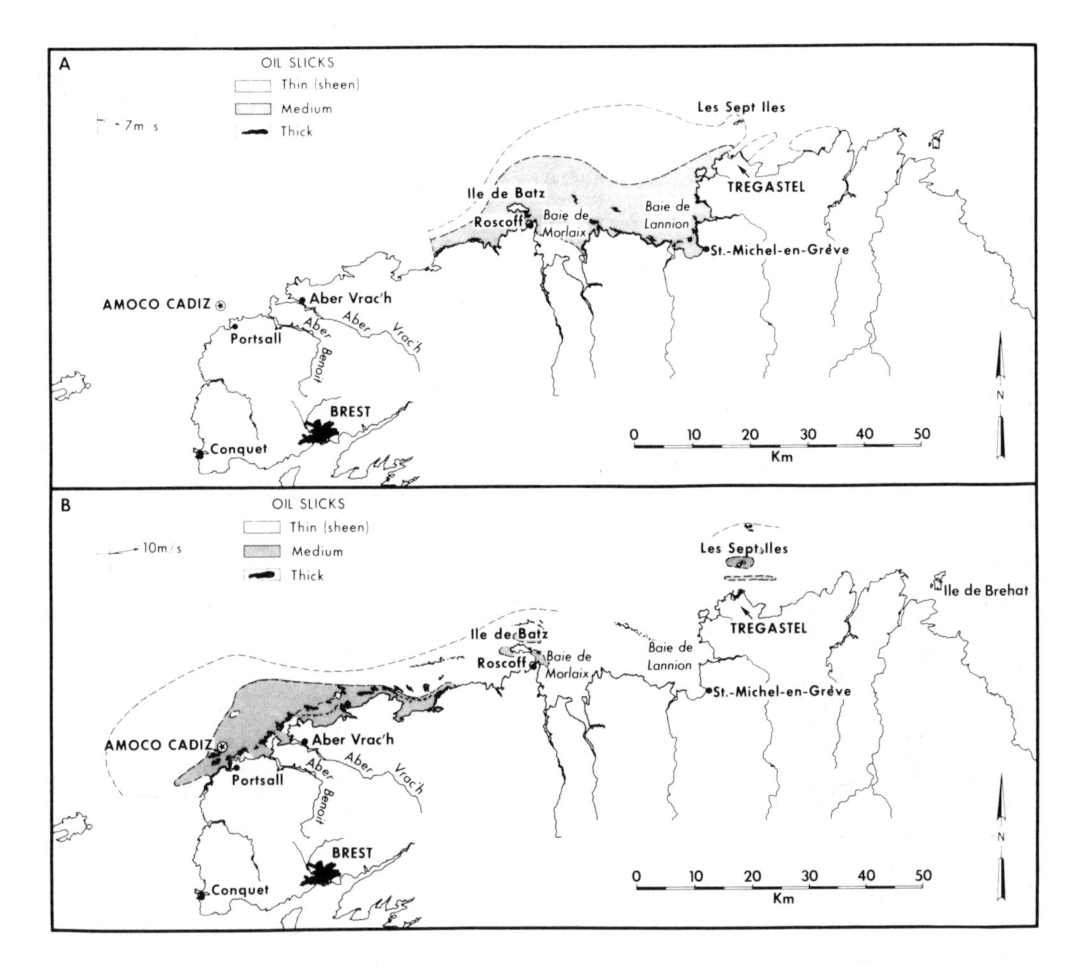

FIGURE 28. (A) The eastern portion of the *Amoco Cadiz* oil slick on March 24, 1978. (B) The oil slick on March 26, 9 days after the breakup of the ship. (All slick boundaries in Figures 27 and 28 after CNEXCO et al.[64])

Clearly the coastal current field and the streaming of oil are strongly influenced by the rugged indented topography, whose complexities have acted as side-wall roughness elements. Three days later (Figure 28A) the oil has filled the much larger Bay of Lannion, which has acted as an impoundment basin, holding the oil until it filled sufficiently to spill over and join the coastal current stream to the east. The average speed of the oil front of only 13 cm/sec between Figures 27B and 28A reflects this topographically induced deceleration. Two days later, on March 26 (Figure 28B), heavy concentrations of oil have rounded the Tregastel Peninsula, and oil is reported visually as far east as Ile de Brehat, indicating an oil migration speed of 17 cm/sec. Speeds of the leading edge of the oil have clearly decreased in the eastern region, probably due both to the large-scale impoundment and the depletion of the oil reservoir in the *Amoco Cadiz.*

2. Coastal Boundary Layer

By March 31 the thermal scanner shows relatively little fresh oil near the tanker, but large patches are seen all along the coast to the east, moved about not only by the wind and tidal currents previously discussed but also by other types of nearshore processes. For example, the *Amoco Cadiz* released its oil inside of what is referred to as the coastal boundary layer,[65] a wedge of water, often highly stratified, that generally

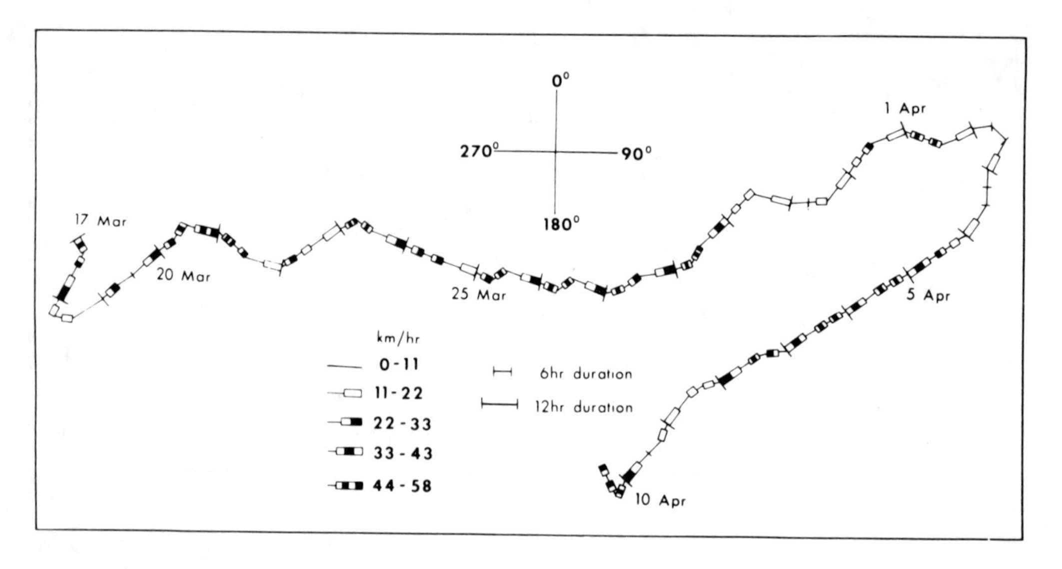

FIGURE 29. The wind velocities at Aber Vrac'h during the *Amoco Cadiz* oil spill. (From French meteorological sources, reproduced in Gundlach and Hays.[66])

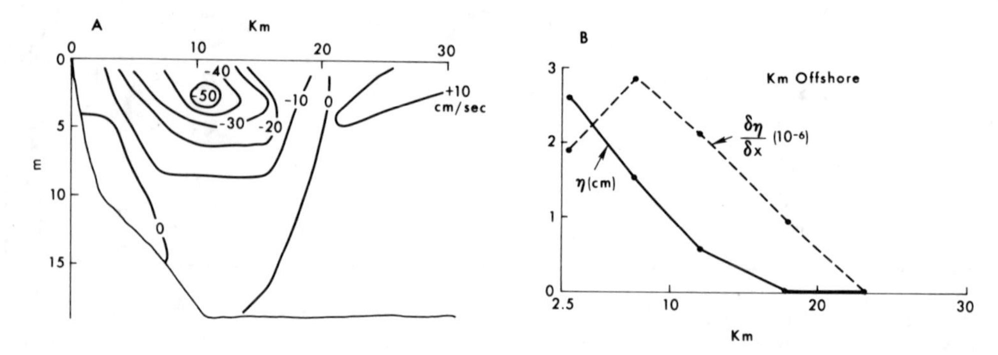

FIGURE 30. (A) The longshore current speed (cm/sec) in the coastal boundary current off Nicaragua. A similar feature likely contributed to movement of oil from the *Amoco Cadiz*. (B) Setup (η) and surface slope ($\delta\eta/\delta x$) across the coastal boundary current seen in (A).

extends out from the coast only a few tens of kilometers. The strong effect of the wind on these stably stratified waters, combined with tidal oscillations, strong density gradients, and the blocking action of the coast, make this a dynamic zone that is often uncoupled from the general circulation farther offshore. Winds during the first 2 weeks of the spill were generally westerly (Figure 29), a situation that, due to Ekman effects, tends to produce a setup of water along the coast. This zone of setup is associated with driving of a strong nearshore current directed to the right of the pressure gradient, in this case the current moving along the coast to the east. Under similar wind angles with respect to the coast, Crout and Murray[67] calculated, from finite-difference forms of Equations 8A and 8B, a 3-cm setup against the coast, which was instrumental in driving a 50-cm/sec current downwind along the coast. The surface slope calculated from the field observations and the well-defined coastal boundary current that they measured along the Nicaragua coast are shown in Figure 30. It is quite likely that the *Amoco Cadiz* oil was moving along the coast in a similar albeit smaller current feature.

Note in Figures 27 and 28 that heavy concentrations of oil have moved directly into at least six estuaries between Portsall and Tregastel. The Aber Vrac'h and Aber Benoit estuaries were especially heavily impacted, the oil pollution becoming entrained in the

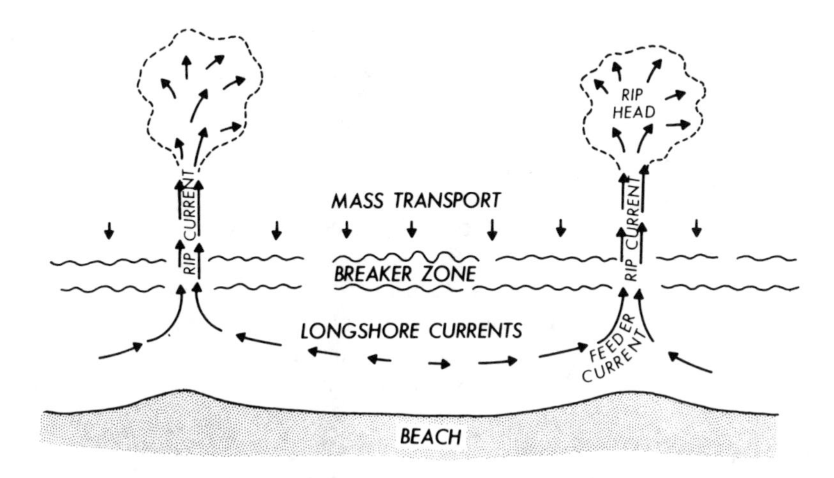

FIGURE 31. Schematic diagram of the littoral (longshore) current, feeder current, and rip current system probably influential during the *Amoco Cadiz* spill. (After Shepard and Inman.[70])

gravitational convection (two-layer mean flow) present in these channels. The effect of the salt-wedge type of circulation present in the Aber Benoit on the distribution of *Amoco Cadiz* oil is discussed by Allen et al.[68] A detailed discussion of estuarine dynamics is presented elsewhere in this volume (Chapter 8).

3. Littoral Current

Once driven within the zone of breaking waves, the *Amoco Cadiz* oil came under the influence of another current regime, usually referred to as the littoral current. Komar[69] emphasizes the two main contributions to the littoral current as (1) those associated with the nearshore circulation cell and (2) those generated by the oblique approach of shoaling waves.

According to theory, the distribution of momentum flux characteristic of shoaling waves must be balanced by a setdown and setup of the mean water level just outside and inside, respectively, the point of breaking. Variations in wave height along the coast due to topographic irregularities (e.g., offshore bars or rocks) will in turn produce longshore gradients in the mean water level that generate longshore currents. Shoreward of the largest waves the water level will be highest and the water will flow along shore toward the next adjacent point of lowest sea level, where it will turn seaward in a narrow band of current referred to as a rip current. The typical circulation cell, which can develop even with wave crests approaching parallel to the beach, is shown schematically in Figure 31.

Longuet-Higgins,[71] on the other hand, considered the case of waves approaching oblique to the beach, and by balancing the wave momentum flux with the frictional drag of the current he obtained the expression

$$U_1 = \frac{5\pi}{8}\frac{\tan\beta}{C_f}u_m \sin\alpha_b \cos\alpha_b \qquad (17)$$

where U_1 is the longshore current speed in the midpoint of the surf zone, β is the beach slope, C_f is a drag coefficient, U_m is the maximum orbital velocity at the breaking wave position, and α_b is the angle of wave approach. With the analysis of much field data Komar has shown that $5\pi/8 \tan\beta/C_f$ has approximately a constant value = 2.7, and Equation 17 then gives very good predictions of longshore currents. Of course in nature currents driven by water level gradients and obliquely shoaling waves nearly always coexist. With values typical of the *Amoco Cadiz* case, we may expect from Equation 17 littoral currents of 0.5 to 1.0 m/sec to have played an important role in the rapid eastward migration of *Amoco Cadiz* oil.

4. Coastal Geomorphology

With an oil spill this close to the shore the local coastal geomorphology may become equal in importance to the physical forces in determining the eventual distribution of oil on the beach. This was certainly true in the *Amoco Cadiz* case, where standard trajectory and forecast models designed for open-ocean conditions were useless.[20] The rugged, heavily indented topography of the Breton coast did indeed play a major areal role in determining where the oil collected.[66] During the first few weeks of westerly winds the west-facing shorelines were severely polluted, while those facing east, particularly within the larger embayments, were barely affected. After the sharp wind shift of April 2 (Figure 29) oil was forced into the mouths of these embayments, where the unusually high tides due to Kelvin wave effects in the English Channel, as explained earlier, then spread the oil high up the beach face and far inland. Color IR aerial photography[72] clearly shows zones of flow separation producing oil-free zones in the lee of headlands and large rocks. While not conclusive, the presence of headland fronts, as discussed earlier, is suggested by the photography to be playing a role in the oil distribution. Thus, as illustrated in Figure 32, several weeks after the spill, coastal areas previously free of oil were subjected to pollution. In fact, this sudden turn of the wind to the northeast reversed the large-scale longshore flow of oil to the east and polluted previously clean areas west of the wreck site all the way south to the Pointe du Raz (southwest of Brest).

Other small-scale geomorphic features common along this coast, emphasized by Gundlach and Hayes,[66] were the tombolos or ridges of sand that connect the mainland with offshore rocks or islands. Convergence of wave rays due to refraction effects over the shallow tombolos apparently led to the systematic trapping of oil behind the rocks and/or islands. Scour pits behind boulders and jointing and bedding patterns along this rocky coast were likewise noted as local oil traps.

5. Oil Pooling

Galt[20] emphasizes the role of pooling of oil against the coast, as observed in nearshore spills such as the *Amoco Cadiz.* Figure 33 illustrates the concept. As winds and tides push oil onto the shore the coast acts as a restraining barrier as new oil accumulates. The trapped oil accumulates into a seaward-thinning lens or pool. As the pool grows the amount of beach face wetted by the oil increases. This oil-wetted beach face is of prime importance, as it is here that the major deposition of pollutants will occur.

A coast consisting of a series of small bays and coves will become a feeder train, each small embayment acting as a pool or oil holding pond until it reaches its capacity and overflows into the next embayment downstream. Galt makes the interesting point that at high tide along the Brittany coast the oil pool is subject to the full effect of both the mean wave-driven longshore current and the flooding (eastward-setting) tidal current. As the tide recedes the oil in contact with the beach tends to be stranded at the high-tide mark, while oil seaward of this zone is "floated" off by the underlying water. A significant portion of the oil is thus removed from the westward-setting ebb tidal currents, in effect adding another component to the mean easterly drift of the oil caused by the large-scale wind. The phase relationships between tide level and tidal current (i.e., progressive vs. standing wave characteristics) then becomes an important process to model in terms of net oil transport in this type of spill.

It is interesting to note that oil pooling also played a major role on the oil spill at the West Hackberry oil storage spill.[73] On September 21, 1978, an accident occurred

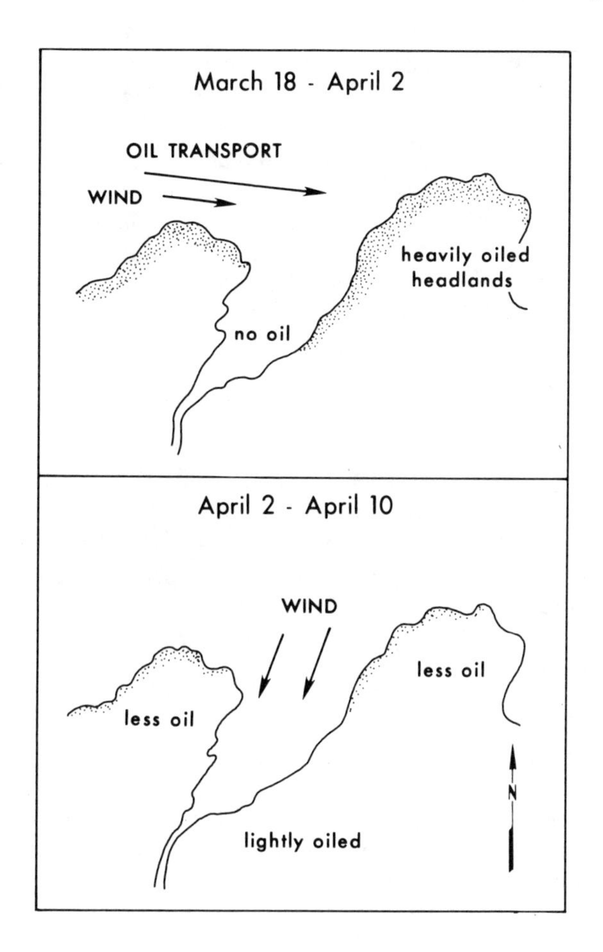

FIGURE 32. Schematic diagram showing how correlation between wind direction and dominant orientation of coastal lineations affected the spread of oil during the *Amoco Cadiz* spill. (After Gundlach and Hayes.[66])

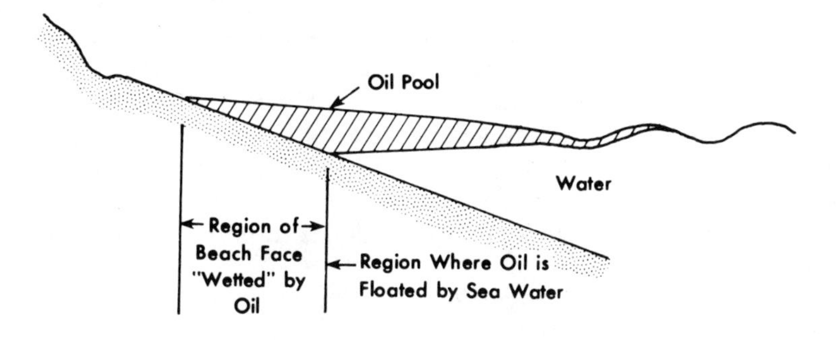

FIGURE 33. Diagram illustrating geometry of oil pool commonly seen during the *Amoco Cadiz* oil spill. (After Galt.[20])

at an oil rig involved in preparing a subsurface salt dome for oil storage in the Strategic Petroleum Reserve Program of the U.S. Department of Energy. An estimated 72,000 bbl (Table 1) of oil spilled to the surface, half of which escaped into adjacent biologically sensitive Black Lake. As oil escaped over a containment wall into the lake, a 5 to 7-m/sec onshore wind stress tended to keep the oil concentrated up against the coast, forming a pool that reached 0.5 m in depth at the shore and 0.1 m at a containment boom moored several hundred meters offshore. The pooling effect helped considerably in allowing nearly total recovery of the oil that escaped into the lake.

Some attempt has been made to study the dynamics of this phenomenon by Sorensen,[74] who studied both analytically and experimentally the pooling of oil behind a barrier such as an oil-collecting boom. He derived expressions for the thickness of the oil lens both for the case with waves and the case without waves. Without waves the dimensionless setup d_o/L, where d_o is the oil pool thickness at the barrier and L is slick length normal to the boom, is given by

$$\frac{d_0}{L} = 2.3 \times 10^{-3} \sqrt{\frac{\rho_W}{\rho_0}} \; \frac{W}{\sqrt{g'L}} \qquad (18)$$

where ϱ_w and ϱ_o are water and oil densities, respectively, W is the reference wind speed, and g′ is the densimetric acceleration of gravity $= g\,(1 - \varrho_o/\varrho_w)$. Experimental tests showed that Equation 18 underestimated the oil pool thickness. Sorensen attributed this to an increase in surface wind stress due to form drag on wind waves and formulated a new expression

$$d_0 = \sqrt{\frac{2BL\rho_W}{g'\rho_0}(W - W_c)} \qquad (19)$$

where the constants B and W_c were determined experimentally as 7.7×10^{-6} and 3.63 m/sec, respectively. W_c is the wind speed at which wave effects become important. While these expressions describe the laboratory data quite well, trial calculations for the *Amoco Cadiz* and Hackberry spills indicate that they underpredicted the oil pool thickness by a factor of 5 or more. Further work is clearly needed on this important aspect of the air-oil-water interface problem.

E. The Gulf of Campeche (Mexico) Spill

The role of large-scale wind and current fields has again come into focus with the largest oil spill in history, which began June 3, 1979, when the Ixtoc I test well blew out (see Figure 34) on the Campeche Bank in the Gulf of Mexico. The dominant dynamic feature of the Gulf of Mexico is the Loop Current, which enters the Gulf of Mexico at speeds of 50 to 75 cm/sec through the channel between the Yucatan Peninsula and Cuba, runs northerly, undergoes a tight 180° turn, and exits the Gulf through the Florida Strait between Cuba and Florida. As shown on the figure, some of the current coming through the Yucatan Passage does turn westerly across the Campeche Bank, toward the spill site, and then turns clockwise northerly with the coastline. A local wind rose shown on the figure indicates that 80% of the time winds are from the southeastern quadrant during July, also acting so as to push the oil northerly along the coast.

The distribution of oil as of August 6, after 53 million gallons (Table 1) of crude had spilled, is seen in Figure 35 to be closely controlled by the wind direction. Massive beach pollution was reported all along the Mexican coast, and oil started to come ashore between Brownsville and Corpus Christi about August 4. By August 5 the spill equalled the volume of the *Amoco Cadiz* and doubled it some time in October. The well remains uncapped at the time of this writing, on December 13, 1979.

Pollution of beaches north of Corpus Christi has not been reported in the press, indicating a halt in the northward progression of the oil between Brownsville and Corpus Christi. This is perhaps due to the convergence in the shelf current field thought to be present in this area and indicated in Figure 34. More likely, the expected increase

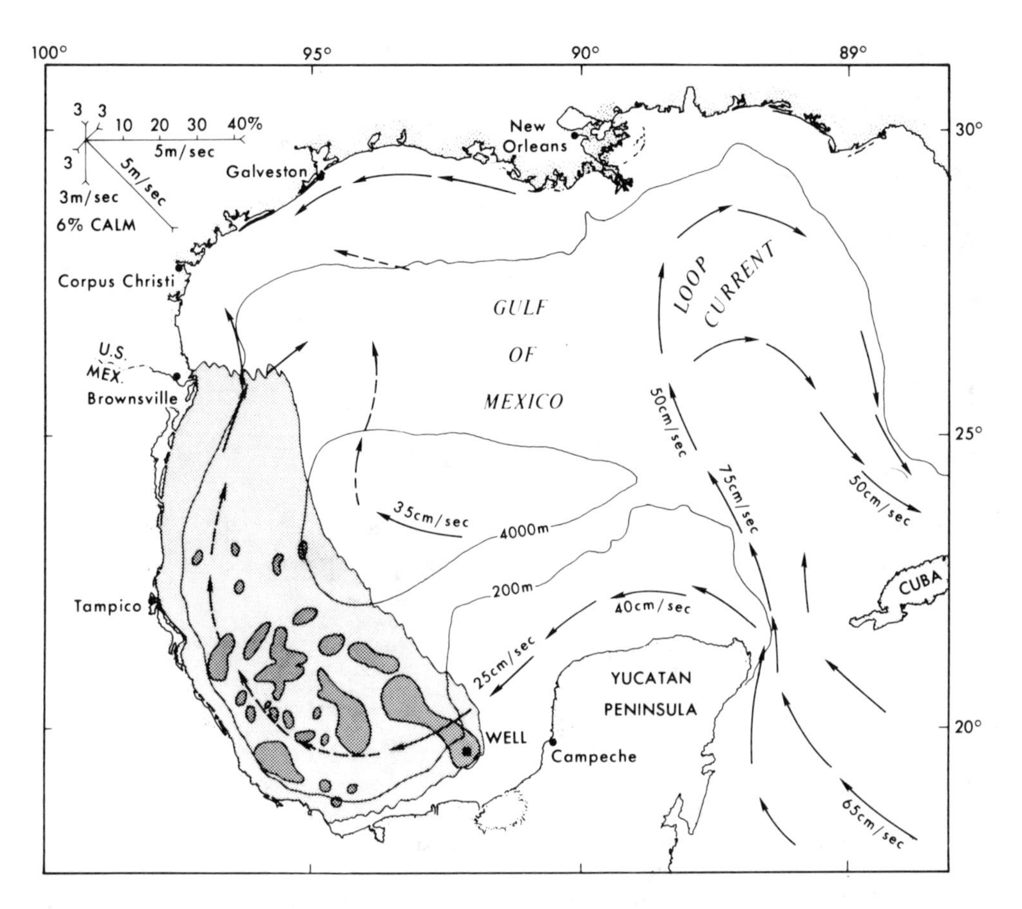

FIGURE 34. The oil spill from the Ixtoc I drilling rig on the Campeche Bank of Mexico on August 6, 1979. The spill, which began on June 3, 1979, is the longest and largest in history. The wind rose gives winds for an average July offshore of Tampico. (Oil distribution from United Press and Associated Press maps, currents from Oceanographic Office.[75])

of winds with northerly components off Brownsville from 16% in July to 30% in August to 60% in September resulted in the halt and/or reversal of the northward movement of the oil. These northerly winds are of course associated with the southerly encroachment of migrating cyclones and anticyclones, signaling the approach of winter.

VI. MODELING

The various processes discussed to this point may all, at one stage or another, come into play in determining the behavior of an oil slick. The time and length scales of each slick are good indices as to which processes will be important at a specific time in a spill history, as illustrated in Figure 35. During the first few hours, spreading, evaporation, and advection by tidal currents are dominant processes in nearly all marine spills, as seen, for example, in the *Argo Merchant* and *Amoco Cadiz* incidents. If the spill is close to shore, as with *Argo Merchant,* littoral currents may also come into play almost immediately. The approximate increases in the length scale of a spill as the duration of the spill increases can be estimated by following the slick growth curve given in the figure. Advection by the wind-driven current of local weather systems is followed in turn by the effects of migrating high- and low-pressure systems so important in the Chevron, *Argo Merchant, Torrey Canyon,* and *Amoco Cadiz* spills. Again,

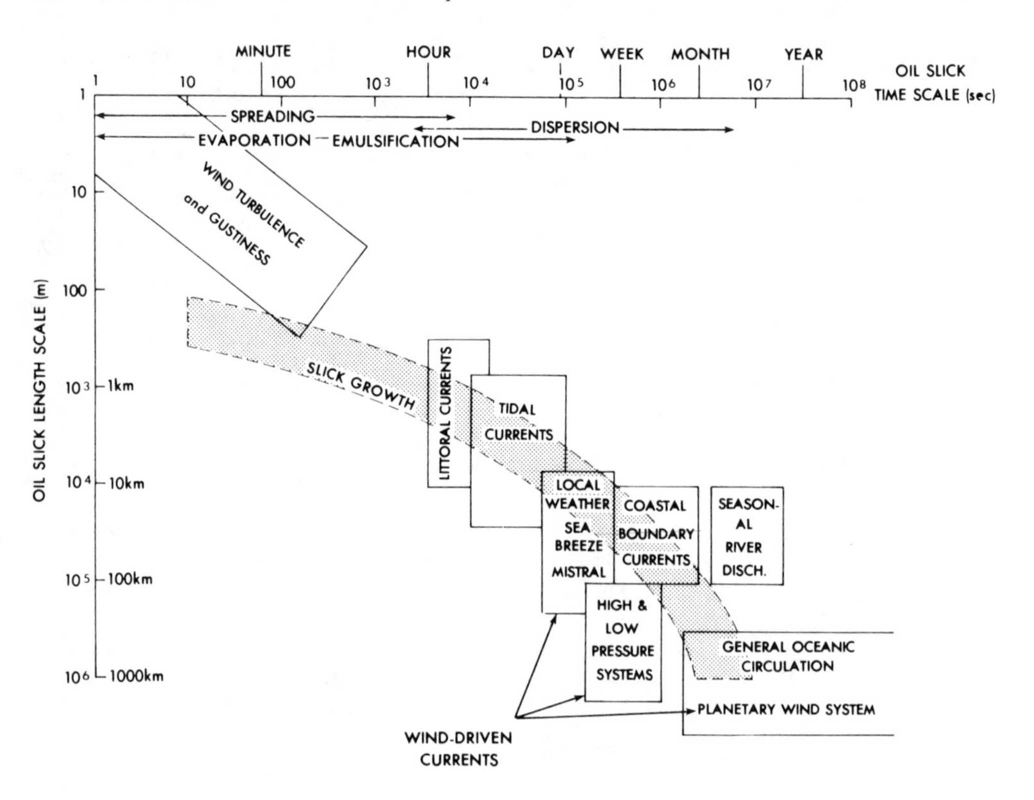

FIGURE 35. Diagram illustrating the time and length dependence of oil slick growth on the physical processes discussed in the text. (Extensively modified from Stolzenbach et al.[23])

if close to the coast, spills from a few days to a few weeks of age may experience significant advection from coastal boundary layer currents. The effects of seasonal river discharge were of enormous influence in the Chevron spill. A few months after the inception of a spill we can expect it to reach a size of 1000 km or more and to be influenced by the general circulation of the ocean and the planetary wind field and its fluctuations; the Campeche Bank oil spill, off Mexico, is an excellent example.

It is important to understand that processes operating at scales larger than that exhibited by a given oil slick at a given time act to displace the center of mass of the slick without deforming its shape, i.e., they advect the entire slick but contribute little to its dispersion. For example, a small spill of a few barrels of a highly volatile crude will be advected by a tidal current and may be completely evaporated before the tidal current reverses and begins to disperse the oil. Similarly, processes acting at length and time scales smaller than those reached by the oil slick act as dispersive processes, rearranging the distribution of the oil in space (spreading it over a larger area) but not providing a net displacement of the center of mass. In the *Amoco Cadiz* spill, for example, after several weeks the length scale reached several hundred kilometers, a magnitude produced by wind-driven currents arising from atmospheric pressure systems. Tidal currents, littoral currents, and coastal boundary layer currents are all still active, but they are embedded in the larger scale flow pattern and are now acting as mixing agents, rearranging or spreading the oil inside the larger slick.

A complete model of the temporal and spatial evolution of an oil spill will, then, not only have to deal with all these dynamic processes, summarized in Figure 35, but also must include the myriad number of local boundary conditions that characterize a real situation in nature, e.g., temporal and spatial variability of density fronts, flow separation around headlands and capes, and thermohaline driving in estuaries and arid regions of the world. The list is nearly endless. Realistically, then, this author believes that the best we can hope for is a series of models that include clusters of the driving forces that are typical of certain stretches of coast identified as critical for biological, recreational, or trafficability reasons. An alternate approach would be to model oil slick movement along the dozen or so types of shorelines[76] typical of the coasts of the world with the oil driven by the subset of forces or dynamical processes characteristic of that climatic region. Stolzenbach et al.[23] take an opposing viewpoint and emphasize the need to learn more of the hydrodynamical and thermodynamic details of the behavior of oil on water.

A. Spreading and Dispersion Models

The first generation of models has incorporated only one or more of the principal driving forces. For example, Fay's[25] work, discussed earlier, was limited to the spreading process (i.e., oil slick growth driven by gravity, buoyancy effects, and, most importantly, surface tension) and thus is restricted to rather short time and length scales during which, as we have seen, dispersion and evaporation are also of major importance. In contrast to Fay's approach, Murray's[27] analysis of selected oil slicks during the Chevron spill in terms of turbulent diffusion theory indicated that dispersive processes were far more important than spreading in a continuous oil spill. In his model slick geometry and total area can be predicted as a function of current speed, horizontal eddy diffusivity, and oil discharge rate, neglecting both evaporation and spreading processes.

Evaporation and spreading, as well as dispersion (slick growth induced by turbulent motions of underlying water), are incorporated in a model by Stolzenbach et al.[23] that is especially interesting in that it treats a two-component oil slick. Evaporation is calculated as a function of wind speed and the vapor pressure of each of the oil fractions in the mixture, spreading rate is obtained from a slightly more elaborate version of the Fay[25] approach, and the growth rate from dispersion is obtained from an Okubo type power law

$$\left.\frac{dR}{dt}\right|_{disp} = \frac{\kappa t^n}{R} \quad (\kappa = 0.01, n = 1.3) \tag{20}$$

of the type discussed in detail elsewhere in this volume. In Equation 20 R is slick radius and κ and n are empirical constants. Trial computations showed that for times less than a day spreading is clearly dominant, while for times greater than a week dispersive processes dominate. At intermediate time and length scales both processes must be considered. Contrary to the observations cited earlier, which emphasized its effect within the first 2 hr, this model suggests that evaporation is important in the interval between 3 hr and 1 day after spill inception.

B. Advective Models

Another class of simple models is completely advective in nature, incorporating no processes other than the displacement of oil parcels by surface currents in exactly the same manner that Figures 5 and 22B were constructed. A recent example of this approach is given in Lissauer and Welsh,[77] where a simple vector addition of 3.5% of the wind speed and the tidal current were used to predict the extreme outer limits of slicks emanating from the *Argo Merchant*. The rotary movement of the tidal current is used to estimate the lateral extent of the oil slick rather than employing a spreading or dispersive relationship. As the model does not allow degrading of the slick through evaporation or dispersive processes, it does overpredict the areal extent of slicks. As the authors note, however, for a quick and reasonably accurate estimate of where the oil will go and how fast it will move, this is an effective method. While useful in terms of the *Argo Merchant* spill, which had no coastal boundaries to complicate it, this simple vectorial addition approach will be of very limited use in realistic coastal spills such as the *Amoco Cadiz.*

A more sophisticated advective model has been advanced by Warner et al.[78] to hindcast the oil slicks from the tanker *Arrow,* wrecked off Nova Scotia in 1970. It is essentially a time-dependent Ekman-type model, being based upon an integral solution to the differential Equations 8A and 8B, including only terms 1, 6, and 7, the local acceleration, the Coriolis force, and the frictional stress, parameterized as the eddy viscosity, respectively. It has the distinct advantages of being theoretically more sound than the wind factor approach and capable of handling a fluctuating wind to drive the slick movement. Disadvantages include the assumption of an eddy viscosity constant with depth and its restriction to offshore areas free from coastal boundary effects. Shelf currents other than direct wind-driven, however, are neglected in the model (e.g., geostrophic, tidal, density gradient). Nonetheless, after fitting a vertical eddy viscosity the model did produce a trajectory that closely simulated observed oil slick motion, whereas the 3% rule proved to be entirely unsatisfactory.

C. Sea Dock Model

A newer class of models has now begun to emerge which include a combination of the processes summarized in Figure 35, of which the Sea Dock model[79,80] is a good example. Sea Dock is a proposed offshore oil unloading facility to be located about 40 km south of Galveston, Texas, a port city located on Figure 34. This model includes wind-driven currents, spreading, evaporation, dissolution, and sedimentation. Biodegradation is discussed but omitted on time-scale considerations. Spreading is accounted for with the usual Fay relationships, but dispersion, clearly more important at the length scale between Sea Dock and the coast, as indicated by the Chevron spill studies[27] and Figure 35, is unaccountably ignored. Exponential decay functions are assumed for the evaporative flux and the loss of hydrocarbons in solution to the water. Sedimentation and emulsification were crudely modeled with a 1% reduction in the oil slick, a computation performed only when the slick is within the "inshore area" (explained below) and wind speeds exceed 32 km/hr, i.e., they are capable of causing the necessary turbidity and agitation to sink and emulsify the oil.

Two sets of wind data were used to derive the model; the first set, representing offshore conditions, was measured over a 23-month period at an offshore oil platform, and the second set, representing coastal conditions, was obtained from the National Weather Bureau station at Galveston. Wind speeds were reduced from the observed values at the 30-m level to the standard 10-m level using a 1/7 power law profile. More sophisticated techniques to forecast the 10-m wind from meteorological observations are discussed extensively in Stolzenbach et al.[23] Currents 3 m below the surface at the offshore platform, measured simultaneously with the wind velocities, were also used in the model.

Showing some insight into the spatial variability of the forcing processes, the author divided the model area into three zones: (1) a "gross zone", enclosing the water seaward of a line 9 km offshore out to the seaward boundary of the model, roughly 50 km offshore; (2) a "nearshore zone", enclosed by lines 3 and 9 km offshore; and (3) an "inshore zone", between the coast and the line 3 km offshore. In the offshore or gross area the movement of the oil slick in the ith time step $\overrightarrow{\Delta P}_i$ is computed as a sum of 3% of the offshore wind velocity $\overrightarrow{W}_o$ plus the subsurface current at the offshore platform $\overrightarrow{C}_o$, i.e.,

$$\Delta\vec{P}_i = (0.03\ \vec{W}_{oi} + \vec{C}_{oi})\ \Delta T_i \qquad (21)$$

The wind-driven current vector 0.03 $\vec{W}_{oi}$ is arbitrarily rotated 15° clockwise to account for the Coriolis deflection in shallow water. In the nearshore zone the expression

$$\vec{\Delta P_i} = (0.03\ (a\ \vec{W}_{ci} + b\ \vec{W}_{oi}) + \vec{C}_{ci})\ \Delta T_i \tag{22}$$

utilized a coastal current $\vec{C}_{ci}$ interpolated from current charts and $\vec{W}_{ci}$ is the wind from the coastal weather bureau site. The weighting terms a and b (a + b = 1) represent the relative distance to the nearshore and inshore boundaries. This is a first attempt to account for a local wind system, the sea breeze, known to be active in this area. Coastal boundary layer currents, which would also be active in this zone, were not treated, however. When the slick moves within the inshore zone its movement is calculated from

$$\vec{\Delta P_i} = (0.03\ \vec{W}_{ci} + \vec{U}_{\ell})\ \Delta T_i \tag{23}$$

where $\vec{U}_{\ell}$ is the longshore or littoral current obtained from a lookup table. In this zone it would be sounder to utilize the data in Figure 6 instead of 0.03 $\vec{W}_{ci}$ and an expression such as Equation 17 instead of a lookup table. In summary, this model, while of necessity still using empirical rules, is notable for its attempt to include the combined effects of many of the processes affecting oil slick movement and its recognition of the importance of different processes in different regions of the coastal ocean.

D. Galt-Pease Diagnostic Model

A considerably more sophisticated approach to predicting oil spill trajectories in shelf waters has recently been advanced by Galt and Pease.[81] In essence this is a linear combination of a geostrophic current field and a wind-driven (Ekman) layer. The geostrophic equations are merely a balance between the Coriolis force and the pressure gradient forces, i.e., term 6 = term 4 + term 5 in Equations 8A and 8B. Substituting the hydrostatic approximation for the pressure increment dp = $-\varrho$g dz and differentiating with respect to z, the geostrophic relations become

$$f\ \frac{\partial v}{\partial z} = \frac{1}{\rho}\ \frac{\partial p}{\partial x} \tag{24A}$$

$$f\ \frac{\partial u}{\partial z} = \frac{1}{\rho}\ \frac{\partial p}{\partial y} \tag{24B}$$

where the pressure gradient term on the right is identically the sum of terms 4 and 5 in Equations 8A and 8B. From this equation we see that the geostrophic velocity can be calculated to within a constant of integration from a known density field, as term 4, Equations 8A and 8B, is independent of depth. Equations such as Equation 24, utilizing known values for otherwise dependent variables, are referred to as diagnostic equations. The Ekman layer equations

$$fv = \frac{1}{\rho}\ \frac{\partial}{\partial z}\ \tau_x = \frac{1}{\rho}\ \frac{\partial}{\partial z}\left(A_x\ \frac{\partial u}{\partial z}\right) \tag{25A}$$

$$fu = \frac{1}{\rho}\ \frac{\partial}{\partial z}\ \tau_y = \frac{1}{\rho}\ \frac{\partial}{\partial z}\left(A_y\ \frac{\partial v}{\partial z}\right) \tag{25B}$$

are then combined with the geostrophic equations and the appropriate boundary conditions to form a numerical diagnostic model. Boundary conditions include the usual surface wind stress relation (Equation 3), a stress condition on the bottom, and a continuity constraint on the transport, i.e., the divergence of total horizontal flow must be zero. According to Galt, from this relatively simple theory "... a diagnostic model can be developed that includes wind-driven current, geostrophic flow (both barotropic and baroclinic modes), frictionally controlled currents along the bottom, and the effects of complex bathymetry." He goes on to demonstrate the utility of the model with a computation of the coastal and shelf circulation along the Alaskan coast. While still strictly only an advective model, its sound theoretical basis indicates that future progress must proceed along these lines rather than attempting to further refine empirical wind factor approaches. The future addition of spreading, dispersion, and evaporation will make this a powerful tool indeed.

E. Hess-Kerr Model

Another sophisticated model based on sound theoretical principles and more complete than Galt's was recently advanced by Hess and Kerr.[82] Their model is actually a composite of three separate models for the motion of oil, the motion of water, and the motion of air. The oil movement model is based upon the two-dimensional mass and linear momentum equations applied to oil as a homogeneous fluid. Terms included in the analysis are the local acceleration of oil, the gravitational buoyancy effects of oil, the frictional stresses at the upper (air) and lower (water) boundaries, and the surface tension effects at the upper and lower boundaries. Terms neglected in the oil model represent mass exchanges (e.g., evaporation) at the free surface boundaries, the Coriolis acceleration of the oil, and turbulent and molecular stresses within the oil. The surface and bottom stresses on the oil slick are taken as the usual quadratic relations with the 10-m velocity and the surface water velocity, respectively. The oil equations are then solved numerically for the oil thickness and the horizontal oil velocity as functions of both time elapsed and displacement on the water surface. Numerical tests were run to compare the model results with the analytical results of Fay, discussed earlier, for the instantaneous release of a large quantity of oil on a calm sea. Agreement was reasonably good, and such tests are continuing.

The model for water motion involves Ekman dynamics coupled with continuity to give a three-dimensional current velocity structure. It is restricted to the surface mixed layer above the thermocline and assumes that the water below the thermocline is motionless and that the eddy viscosity is a function of wind speed. The eddy viscosity is in fact chosen so that the surface current is about 3% of the wind speed. The open-sea boundary condition is set with the inverse barometer effect, and the normal gradient of the flow at the lateral boundaries is forced to zero. The surface current speed computed from the model is then used to calculate the shear stress on the lower surface of the oil slick. Calibration of the water velocity model was said to be only in the preliminary stage.

The model of the wind velocity field is based upon existing operational National Weather Service models of the lower atmosphere which compute wind velocity, pressure, and temperature at set heights above the surface of the earth. Using various relationships from boundary layer theory for the Monin-Obukov length, the sea surface temperature, and the surface roughness z_o, the authors are able to solve for the friction velocity u^*, which then allows computation of the surface wind stress τ_o. This is then used as the stress boundary condition on the upper surface of the oil slick.

In essence, then, the Hess-Kerr model includes only spreading and advective effects arising from large-scale atmospheric pressure and wind effects. Turbulent dispersive effects, tidal currents, pressure gradients arising from density gradients in the water and mass loss due to evaporation, sedimentation, and emulsification are all neglected. Nonlinear effects that would arise from coastal headlands and embayments are like-

Table 2
SUGGESTED CATEGORIES FOR OIL-SPILL MODELS RESEARCH

Environmental category	Processes
Coastal	Littoral currents
	Rip currents
	Flow around headlands
	Flow in bays
	River-mouth effluents
	Local weather, sea breeze, mistral
	Wind effects in estuaries
	Tidal inlets
	Storm surge
	Pooling
Shelf	Coastal boundary layer
	Upwelling
	Tidal driving
	Atmospheric pressure systems
	Storm surge
	Ekman effects on wind-driven currents
Oceanic	General circulation
	Western boundary currents
	Hurricanes-typhoons
	Mesoscale eddies

wise not treated in this first version. It is, then, best suited for continental shelf application, and appropriately the authors make a computational comparison with observed slicks from the *Argo Merchant* spill discussed earlier. The agreement between model and field observations is extremely encouraging. Reproduction of the patchiness or local concentrations of oil in these slicks and its ability to predict oil thickness indicate that this model will eventually be quite superior to the simple vector addition technique currently in use.

F. Further Work

After comparing the types of oil spill models in the literature with the locations of and problems associated with real oil spills in the field, this author concludes that further oil-spill modeling should be encouraged to deal with processes in specific environmental categories. We must begin to encourage development of models that will fill gaps in our knowledge of how oil spills respond to driving forces in specific coastal and marine environments. A simple such scheme is outlined in Table 2, where I have identified coastal, shelf, and oceanic waters as the principal environmental categories and have listed some important processes that will affect oil-spill movement in each category. Galt and Pease's[81] model is a good example of a shelf-oriented model, while Hess and Kerr's[82] model is most appropriate for oceanic conditions. Paradoxically, the coastal area, where most of the environmental damage occurs and which displays the greatest variety of pertinent physical processes, has been largely ignored from the viewpoint of quantitative oil-spill models. For example, a research project on the effects of littoral currents and rip currents on oil advection and diffusion is clearly a high priority.

VII. SUMMARY

The introduction of petroleum products and crude oil from ship accidents and damaged platforms into the ocean is clearly on the rise. Recent research on continental shelves and coastal waters has shown the dominant role of winds and weather systems on forcing the motion of these waters. While wind effects are well recognized as a major factor in oil slick motion, their role as an element in a coherent predictable weather system is largely unappreciated.

Weather systems of nearly all sizes and time scales may have strong effects on oil slick movement and dispersal. Thunderstorms, local weather systems characteristic of certain regions of the world, mid-latitude high- and low-pressure systems, tropical cyclones, and the trade winds and prevailing westerlies of the planetary wind system are all potentially important agents in the movement and dispersal of oil slicks. Currents driven by these wind systems are influenced by the rotation of the earth, which causes them to veer to the right of the wind in the northern hemisphere and to the left of the wind in the southern hemisphere. Wind shifts or sudden decreases in wind stress induce circular or inertial oscillations whose period varies with latitude. Near the shore these effects are severely damped by the blocking action of the coast, causing the flow to run more or less parallel to the coastal boundary. All these effects will in turn exert significant control over the movement of entrained oil slicks.

Evidence is accumulating that energy is being transferred from surface waves directly into moving oil slicks, imparting to them a speed slightly greater than that of the wind-driven current entraining the oil slick. The magnitude of this excess speed is close to that expected from the Stokes drift of waves in shallow water.

In the near-field region of an oil spill tidal currents can also be of considerable importance. Rotary currents, characteristic of open-shelf waters and effective dispersal agents of oil, arise from the influence of the rotation of the earth on the tidal current. Another such interaction between rotation of the earth and the tide produces Kelvin waves, which result in unusually high tidal ranges along the coast to the right of the tidal wave propagation. Both effects have been important in recent oil spills.

As oil is advected along in the current stream another suite of processes is active in altering its chemical and physical characteristics. Evaporation, emulsification, turbulence, surface tension effects, sedimentation, and microbial activity are all active in dispersing or degrading the oil slick. Somewhat paradoxically, emulsification by augmenting the slick volume has substantially increased the impact of oil pollution in certain spill episodes.

All these oceanographic processes and others have played key roles in major spills over the last 10 years. In the *Torrey Canyon* spill off the west coast of England large-scale winds were the controlling factor, while in the Chevron Platform spill in the Gulf of Mexico rapidly migrating high- and low-pressure systems as well as density fronts associated with seasonal high river discharge played determinative roles. The *Argo Merchant* spill, thanks to the passage of four successive low-pressure cells to the north of the wreck site, was largely controlled by oceanic processes. Strong tidal currents were important only in the immediate vicinity of the wreck. The *Amoco Cadiz,* conversely, was a coastal spill. Coastal boundary layer currents, littoral currents, estuarine processes, and coastal topographic effects such as flow separation around headlands were by far the most critical factors in determining the fate of this spill. The largest oil spill in the world to date from a damaged well on the Campeche Bank of Mexico, has shown again that spills of long time and length scales will be influenced by oceanic scale circulation and large-scale wind patterns.

Modeling of oil spill movement has only recently progressed beyond the elementary stage of development. The most effective models still utilize empirical rules to predict oil movement, such as the 3% of the wind speed rule for surface currents. Evaporation, spreading due to surface tension effects, and dispersion are incorporated in some models, but not in others. A complex model incorporating all the processes and coastal boundary effects active in the *Amoco Cadiz* spill remains far in the future. Galt's[20] numerical diagnostic model, while still strictly advective in nature and not including effects due to the presence of oil, has the distinct advantage of (1) allowing for complex coastal boundaries, and (2) incorporating baroclinic pressure gradients, i.e., currents that arise from the water density field on the shelf. The model of Hess and Kerr[82] is notable in that it solves the oil momentum equations directly for the thickness of the oil slick as well as the velocity of the oil in space and time. It allows for spreading driven by surface tension effects but not for evaporation.

Surveying the effect of physical oceanographic processes on oil spills over the last 10 years, this author considers the outstanding feature to be the great variability of controlling processes from one spill to the next. An offshore spill driven by oceanic circulation and large-scale field is followed by a coastal spill in which littoral currents, shoaling waves, and coastal morphology play dominant roles. In other cases estuarine dynamics and density frontal dynamics are of paramount importance. In short, it appears that sooner or later every dynamical phenomenon studied in physical oceanography will come into play in an oil spill incident. The modeling task, then, is enormous, and the best we can plan for is to model suites of processes representative of certain areas considered critical on the basis of biological, recreational, or trafficability considerations. The next major oil spill might well occur in some obscure corner of the Middle East where extreme evaporation rates, hypersaline water masses, and strong local weather effects will produce effects quite unknown in European and American coastal waters.

ACKNOWLEDGMENTS

The author's research on the dynamics of coastal waters and the movement of oil slicks in the coastal ocean has been sponsored by the Coastal Sciences Programs of the Office of Naval Research, Arlington, Virginia, under a contract with the Coastal Studies Institute, Louisiana State University. Their support during the preparation of this manuscript is also gratefully acknowledged. Mr. Ken Biglane, of the U.S. Environmental Protection Agency, was especially helpful in obtaining source material. Special thanks also go to Mrs. G. Dunn for her outstanding efforts in producing the illustrations.

NOTATIONS

A_x, A_y	Horizontal components of the eddy viscosity
B	A constant in oil pool thickness theory
b	Channel width
$\vec{C}$	Constant vector in oil slick analysis
C_d	Drag coefficient relating wind stress to wind speed
C_f	Drag coefficient
D	Oil slick diameter in spreading theory
d	Diameter of inertial oscillation
d_o	Thickness of oil spill
f	Coriolis parameter
g	Acceleration resulting from gravity
g'	Densimetric acceleration of gravity
h	Local water depth
h_o	Initial height of oil in spreading theory
k	Wind factor, a coefficient relating surface current speed to wind velocity
k_3	A constant equal to about 1.6 in spreading theory
L	Length of oil slick

n	Constant in Stolzenbach's slick dispersion law
P	Position of oil slick in Sea Dock model
p	Pressure
R	Radius of oil patch
Ri	Richardson number
$\vec{S}$	Oil slick vector with components S_x and S_y
$S(\omega)$	Spectra of surface waves as a function of wave frequency
T_p	Period of inertial oscillation
t	Time
U_1	Longshore current speed at midpoint of breaker zone
$\overline{U}_\ell$	Longshore current from lookup table used in Sea Dock model
U_m	Maximum orbital velocity under a breaking wave
U_s	Net surface drift caused by wave motion
U_T	Amplitude of tidal current
$\overline{U}_T$	Cross-sectional average of tidal current amplitude
u	East-west component of water velocity positive to the east
u*	Friction velocity
V_o	Surface current speed
V_T	Inertial current velocity
v	North-south component of water velocity, positive to the north
$\vec{W}$	Wind velocity
W_c	A critical wind speed at which wave effects become important
$\overline{W}$	In complex notation, the steady current speed
W_o	In complex notation, the strength of the inertial current
w	Vertical component of water velocity, positive up
x	East-west coordinate, positive to the east
y	North-south coordinate, positive to the north
$\mathcal{Z}$	In complex notation, the displacement from the initial position
$\mathcal{Z}_o$	In complex notation, the initial position
z	Vertical coordinate, positive up
z_o	Parameter for length of surface roughness
α_b	Angle of wave approach to shore
β	Angle of beach slope
η	Elevation of sea surface above an undisturbed mean water level; η is taken positive up
$\varkappa$	Constant in Stolzenbach's slick dispersion law
ν_w	Viscosity of water
Ω	Angular velocity of the earth
ω	Wave frequency
ϕ	Latitude
ϱ	Density of water
ϱ_a	Density of air
ϱ_o	Density of oil
ϱ_w	Density of water
σ	Surface tension of oil on water
τ	Frictional or eddy stresses arising from turbulence in the water
$\vec{\tau}_o$	Stress of wind on the surface of water

REFERENCES

1. **Reed, A. W.,** Ocean Waste Disposal Practices, *Pollution Technology Review,* No. 23, Noyes Data Corporation, Park Ridge, N.J., 1975, 336.
2. **Grose, P. L. and Mattson, J. S.,** The *Argo Merchant* Oil Spill, A Preliminary Scientific Report, National Oceanic and Atmospheric Administration, Environmental Research Laboratories, Boulder, Colorado, 1977, 133 and 8 appendices.
3. **Niiler, P.,** A report on the continental shelf circulation and coastal upwelling, *Rev. Geophys. Space Phys.,* 13, 609, 1975.
4. U.S. Weather Bureau, *The Thunderstorm,* U.S. Government Printing Office, Washington, D.C., 1949.
5. **Hsu, S. A.,** Acoustic sounding of the atmospheric boundary layer over a tropical windward coast, in *Proc. 4th Symp. Meteorological Observations and Instrumentation,* American Meteorological Society, Boston, Massachusetts, 1978, 333.
6. **Johnson, A., Jr. and O'Brien, J. J.,** A study on an Oregon sea breeze event, *J. Appl. Meteorol.,* 18, 1267, 1973.
7. **Hsu, S. A.,** Coastal air circulation system: observations and empirical model, *Mon. Weather Rev.,* 98, 487, 1970.
8. **Sonu, C. J., Murray, S. P., Hsu, S. A., Suhayda, J. N., and Waddell, E.,** Sea breeze and coastal processes, *EOS Trans.,* 54, 820, 1973.
9. **Atkinson, G. P.,** Forecasters' Guide to Tropical Meteorology, Tech. Rept. 240, U.S. Air Force, Air Weather Service, Washington, D.C., 1971.
10. **Ichiye, T.,** Circulation changes caused by hurricanes, in *Contributions on the Physical Oceanography of the Gulf of Mexico,* Capurro, L. and Reid, J. L., Eds., Gulf Publishing Company, Houston, 1972, 229.
11. **Forristall, G. Z.,** Three dimensional structure of storm generated currents, *J. Geophys. Res.,* 79, 2721, 1974.
12. **Murray, S. P.,** Bottom currents near the coast during Hurricane Camille, *J. Geophys. Res.,* 75, 4579, 1970.
13. **Murray, S. P.,** Turbulence in hurricane generated coastal currents, in *Proc. 12th Coastal Engineering Conf.,* American Society of Civil Engineers, New York, 1971, 2051.
14. **Stommel, H., Voorhis, A., and Webb, D.,** Submarine clouds in the deep ocean, *Am. Sci.,* 59, 716, 1971.
15. **Byers, H.,** *General Meteorology,* McGraw-Hill, New York, 1959, 540.
16. **FAO,** *Impact of Oil in the Marine Environment,* Reports and Studies, No. 6 Food and Agricultural Organization, United Nations, New York, 1977, 250.
17. **Milgram, J. H.,** The role of physical studies before, during and after oil spills, in *In The Wake of The Argo Merchant, A Symposium,* Center for Ocean Management Studies, University of Rhode Island, 1978, 5.
18. **Harrison, W., Winnick, M. A., Kwang, P., and McKay, D.,** Crude oil spills, disappearance of aromatic and aliphatic components from small sea surface slicks, *Environ. Sci. Technol.,* 9, 231, 1975.
19. **Sivadier, H. O. and Mikolaj, P. G.,** Measurement of evaporation rates from oil slicks on the open sea, in *Proc. Conf. Prevention and Control of Oil Spills,* American Petroleum Institute, Washington, D.C., 1973, 475.
20. **Galt, J. A.,** Investigation of physical processes, in The *Amoco Cadiz* Oil Spill — A Preliminary Scientific Report, Hess, W. N., Ed., National Oceanic and Atmospheric Administration/Environmental Protection Agency Spec. Rept., Washington, D.C., 1978, 7.
21. **Smith, J. E.,** *Torrey Canyon Pollution and Marine Life,* Cambridge University Press, Cambridge, 1968.
22. **Berridge, S. A., Thew, M. T., and Loristen-Clarke, A. G.,** Formation and stability of emulsions of water in crude petroleum and similar stocks, *J. Inst. Petrol.,* 54, 333, 1969.
23. **Stolzenbach, K. L., Madsen, O., Adams, E., Pollack, A., and Cooper, C.,** A Review and Evaluation of Basic Techniques for Predicting the Behavior of Surface Oil Slicks, Report No. MIT SG77-8, Massachusetts Institute of Technology, Cambridge, 1977.
24. **Blumer, M. and Sass, J.,** Oil pollution, persistence and degradation of spilled fuel oil, *Science,* 176, 1120, 1972.
25. **Fay, J. A.,** The spread of oil slicks on a calm sea, in *Oil on the Sea,* Hoult, D. P., Ed., Plenum Press, New York, 1969, 53.
26. **Hoult, D. P.,** Oil Spreading on the sea, in *Annu. Rev. Fluid Mech.,* 341, 1972.
27. **Murray, S. P.,** Turbulent diffusion of oil in the ocean, *Limnol. Oceanogr.,* 17, 651, 1972a.
28. **Neumann, G. and Pierson, W. J., Jr.,** *Principles of Physical Oceanography,* Prentice Hall, Englewood Cliffs, New Jersey, 1966, 545.

29. **Hsu, S. A.,** Determination of the momentum flux at the air-sea interface under variable meteorological and oceanographic conditions: further application of the wind-wave interaction methods, *Boundary Layer Meteorol.*, 10, 1976, 221.
30. **Wu, J.,** Wind stress and surface roughness at air-sea interface, *J. Geophys. Res.*, 74, 444, 1969.
31. **Chang, M. S.,** Mass transport in deep water long crested random gravity waves, *J. Geophys. Res.*, 74, 1515, 1969.
32. **Munday, J. C., Jr., Harrison, W., and MacIntyre, W. G.,** Oil slick motion near Chesapeake Bay entrance, *Water Resour. Bull.*, 6, 879, 1970.
33. **Ekman, V. W.,** On the influence of the earth's rotation on ocean currents, *Ark. Mat. Astron. Fys.*, 2, 1, 1905.
34. **Sverdrup, H. U., Johnson, M. W., and Fleming, R. H.,** *The Oceans, Their Physics, Chemistry, and General Biology,* Prentice Hall, Englewood Cliffs, N.J., 1942, 1087.
35. **Hughes, P.,** A determination of the relation between wind and sea surface drift, *Q. J. R. Meteorol., Soc.*, 82, 494, 1956.
36. **Neumann, H.,** The relation between wind and surface current derived from drift card investigations, *Dtsche. Hydrogr. Z.*, 19, 253, 1966.
37. **Jeffreys, H.,** The effect of a steady wind on the sea level near a straight shore, *Philos. Mag.*, 46, 115, 1929.
38. **Murray, S. P.,** Speeds and trajectories of currents near the coast, *J. Phys. Oceanogr.*, 5, 347, 1975a.
39. **Madsen, O.,** A realistic model of the wind induced Ekman boundary layer, *J. Phys. Oceanogr.*, 7, 248, 1977.
40. **Pollard, R. T. and Millard, R. C., Jr.,** Comparisons between observed and simulated wind generated inertial oscillations, *Deep Sea Res.*, 17, 813, 1970.
41. **Pollard, R. T.,** On the generation by winds of inertial waves in the ocean, *Deep Sea Res.*, 17, 795, 1970.
42. **Daddio, E., Wiseman, W. J., Jr., and Murray, S. P.,** Inertial currents over the inner shelf near 30°N, *J. Phys. Oceanogr.*, 8, 728, 1978.
43. **Murray, S. P., Roberts, H. H., Wiseman, W. J., Jr., Tornatore, H. G., and Whelan, W. T.,** An over the horizon radio direction-finding system for tracking coastal and shelf currents, *Geophys. Res. Lett.*, 2, 211, 1975.
44. **Wiseman, W. J., Jr., Murray, S. P., and Roberts, H. H.,** High frequency techniques and over-the-horizon radar in coastal research, in Proc. Russell Symp. Coastal Research, Louisiana State University, Baton Rouge, 1977.
45. **Proudman, J.,** *Dynamical Oceanography,* John Wiley & Sons, New York, 1953, 409.
46. **Hansen, D. and Rattray, M ., Jr.,** Gravitational circulation in estuaries, *J. Mar. Res.*, 23, 104, 1965.
47. **Petrow, R.,** *In the Wake of Torrey Canyon,* David McKay Inc., New York, 1968, 256.
48. **Admiralty,** *The English and Bristol Channels Tidal Stream* Atlas, Hydrographer of the Navy, Taunton, Somerset, 1961, 16.
49. **Wright, L. D. and Coleman, J. M.,** Effluent expansion and interfacial mixing in the presence of a salt wedge, Mississippi River delta, *J. Geophys. Res.*, 76, 8649, 1971.
50. **Murray, S. P.,** Observations on wind, tidal and density driven currents in the vicinity of the Mississippi River delta, in *Shelf Sediment Transport,* Swift, D. J. P., Duane, D. B., and Pilkey, O. H., Eds., Dowden, Hutchinson, and Ross, Stroudsberg, Pa., 1972b, 127.
51. **Murray, S. P.,** Wind and current effects on large-scale oil slicks, in *Proc. 7th Annu. Offshore Technology Conf.*, Paper no. OTC 2389, Offshore Technology Conference, Dallas, 1975b, 523.
52. **Bowden, K. F. and Hughes, P.,** The flow of water through the Irish Sea and its relation to wind, *Geophys. J. R. Astron. Soc.*, 5, 265, 1961.
53. **Garvine, R. W.,** Physical features of the Connecticut River outflow during high discharge, *J. Geophys. Res.*, 79, 831, 1974a.
54. **Garvine, R. W.,** Dynamics of small-scale oceanic fronts, *J. Phys. Oceanogr.*, 4, 557, 1974b.
55. **Garvine, R. W.,** Observation of the motion field of the Connecticut River plume, *J. Geophys. Res.*, 82, 441, 1977.
56. **Garvine, R. W. and Monk, J. D.,** Frontal structure of a river plume, *J. Geophys. Res.*, 79, 2251, 1974.
57. **Horne, E.,** Physical aspects of the Nova Scotian shelf break fronts, in *Oceanic Fronts in Coastal Processes,* Bowman, M. J. and Esaias, W. E., Eds., Spec. Rept. No. 10, Marine Science Research Center, State University of New York, Stony Brook, 1977, chap. 7.
58. **Mooers, C., Flagg, C., and Boicourt, W.,** Prograde and retrograde fronts, in *Ocean Fronts in Coastal Processes,* Bowman, M. J., and Esaias, W. E., Eds., Spec. Rept. No. 10, Marine Science Research Center, State University of New York, Stony Brook, 1977, chap. 6.
59. **Simpson, J. and Pingree, R.,** Shallow sea fronts produced by tidal stirring, in *Oceanic Fronts in Coastal Processes,* Bowman, M. J. and Esaias, W. E., Eds., Spec. Rept. No. 10, Marine Science Research Center, State University of New York, Stony Brook, 1977, chap. 5.

60. **Pingree, R., Bowman, M., and Esaias, W. E.,** Headland fronts, in *Oceanic Fronts in Coastal Processes,* Bowman, M. J. and Esaias, W. E., Eds., Spec. Rept. No. 10, Marine Science Research Center, State University of New York, Stony Brook, 1977, chap. 9.
61. **Murray, S. P., Smith, W. G., and Sonu, C. J.,** *Oceanographic Observations and Theoretical Analysis of Oil Slicks during the Chevron Spill, March, 1970,* Tech. Rept. No. 87, Coastal Studies Institute, Louisiana State University, Baton Rouge, 1970, 106.
62. **Grose, P. L.,** The behavior of floating oil from the *Argo Merchant,* in *In the Wake of the Argo Merchant, a Symposium,* Center for Ocean Management Studies, University of Rhode Island, Providence, 1978, 19.
63. **Hess, W. N.,** 1978, Executive Summary, in The *Amoco Cadiz* Oil Spill — A Preliminary Scientific Report, Hess, W. N., Ed., National Oceanic and Atmospheric Administration/Environmental Protection Agency Spec. Rept., Washington, D.C., 1978, 1.
64. CNEXCO, Remote Sensing of Oil Spills, Preliminary Report, Centre National pour L'Exploration des Oceans, Institut Francais du Petrole and Institut Geographique National, Paris, 1978.
65. **Csanady, G. T.,** The coastal boundary layer, in *Estuaries, Geophysics, and the Environment,* Studies in Geophysics Series, National Academy of Sciences, Washington, D.C., 1977, 57.
66. **Gundlach, E. R. and Hayes, M. O.,** Investigation of beach processes, in The *Amoco Cadiz* Oil Spill — A Preliminary Scientific Report, Hess, W. N., Ed., National Oceanic and Atmospheric Administration/Environmental Protection Agency, Washington, D.C., Spec. Rept., 1978, 85.
67. **Crout, R. L. and Murray, S. P.,** Shelf and coastal boundary layer currents, Miskito Bank of Nicaragua, in *Proc. 16th Coastal Engineering Conf.,* American Society of Civil Engineers, New York, 1979, 2715.
68. **Allen, G., D'Ozouville, L., and L'Yavanc, J.,** Etat de la pollution par les hydrocarbures dans l'Aber Benoit, in *Amoco Cadiz, Primieres Observations sur la Pollution Par Les Hydrocarbures,* Actes de Colloques, No. 6, Centre National Pour L'Exploitation des Oceans, Centre Oceanologique de Bretagne, Brest, 1978.
69. **Komar, P. D.,** Nearshore currents and sediment transport and the resulting beach, in *Marine Sediment Transport and Environmental Management,* Stanley, D. J. and Swift, D. J. P., Eds., Interscience, New York, 1976, 241.
70. **Shepard, F. P. and Inman, D. L.,** Nearshore circulation related to bottom topography and wave refraction, *Trans. Am. Geophys. Un.,* 31, 555, 1950.
71. **Longuet-Higgins, M. S.,** Longshore currents generated by obliquely incident sea waves, *J. Geophys. Res.,* 75, 6778, 1970.
72. **Berne, S., Brossier, R., Fontanel, A., D'Ozouville, L., Serriere, J., and Wadsworth, A.,** Teledetection des pollutions par hydrocarbures de *l'Amoco Cadiz,* in *Amoco Cadiz, Primieres Observations sur La Pollution Par les Hydrocarbures,* Actes de Colloques, No. 6, Centre National Pour L'Exploitation des Oceans, Centre Oceanologique de Bretagne, Brest, 1978.
73. DOE, Strategic Petroleum Reserve, West Hackberry Oil Storage Cavern Fire and Spill, September 21, 1978, DOE-OSC After Action Report, Department of Energy, New Orleans, 1978, 99.
74. **Sorenson, R. M. and Spencer, E. B.,** Two dimensional wind set up of oil on water, *J. Waterways, Harbors Coastal Eng. Div.,* WW3, 517, 1971.
75. Atlas of Pilot Charts, Central American Waters and South Atlantic Ocean H.O. Publ. no. 576, Oceanographic Office, U.S. Navy, Washington, D.C., 1955.
76. **Coleman, J. M. and Murray, S. P.,** Coastal sciences: recent advances and future outlook, in *Science, Technology, and the Modern Navy,* Salkowitz, E. I., Ed., Department of the Navy, Office of Naval Research, Arlington, Va., 1976, 347.
77. **Lissauer, I. and Welsh, P.,** Can oil spill movement be predicted?, in *In the Wake of the Argo Merchant, A Symposium,* Center for Ocean Management Studies, University of Rhode Island, Providence, 1978, 22.
78. **Warner, J. L., Graham, J. W., and Dean, R. G.,** Prediction of the movement of an oil spill on the surface of the water, in Proc. Offshore Technology Conf., paper no. OTC 1550, Offshore Technology Conference, Dallas, 1972.
79. **Williams, G. N., Hann, R., and James, W. P.,** Predicting the fate of oil in the marine environment, in *Proc. Joint Conf. Prevention and Control of Oil Spills,* American Petroleum Institute, Washington, D.C., 1975.
80. Projected Movement of Oil Slicks, Environmental Report Vol. II, Sea Dock, Houston, 1975, 48.
81. **Galt, J. A. and Pease, C. H.,** The use of a diagnostic circulation model for oil trajectory analysis, in *Proc. Oil Spill Conf. (Prevention, Behavior, Control, Clean Up),* Publ. no. 484, American Petroleum Institute, Washington, D.C., 1977, 447.
82. **Hess, K. W. and Kerr, C. L.,** A model to forecast the motion of oil on the sea, in *Proc. Oil Spill Conf. (Prevention, Behavior, Control, Cleanup),* American Petroleum Institute, Washington, D.C., 1979, 27.

INDEX

C

D

E

O

P

Q

R

S

T

U

V

W

X

Y

Z